An understanding of organic reaction mechanisms is an essential part of any undergraduate chemistry course. This book describes the principles that govern chemical reactivity, and shows how these principles can be used to make predictions about the mechanisms and outcomes of chemical reactions.

Molecular orbital theory is used to provide up-to-date explanations of chemical reactivity, in an entirely non-mathematical approach suited to organic chemists. A valuable section explains the use of curly arrows, vital for describing reaction mechanisms. A whole chapter is devoted to exploring the thought processes involved in predicting the mechanisms of unfamiliar reactions. Each chapter is followed by a summary of the important points and a selection of problems to help the reader make sure that the material in that chapter has been assimilated. The book concludes with a comprehensive glossary of technical terms.

Understanding organic reaction mechanisms

Understanding
organic reaction mechanisms

A. Jacobs

CAMBRIDGE
UNIVERSITY PRESS

PUBLISHED BY THE PRESS SYNDICATE OF THE UNIVERSITY OF CAMBRIDGE
The Pitt Building, Trumpington Street, Cambridge CB2 1RP,
United Kingdom

CAMBRIDGE UNIVERSITY PRESS
The Edinburgh Building, Cambridge CB2 2RU, United Kingdom
40 West 20th Street, New York, NY 10011–4211, USA
10 Stamford Road, Oakleigh, Melbourne 3166, Australia

First published 1997

Printed in the United Kingdom at the University Press, Cambridge

Typeset in Monotype Times 10/13pt

A catalogue record for this book is available from the British Library

Library of Congress Cataloguing in Publication data

Jacobs, A. (Adam), 1966–
Understanding organic reaction mechanisms / A. Jacobs.
p. cm.
Includes index.
ISBN 0-521-46217-7 (hc). – ISBN 0-521-46776-4 (pb)
1. Chemistry, Organic. 2. Chemical reactions. I. Title.
QD251.2.J33 1997
547.1′39–dc20 96-44742 CIP

ISBN 0 521 46217 7 hardback
ISBN 0 521 46776 4 paperback

Contents

CONTENTS

Preface

Reaction mechanisms are a fundamental part of the study of organic chemistry, and the aim of this book is to help you to understand them. Organic reaction mechanisms are sometimes perceived to be an incoherent and difficult subject, but in fact there are principles underlying them that make them much easier to grasp. Reading this book should give you a mastery of these principles.

Chapter 1 introduces the basics of chemical bonding. This contains a discussion of frontier orbitals (HOMOs and LUMOs), which are an important part of understanding chemical reactivity. Chapters 2 and 3 describe more of the background to understanding reaction mechanisms, namely the nature of ionic species, which are found in the vast majority of reactions, and the driving forces behind reactions. Chapters 4 and 5 look more closely at the molecules that take part in organic reactions, Chapter 4 dealing with species whose reactivity is centred on carbon, and Chapter 5 addressing molecules with other atoms. Chapter 6 describes the reactions themselves. By this stage in the book, most of the principles behind chemical reactivity have already been explained, so the reactions can be seen to be no more than logical consequences of these. Chapter 7 is something of an aside, and looks at how we know about reaction mechanisms from experimental evidence. Chapter 8 draws on the material presented earlier in the book to help you to suggest mechanisms for unknown reactions, a vital part of any undergraduate chemistry course. Finally, Chapter 9 looks at how knowledge of reaction mechanisms has been used in practice.

This book is primarily designed to be read sequentially, as each chapter makes use of principles explained in previous ones. However, you can also turn to any part of the book to look up a particular topic, as material explained earlier in the book is indicated by cross references.

Organic chemistry is a very rewarding subject, and I hope this book will allow you to pursue your studies more easily. Although organic reaction mechanisms can seem something of a mystery at first, time spent studying them will be richly rewarded in a greater understanding of organic chemistry as a whole.

A. Jacobs   1997

Acknowledgements

A book such as this requires the work of many more people than just the author. I should like to express my thanks to the staff of Cambridge University Press for all their help in producing this book, and in particular Fiona Thomson, without whom this book would never have been written in the first place. I am also very grateful to Dr Linda Lazarus and Dr Karen Jonsen, both of whom read draft versions of this book and were responsible for removing many of my errors and incomprehensible sentences.

Finally, I should like to thank Carolyn Jones, who not only made many helpful suggestions on a draft of the book, but also bore her book widowhood with fortitude and helped to ensure that all the time I spent slaving over a word processor did not (entirely) turn me into a gibbering wreck.

1

Chemical structure

As organic chemists, we are primarily interested in the covalent bonds formed by the elements in the first row of the periodic table, in particular those of carbon. The study of organic chemistry is about the making and breaking of these bonds. However, before we consider the mechanisms by which these processes take place, it is worth spending a little time to look at the structures of organic molecules and the nature of the bonds which hold them together.

In this chapter we will look at the behaviour of electrons in atomic orbitals and the way these atomic orbitals can combine to form bonds. We will look at the movement of electrons in reactions, and how we can write this down so that we can see clearly what is happening. We will pay particular attention to those orbitals that take part directly in reactions, namely the frontier orbitals.

1.1 Chemical bonds

1.1.1 Atomic orbitals

Chemical bonds are formed when the atomic orbitals on one atom interact with those on another to form molecular orbitals. The electrons that each of the atoms brings to the bond are shared between them, occupying the resulting molecular orbital. Let us first look at the atomic orbitals that contribute to these bonds.

The behaviour of electrons in atoms is governed by the Schrödinger equation. Solving this equation[1] gives an infinite number of wavefunctions, each of which is a mathematical description of the energy of the electron and the probability of finding it in a certain place, and is defined by a set of four quantum numbers. Fortunately, although the number of mathematically possible wavefunctions is infinite, we need

---

[1] The Schrödinger equation may be solved exactly only for the hydrogen atom, and must be solved by numerical methods for more complicated atoms or molecules.

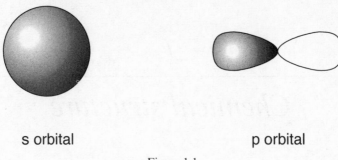

s orbital   p orbital

Figure 1.1

consider only a small number of them, as all the others are of too high energy (i.e. energetically unfavourable) to be chemically relevant. Wavefunctions are part of the language of the theoretical chemist, but they correspond to orbitals, which are an organic chemist's way of looking at the same thing.

The principal quantum number, n, may have any positive integer value from 1 upwards. This corresponds to the shell that the electrons occupy around the nucleus, thus an orbital with $n=1$ is in the closest shell to the nucleus (and the only occupied shell in a hydrogen or helium atom), $n=2$ describes the next shell outwards, and so on. Organic chemists are seldom interested in any orbitals with n greater than 3; $n=3$ corresponds to the outer orbitals of a second-row element. The higher the value of n, the greater is the average distance of the electron from the nucleus, and consequently the higher the energy.

The angular momentum quantum number, l, is perhaps of more interest to the organic chemist, as it specifies the shape of the orbital. Possible values of l range from 0 to $(n-1)$, thus when $n=1$, l must be 0, when $n=2$, l may be either 0 or 1, and so on. When $l=0$, the resulting orbital is spherically symmetric, and is called an s orbital. When $l=1$, this gives rise to a p orbital, which is dumbbell-shaped. These are the shapes of orbitals that are most relevant to organic chemistry, and are shown in Figure 1.1. Note that the two lobes of the p orbital have opposite numerical signs (the change in sign is shown by shading); the significance of this will become clear when we look at bonding in Section 1.1.2. The signs of wavefunctions are particularly important in pericyclic reactions, as we shall see in Section 6.5. Sometimes we may also be interested in d orbitals, for which $l=2$, but extremely rarely in any higher values of l.

The magnetic quantum number, m, may take any integer value between l and $-l$. This represents the spatial orientation of the orbital. Thus for a p orbital there are three possible values of m, and hence three possible orientations of the orbital. These are at right angles to each other, and are known as the p_x, p_y, and p_z orbitals.[2]

[2] The p_x, p_y, and p_z orbitals are not in fact those orbitals with m values of 1, 0, and -1. You need not, however, be concerned with this. The important point is that there are three possible values of m, and three differently oriented p orbitals. There are many good theoretical chemistry textbooks that go into this in more detail for those who are interested.

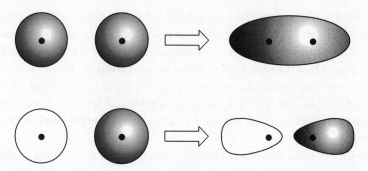

Figure 1.2 The bonding (top) and antibonding (bottom) orbitals of the hydrogen molecule.

The spin quantum number, s, has no effect on the size, shape, or orientation of the orbital, but refers to the electron. It may take the values $+1/2$ or $-1/2$, irrespective of the nature of the orbital. Electron spin is a purely quantum mechanical phenomenon, which has no counterpart in the macroscopic world, and cannot be described other than in mathematical terms. From the organic chemist's point of view, its consequence is this. The Pauli exclusion principle states that no two electrons in any atom or molecule may have the same set of quantum numbers. If two electrons occupy the same orbital, then they must have different values of s, as their other quantum numbers will be the same. As there are only two values of s allowed for the electron, it is not possible to have more than two electrons in any orbital.

Although these orbitals may be given an accurate mathematical specification only for the hydrogen atom, the above description applies just as well to other atoms in qualitative terms. Organic chemists are not interested in a precise mathematical description of orbitals, only in their approximate sizes and shapes, and hence how they influence chemical reactivity.

1.1.2 Simple bonds

We will begin by considering the simplest stable molecule of all: the hydrogen molecule. Each isolated hydrogen atom has one electron, contained in a 1s orbital. As the two atoms are brought together, their 1s orbitals interact, which they may do in two ways, either in phase or out of phase (Figure 1.2). When they are in phase, they add together to produce a wavefunction with a higher value between the nuclei. When they are out of phase, they will tend to cancel each other out, so that there is a nodal plane (a plane in which there is zero probability of finding the electron) halfway between the two nuclei. Thus we started with two atomic orbitals, and we now have two molecular orbitals. This is a general phenomenon – whenever there are a total of n atomic orbitals combining, the resultant molecule will always have n molecular orbitals.

3

What exactly do we mean by atomic orbitals adding in phase or out of phase? The wavefunction for an electron at a particular point has a sign, in other words it may be positive or negative. This has no physical meaning by itself, because electron density is proportional to the square of the wavefunction. If, however, we imagine both atomic orbitals to have the same sign, then they will add in phase. Conversely, they will add out of phase if they have opposite signs. The signs of wavefunctions become more important as we consider p orbitals, as these have different signs on the two lobes.

It is obvious that the molecular orbitals from these two types of combination are not the same. The orbital resulting from the in-phase combination of atomic orbitals has a region of relatively high electron density between the two nuclei. This electron density attracts the positively charged nuclei and keeps them together. For this reason, this orbital is known as the bonding orbital. The other orbital has decreased electron density between the nuclei. This means that the two positive charges are not shielded from each other, and so they experience a repulsive force. This orbital is therefore known as the antibonding orbital, and is of higher energy (i.e. less stable) than the bonding orbital.

In the hydrogen molecule there are only two electrons, and these may both be accommodated in the bonding orbital, the antibonding orbital remaining empty. This gives rise to a stable bond, as we would expect. If we imagine a hypothetical helium molecule, He_2, we have essentially the same orbitals, but now four electrons. The bonding orbital can accommodate only two electrons, so the other two must occupy the antibonding orbital. These two electrons in the antibonding orbital cancel out the effect of the two in the bonding orbital, so there is no net bonding interaction.[3] This is why the helium molecule does not exist.

Combining atomic orbitals in this way is a mathematical convenience rather than a real physical process. An alternative way of arriving at the above molecular orbitals would be to solve the Schrödinger equation for the hydrogen molecule. Were we to do this (impossible to do exactly, although excellent approximations can be made), we would find the same two molecular orbitals, one bonding and one anti-bonding. (We would also find an infinite number of higher-energy orbitals, as indeed we would if we were to combine the higher-energy atomic orbitals in the way we have done for the 1s orbitals, but we can cheerfully ignore these because they contain no electrons.) Although it is mathematically more correct to think of molecular orbitals as being solutions to the Schrödinger equation for a whole molecule, it is usually more convenient to think of them as combinations of atomic orbitals, as we have done for hydrogen.

In principle, we can solve the Schrödinger equation for any molecule, and the wavefunctions that we obtain will describe the distribution of electrons, and hence

---

[3] There is in fact a small net antibonding interaction, as the antibonding effect of an antibonding orbital is slightly greater than the bonding effect of a bonding orbital.

the bonding in the molecule. In practice, however, we will usually consider bonding in molecules as combinations of atomic orbitals, as we have done here. A precise mathematical description of electrons is not important in organic chemistry; we are more interested in how they take part in reactions. It is easier to predict this from a consideration of the properties of atoms and functional groups than from mathematical equations.

1.1.3 More complicated bonding and hybridization

The above description gives a good idea of what a chemical bond is at its simplest level, but as organic chemists we are not much concerned with hydrogen molecules. We are more interested in organic molecules, in other words those containing carbon. Carbon is in the first main row of the periodic table, so its 1s orbital is full, and we may ignore it. We are safe in doing this because orbitals interact significantly with other orbitals only if they are of similar energy. A filled inner orbital is much lower in energy than a valence orbital (the outer orbital of the atom, which takes part in reactions), and so the interaction between the carbon 1s orbital and the valence orbital of any atom with which carbon is bonding will be negligible. Of course, if the other atom also has a filled inner orbital, this may well be of a similar energy to the carbon 1s orbital, but we can still ignore the interaction, because the interaction of two filled orbitals leads to a bonding orbital and an antibonding orbital, which cancel each other out.

Carbon's valence shell consists of the 2s and three 2p orbitals. Carbon has six electrons, two of which occupy the 1s orbital, leaving four for the valence shell. As there are four orbitals and four electrons in the valence shell, we would expect carbon to be able to form four covalent bonds (each orbital has only one electron and so must gain a further electron from a covalent bond with another atom), and indeed we know that most carbon atoms in organic molecules are tetravalent.

Before we consider tetravalent carbons, however, let us first consider the simpler example of the methyl cation, CH_3^+. Although this is extremely reactive and cannot be isolated, it does nonetheless have three stable bonds to hydrogen atoms and can be detected spectroscopically. Its reactivity arises from its vacant 2p orbital, perpendicular to the plane of the paper in Figure 1.3. The two remaining 2p orbitals and the 2s orbital are used in bonding to the hydrogens. If we consider the symmetry properties of the orbitals (this is really a job for theoretical chemists, so do not worry if this seems alien to you) we arrive at three combinations of atomic orbitals leading to bonding molecular orbitals. There are, of course, also three antibonding orbitals. Each of the three bonding orbitals contains two electrons, and so three bonds are formed.

This has already become much more complicated than the hydrogen molecule, and we still have only four atoms. Clearly, it would be advantageous to have some way of simplifying the problem, or we are going to run into great difficulties when

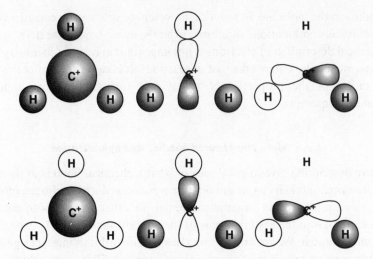

Figure 1.3 The three bonding (top) and three antibonding (bottom) orbitals
of the methyl cation.

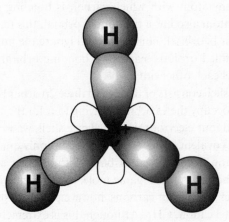

Figure 1.4

considering more grown-up organic molecules. Fortunately, such a trick is available.
This is known as hybridization. Instead of considering the orbitals on the carbon
atom in the way we have done above, we may think of the orbitals involved in the
bonding as being equivalent. In this example, the resulting orbitals will be called sp_2
hybrids, because they are made up of one s orbital and two p orbitals. Each orbital
has one third of the character of an s orbital, and two thirds that of a p orbital. They
are arranged at 120° to each other, with each one pointing to a hydrogen atom
(Figure 1.4). It is now much easier to see how three bonds are formed. Each of these
bonds can be thought of as being localized and separate from the other two, and can
hold two electrons. As before, we have six electrons taking part in bonding.

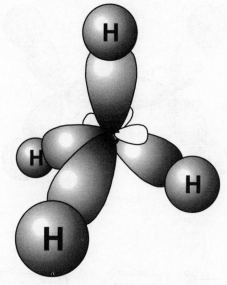

Figure 1.5

Figure 1.6

We should remember that hybridization of orbitals is just a mathematical trick; hybrid orbitals do not really exist. However, the molecular orbitals arrived at by considering hybrid orbitals have the same electron distribution as those derived from combinations of real atomic orbitals, or for that matter, from solving the Schrödinger equation for the whole molecule. It is obviously much more convenient to think of bonds in terms of the appropriate hybrid orbitals.

We may apply the same technique to methane, in which carbon is bonded to four hydrogen atoms at the vertices of a regular tetrahedron (Figure 1.5). All four orbitals participate in bonding, so we now use sp_3 hybrids, derived from one s orbital and three p orbitals. These are arranged tetrahedrally, with each one forming a bond to a hydrogen atom. We therefore have four localized bonds, as we would expect.

The third commonly used possibility is an sp hybrid, as found in acetylene. We shall ignore the bonding between the two carbon atoms in acetylene for the moment; this is explained in Section 1.1.4. The carbon–hydrogen bond is formed from an sp hybrid orbital on the carbon atom, which is at 180° to the carbon–carbon bond (Figure 1.6). This gives the molecule a linear shape.

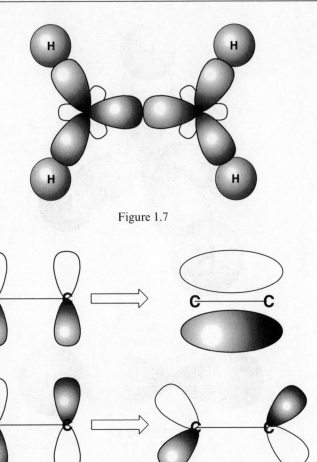

Figure 1.7

Figure 1.8 The bonding (top) and antibonding (bottom) π orbitals of ethylene.

1.1.4 Multiple bonds

Let us now consider the bonding in ethylene, one of the simplest molecules to have a multiple bond. It is easiest to think of the carbon atoms being sp_2 hybridized. We may thus draw the molecule as shown in Figure 1.7, with two of the sp_2 hybrid orbitals pointing towards the hydrogen atoms, and the third towards the other carbon.

This leaves us with a spare p orbital on each carbon atom. The p orbitals are perpendicular to the plane of the molecule. If we look at the molecule edge on (Figure 1.8) we see that there are two ways in which they may interact. This is exactly analogous to the overlap of s orbitals; in-phase overlap leads to a bonding orbital, out-of-phase overlap results in an antibonding orbital.

It is obvious that this is a different type of bond to those we have considered

8

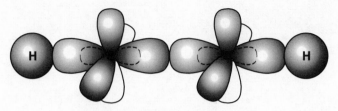

Figure 1.9

hitherto. Previously, we have looked at bonds in which the bonding overlap occurs along the axis joining the two nuclei. Such interactions are known as σ (sigma) bonds. In this new bond, there is no overlap of orbitals along the internuclear axis. The overlap occurs instead above and below it. This is called a π (pi) bond. Ethylene has one σ bond and one π bond joining the two carbon atoms; in other words there are two bonds, which is why ethylene is commonly written with a double bond.

A π bond is not as strong as a σ bond, because the orbitals are overlapping sideways, rather than head on. Consequently, although a double bond is stronger (and shorter) than a single bond, it is less than twice as strong. This is worth remembering, as there are many reactions in which σ bonds are formed at the expense of π bonds, for example the Diels–Alder reaction (Section 6.5.3).

An important feature of π bonds is that because there is no electron density between the two nuclei, it is all concentrated into electron clouds above and below the plane of the bond. These electron clouds readily take part in reactions, as they are held less tightly by the nuclei than electrons in σ bonds. This explains the tendency of olefins (compounds with carbon–carbon double bonds, otherwise known as alkenes) to react with electrophilic reagents, for example bromine (Section 6.1.3).

We can now return to acetylene. The σ framework consists of the sp hybrid orbitals from carbon and the s orbitals from hydrogen. This leaves two p orbitals on each carbon, which are perpendicular to the σ bonds and to each other. These form π bonds in just the same way that the π bond in ethylene is formed, but there are now two such bonds, one above and below the molecule, and one to either side of it (Figure 1.9). Thus the triple bond of acetylene is composed of one σ bond and two π bonds.

1.2 Resonance structures

The types of molecules we have discussed so far can be adequately described by a single structural formula. This simple state of affairs is not always so. Let us consider nitromethane. This may be written as shown in Figure 1.10a. According to this formula, one of the oxygen atoms is joined to the nitrogen with a double bond, and is electrically neutral, while the other is joined by a single bond and carries a negative charge.

Figure 1.10

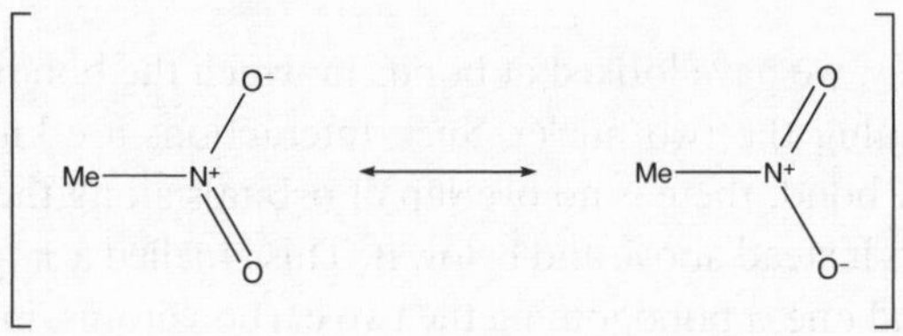

Figure 1.11

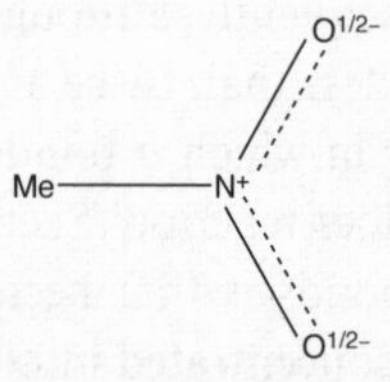

Figure 1.12

If we were to believe this formula, we would think that the two oxygen atoms are different. It is known, however, that the two oxygen atoms of the nitro group are equivalent. Each of the two bonds between oxygen and nitrogen is intermediate in length and strength between a single and a double bond. We could also write the structure of nitromethane as shown in Figure 1.10b. Here, the bonding of the two oxygens is written the other way round. We may think of the real bonding as being intermediate between the two ways of writing the structure.

Neither of these two formulae on their own is sufficient to describe the structure accurately. We may, however, write the structure of nitromethane as being composed of both structures, as in Figure 1.11. These are known as resonance structures. The double-headed arrow is the conventional way of showing that two structures are resonance forms of the same thing. Note also that it is common for resonance structures to be written within square brackets. It is important to realise that although we write the structure as these two separate forms, the real structure is in between the two. A common mistake is to think that the molecule spends some time as one of these structures as drawn, then the electrons rearrange themselves into the other form, and so on. This is not so. The molecule spends its entire time in an intermediate state. It would be more accurate to draw the molecule as in Figure 1.12, with half

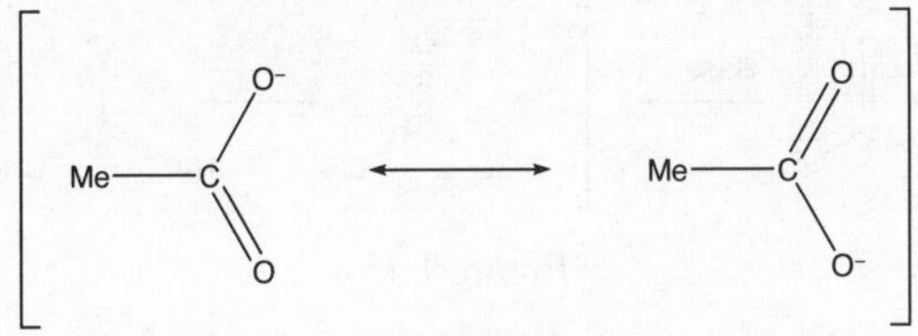

Figure 1.13

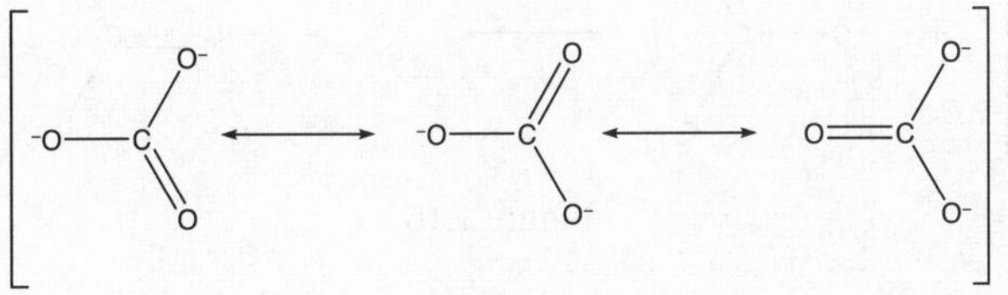

Figure 1.14

a negative charge on each oxygen, and a bond intermediate between a single and a double bond joining each oxygen to the nitrogen. In practice, this is more cumbersome to write, so it is more usual to draw one of the two resonance structures instead. We must never forget that this is only shorthand.

The electrons that give rise to the negative charge and that form the double bond (four in all) are said to be delocalized. The origin of this term should be clear, as they cannot be considered to belong to any one atom or bond, but are spread over a number of atoms and bonds.

The resonance forms that may be written for a particular structure are also known as canonical forms. When taken together, the combination of resonance structures that may be written for a molecule describe it as a resonance hybrid.

Many other molecules have separate resonance forms. The acetate anion (Figure 1.13) is analogous to nitromethane. Each of the two oxygens carries half a negative charge. Another commonly found example of a molecule with resonance forms is benzene (of which more in Section 1.5). Benzene does not have three single and three double bonds, but six equivalent ones.

There are more complex examples, in which two resonance forms are insufficient to describe the structure, and we must write three or more. The carbonate anion (Figure 1.14) is one such molecule, in which the two negative charges are divided equally between the three oxygens.

In all the species we have considered so far, each canonical form contributes equally to the resonance hybrid. This is not always true. Let us take the enolate anion derived from acetone, in other words the structure that we get by treating acetone with base to remove a proton (Figure 1.15). This may be written either with the negative charge on the carbon atom or on the oxygen. In practice, most of the negative charge is found on the oxygen, so we may think of the structure with the

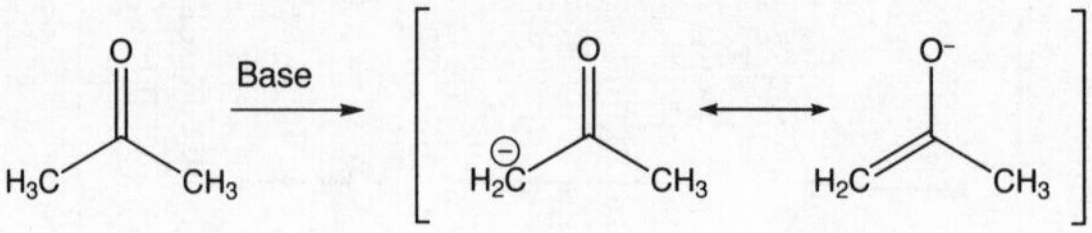

Figure 1.15

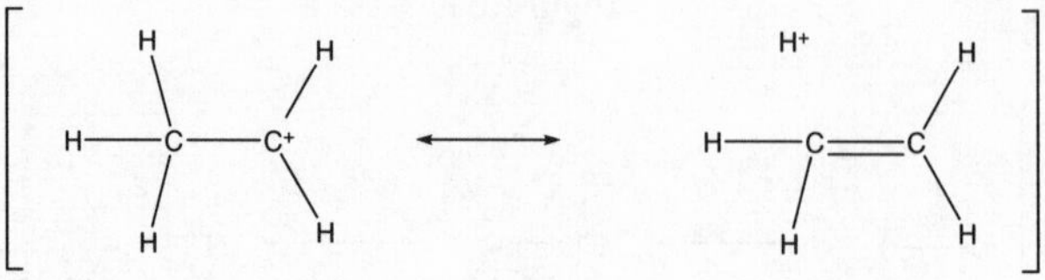

Figure 1.16

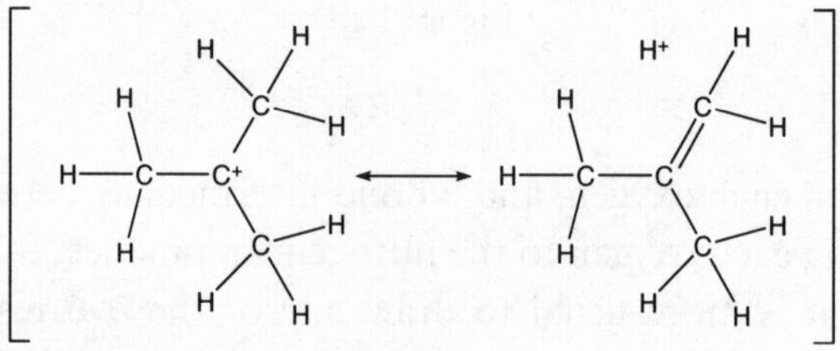

Figure 1.17

negatively charged oxygen as providing a greater contribution to the resonance hybrid than the structure with the negatively charged carbon. This is reflected in the tendency for chemists to write the acetone anion with the negative charge on the oxygen.

A more extreme example is found in the ethyl cation. We may write a resonance structure for this in which a bond to a hydrogen atom is broken altogether, and the positive charge is carried on the resulting free proton (Figure 1.16). This is called hyperconjugation. There are three such forms that can be written for the ethyl cation, each with a different proton becoming free. These forms make only a minor contribution to the resonance hybrid and the protons do not actually become detached from the rest of the molecule. The effect is nonetheless significant, especially when many different hyperconjugated resonance structures can be written, as their effects are additive. There are nine such forms for the t-butyl cation (Figure 1.17); this helps to account for its stability.

1.3 Curly arrows

Curly arrows are of enormous importance in organic chemistry. They are the means by which we show the movement of electrons in reactions, and hence the mechanisms of reactions. To function properly as an organic chemist you must become so

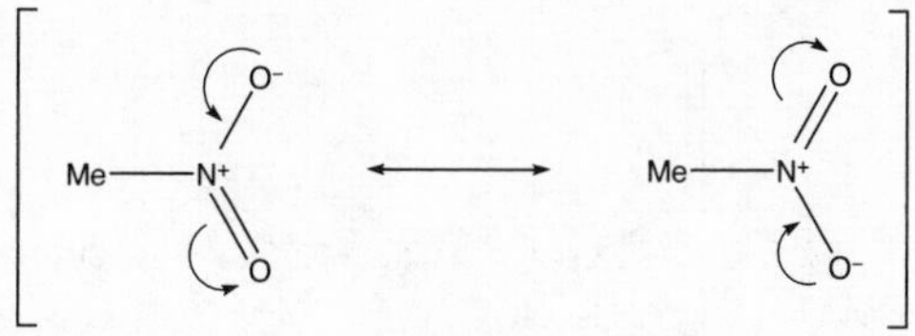

Figure 1.18

familiar with the rules governing curly arrows that their use becomes second nature to you.

It is unfortunate that curly arrows have two distinct purposes: showing the imaginary movement of electrons in resonance structures, and showing the real movement of electrons in reactions. Nonetheless, this is seldom too much of a problem because the context will generally make clear what the curly arrows are showing. We will now look at the two different functions of curly arrows in turn.

1.3.1 Curly arrows and resonance structures

All the resonance structures described in Section 1.2 have one thing in common. Each form may be related to another in a given resonance hybrid by the movement of electrons. This can be shown with curly arrows. Here, the movement of electrons is purely imaginary, but we shall see in Section 1.3.2 that curly arrows may also be used in a different sense to represent real movement of electrons.

Let us return to nitromethane. If we consider each structure to be a real entity (which of course we know it is not), then we can see that one may be related to another by the movement of electrons from one oxygen to the other. We may write this imaginary movement using curly arrows (Figure 1.18). Each curly arrow represents the movement of a pair of electrons. Here, one pair of electrons (a lone pair on the oxygen bearing the negative charge) moves from its original position to form a π bond between the oxygen and the nitrogen. The other pair moves from being a π bond to the other oxygen (which it is forced to do, so that we do end up with a pentavalent nitrogen) to being a lone pair on this oxygen. We can see that this movement of electrons has transferred the negative charge from one oxygen to the other.

This use of curly arrows is purely artificial. As stated previously, the electrons do not actually move, and nitromethane is not accurately represented by one structure or the other. However, as it is common practice to draw such molecules as their individual resonance forms, it is as well to have a way of relating one form to another.

From this simple example we may learn the fundamentals of the correct use of curly arrows. Firstly, each arrow represents the movement of a pair of electrons. This is an important point to remember; in organic chemistry a curly arrow is never used for any other purpose. We start with an oxygen lone pair (two electrons) and end up

13

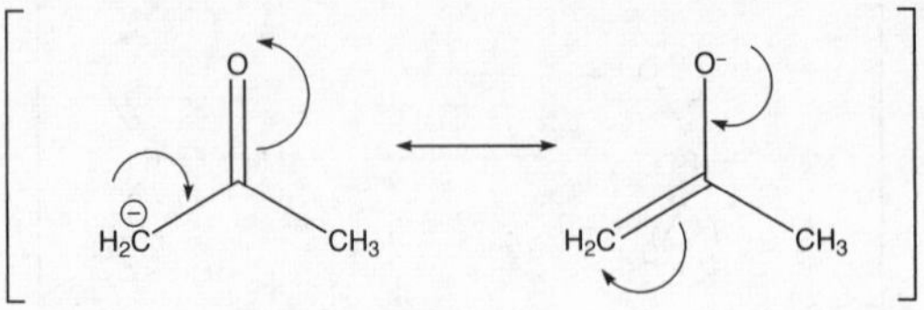

Figure 1.19

with a π bond (also two electrons). In the process the oxygen loses its negative charge by having to share two electrons which it previously had to itself.

We should note in passing that curly arrows can also be used to denote the movement of single electrons, rather than electron pairs, as is found in reactions of free radicals. For this purpose, the arrows are written with a single head rather than a double head. We will meet free radicals again in other places in this book, in particular in Section 4.3.

An important point to notice is that the succession of curly arrows begins at an electron source and ends at an electron sink. The electron source here is the negatively charged oxygen. Electron sources, from which curly arrows start, are frequently negative charges. Lone pairs also commonly fulfil this function, as do (though less often) electron-rich π bonds.

The electron sink is a neutral oxygen, which becomes negatively charged as it accepts the electrons. Atoms which may bear a stable negative charge are one common type of electron sink, positively charged species are another.

Finally, we may note that these two curly arrows are connected. As one of them shows the movement of electrons towards the nitrogen, the other shows the electrons flowing away from it to their ultimate electron sink. In any chain of curly arrows, the arrows will always connect like this. Every atom along the chain thus ends up with as many electrons as it had before, except the atoms at each end of the chain (although sometimes, as we shall see later, the chain has no beginning or end, so all the atoms have the same number of electrons as at the start). The atom at the beginning of the chain ends up with fewer electrons than it had to start with, and that at the end of the chain with more. In other words, a succession of curly arrows represents a flow of electrons from one place (the electron source) to another (the electron sink).

Another example we looked at in Section 1.2 was the acetone enolate ion. Again, we may mentally move the negative charge from carbon to oxygen with curly arrows; and of course we can equally well move it back again (Figure 1.19). It might be thought that we are violating our rule of ending our curly arrows at electron sinks by moving the electrons onto a carbon atom. Carbon atoms are not normally thought of as being able to bear a stable negative charge. We must remember, however, that the structure in which the carbon is negatively charged is only a minor contribution to the resonance hybrid. Although the carbon atom is not a very good electron sink,

14

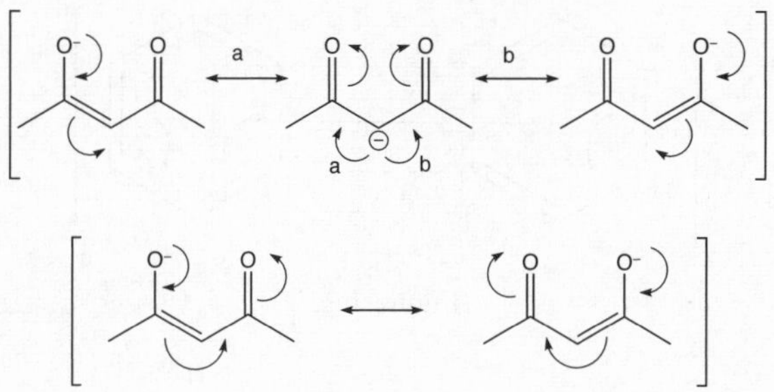

Figure 1.20

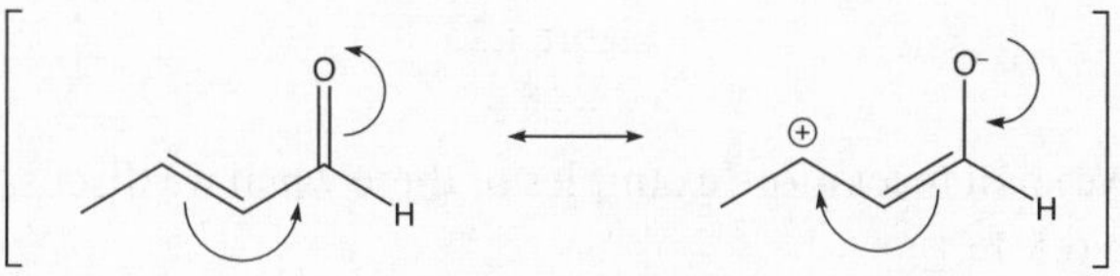

Figure 1.21

it does fulfil the main requirement of an electron sink in that it is able to accept the electrons. As this movement of electrons is, in this instance, purely imaginary, it does not matter that the resulting negative charge is of high energy (i.e. unstable). It is the contribution of the other resonance form that stabilizes this charge.

We may extend this last example by looking at the anion derived from acetylacetone (1,3-pentanedione). The negative charge on the central carbon atom may be delocalized onto either of the two oxygens (Figure 1.20). This gives us three structures that contribute to the resonance hybrid, one with the negative charge on carbon, and two with negatively charged oxygens. We may relate these last two structures to each other with three curly arrows, in which the charge is shown to move from one oxygen to the other through the intervening carbons.

Curly arrows may also be used for the resonance forms of crotonaldehyde (Figure 1.21). One of these structures has charge separation and makes only a small contribution to the resonance hybrid. Nonetheless, we see that we may use the carbon–carbon double bond as an electron source, and move the electrons onto the oxygen, which we know to be a good electron sink. In the reverse direction, the use of the curly arrows is more obvious; the negatively charged oxygen is clearly an electron source and the positively charged carbon is clearly an electron sink.

A variation occurs in interrelating the two canonical forms of benzene (Figure 1.22). Here, the electrons move in a circular fashion, with neither an end nor a beginning to the flow and it is not appropriate to talk of an electron source or sink. All atoms end up with the same number of electrons that they had originally. Situations such as this, when the electrons flow in a closed circuit, are the only exception to the rule that successions of curly arrows must begin at an electron source and end at an

15

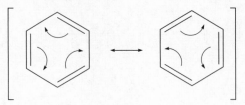

Figure 1.22

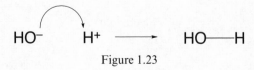

Figure 1.23

electron sink. We shall meet more examples of these when we discuss pericyclic reactions in Section 6.5.1.

1.3.2 Curly arrows and reactions

Let us now turn to the use of curly arrows for describing actual flow of electrons, or in other words, reactions. The rules governing curly arrows here are the same as for their use in describing resonance structures, but now they represent a real physical process.

It is perhaps confusing that the curly arrow should be made to serve this dual purpose. You must bear in mind that the two uses of these arrows are not the same, although they may appear to be on paper. An important point to remember is that for resonance structures, the position of the atoms does not change from one structure to the next; it is only the position of the electrons that differs (on paper, at any rate). When curly arrows are used to describe reactions, chemical bonds are made and broken, and so the position of atoms changes.

A simple example of a chemical reaction is the combination of a hydroxide ion with a proton to form a water molecule (Figure 1.23). One of the electron pairs on the oxygen (the electron source) is donated to the proton (the electron sink), forming a covalent bond. We may write this movement of an electron pair with a curly arrow. Here, one curly arrow is enough to describe the reaction, but more complex examples will often require many arrows.

A mistake students sometimes make when drawing proton transfer reactions such as this is to use the arrow to represent the movement of the proton. This is wrong. Curly arrows are only ever used to show the movement of electrons.

A slightly more complex reaction is that of hydroxide ion with methyl chloride to give methanol (Figure 1.24). This is an example of an S_N2 reaction, of which more in Section 6.3.2. The electron source, as in the last example, is an electron pair from the hydroxide oxygen. The electron sink is the chlorine atom (we know that chlorine

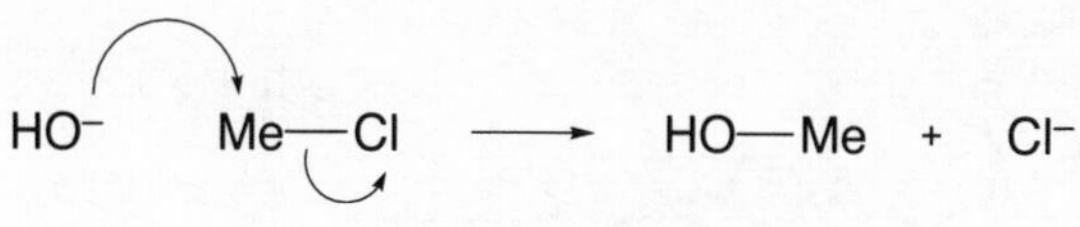

Figure 1.24

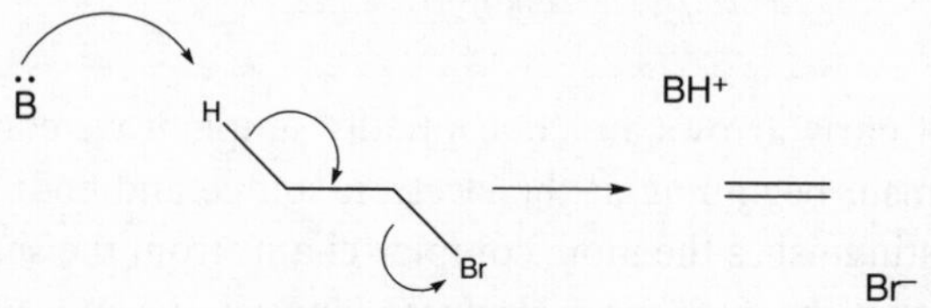

Figure 1.25

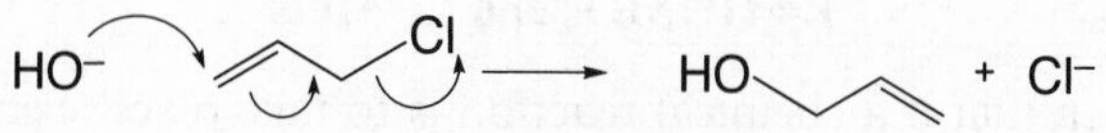

Figure 1.26

atoms can hold a stable negative charge). The electrons are transferred from one to the other through the carbon atom. This process requires two curly arrows; one to show the movement of electrons from oxygen to carbon, and the other to show the breaking of the carbon–chlorine bond, with its electrons ending up on the chlorine atom.

Another simple reaction is the base-catalysed elimination of HBr from ethyl bromide (see Section 6.2.2) (Figure 1.25). An electron pair on the base (the nature of the base is not important) acts as the electron source, and the bromine as the electron sink. This is similar to the last example. Where this example differs is simply that the chain of electron flow is longer. There is now another electron pair between the two ends taking part in the flow, which moves from a carbon–hydrogen σ bond to a carbon–carbon π bond.

Something that commonly happens is that the flow of electrons is transmitted through π bonds. Either of the preceding two examples may be extended in this way. For the displacement of chloride, we may consider allyl chloride (Figure 1.26). The hydroxide ion may attack the chlorinated carbon directly, as it did with methyl chloride. An alternative mode of reaction (we shall not concern ourselves here with which mode is more likely in practice) is that the hydroxide ion may attack the double bond, which in turn is responsible for ejecting the chloride ion. The beginning and end of the chain of electron flow are just the same as for methyl chloride; all that is different here is that the chain is longer.

Similarly, we may extend the elimination reaction by considering crotyl bromide (Figure 1.27). Again, the beginning and end of the flow are the same as before, but we have more electrons making up the chain.

17

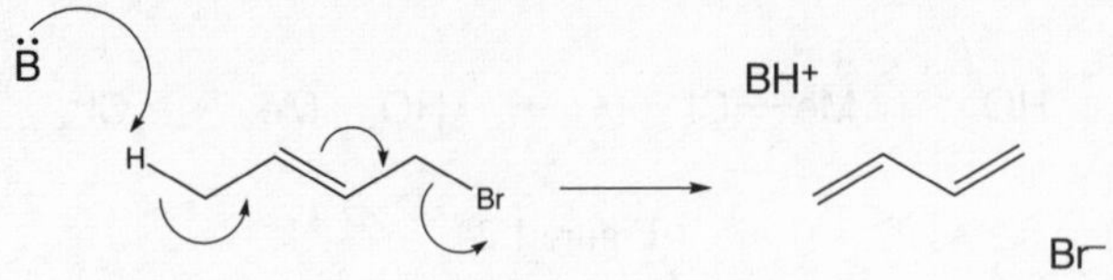

Figure 1.27

Most chains of curly arrows are conceptually simple if we consider only where they start and finish, beginning at the electron source and ending at the electron sink.[4] All that distinguishes the more complex chains from the simple examples we have seen here is that there are more electrons between the ends of the chain.

1.4 HOMOs and LUMOs

We have now seen that if a chemical reaction is to take place, electrons must flow from one molecule to another. We have also seen that electrons occupy orbitals. If a reaction is to take place, therefore, an orbital of one molecule must interact with an orbital of another molecule. The orbitals that take part in these reactions are known as frontier orbitals, because they are at the frontier of the molecules' reactivity. Once we understand the nature of these frontier orbitals, we shall be in a better position to understand many features of chemical reactivity.

The subject of frontier orbitals is often treated rather superficially in many under-graduate organic chemistry courses, because it is felt to be too complicated. This is a pity, because whilst there are undoubtedly some complex points to be understood, once these have been mastered they can greatly simplify the explanation of many reactions. You should therefore study this section carefully, and not be disheartened if at first it seems heavy going. You can always come back to this section after you have seen how frontier orbitals are used in explaining reactions later in the book.

The two main types of frontier orbitals are HOMOs and LUMOs. The acronym HOMO stands for Highest Occupied Molecular Orbital, and LUMO for Lowest Unoccupied Molecular Orbital. This means that the HOMO of a particular mole-cule is the highest energy orbital to contain a pair of electrons, and the LUMO is the lowest energy empty orbital.

An example will make this clear. We have already seen that the hydrogen mole-cule has two molecular orbitals formed from the combination of two 1s orbitals (Figure 1.2). The lower in energy of these two orbitals, the bonding orbital, contains the two electrons. As this is the only orbital with any electrons in it, it must be the highest-energy occupied orbital, or the HOMO. The antibonding orbital formed from the combination of the 1s electrons is higher in energy than the bonding orbital, but lower in energy than any others. There are an infinite number of mole-

---

[4] As already mentioned, pericyclic reactions are an exception to this rule (Section 6.5).

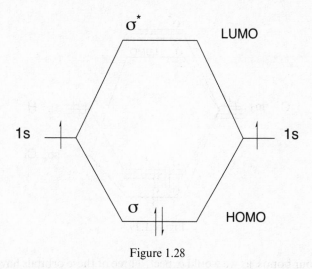

Figure 1.28

cular orbitals on the hydrogen molecule, but they are almost all too high in energy to be chemically significant. As the antibonding orbital derived from the 1s atomic orbitals is the lowest in energy of all these unoccupied orbitals it has a special status, and is designated by the term LUMO.

HOMOs and LUMOs can be described with energy-level diagrams. In these diagrams, we show different types of orbital along the horizontal scale (here, for example, the hydrogen molecular orbitals and the hydrogen 1s orbitals from which they are derived), and the energy of these orbitals on the vertical scale. The higher on the vertical scale a particular orbital is, the higher its energy, and so the less stable it is. Thus orbitals closer to the horizontal axis will be more energetically favourable, and so more likely to contain electrons. We should note that while the vertical axis is a quantitative scale, the horizontal axis is purely qualitative, in other words it does not represent any numerical value.

An energy-level diagram for the hydrogen molecule clearly shows the two orbitals, one (the bonding orbital) lower in energy than the other (Figure 1.28). By convention, the electrons are represented by single-headed arrows, the direction of the arrow corresponding to the spin of the electron (quantum number s). A pair of electrons in any one orbital must necessarily have opposite spins; this is shown by a pair of arrows pointing in opposite directions. Antibonding orbitals are denoted by an asterisk. In this simple molecule the energy-level diagram may be superfluous, but in more complex examples it will make it much easier to see what is happening.

Let us consider an energy-level diagram for methyl chloride (Figure 1.29). Eight molecular orbitals are formed from a combination of the four carbon sp_3 orbitals, the three hydrogen 1s orbitals, and one of the chlorine sp_3 orbitals (the other three chlorine valence orbitals correspond to its lone pairs). Of these eight orbitals, four are bonding and four are antibonding. The four bonding orbitals are all filled, which

19

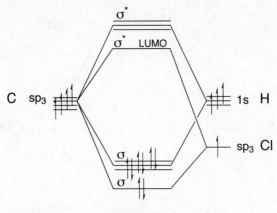

Figure 1.29

gives rise to four bonds, as we would expect. Three of these orbitals have equal energies, although the orbitals are shown at slightly different heights on the diagram for clarity (otherwise we might think there is only one orbital). Orbitals of equal energy are said to be degenerate. The orbitals involving the chlorine atom are lower in energy than the others, because of the electronegativity of chlorine. The LUMO is the lowest in energy of the antibonding orbitals, which, as we can see from the diagram, is that between carbon and chlorine.

At this point, we can see that the concept of HOMOs and LUMOs is not just some esoteric mathematical device, but has a direct bearing on chemical reactivity. In the reaction between methyl chloride and hydroxide, how is the mode of attack of the hydroxide determined? The answer is that the lone pair of the hydroxide (its HOMO) reacts with the LUMO of the methyl chloride. This is the antibonding orbital between the carbon and the chlorine, so the result of this reaction is that the electrons from the hydroxide ion enter this antibonding orbital, causing the carbon–chlorine bond to break. Chemical reactivity cannot always be satisfactorily explained by such a simple use of HOMOs and LUMOs, as there are usually many other factors at work; indeed even here it is doubtful whether the explanation given above is really the whole story. Nonetheless, this example serves as a useful illustration of how a consideration of HOMOs and LUMOs may frequently yield valuable information about the reactivity of various molecules.

This is a recurrent feature in chemical reactions; the HOMO of one reactant donates its electrons to the LUMO of the other. Simple chemical knowledge tells us how methyl chloride will react, but in more complex examples a knowledge of HOMOs and LUMOs allows us to predict the path of chemical reactions which might otherwise be considerably less obvious.

HOMOs and LUMOs are particularly useful when considering reactions of complicated π systems. A molecular orbital diagram for 1,3-butadiene is shown in Figure 1.30. We can ignore the σ framework here because it is lower in energy, and

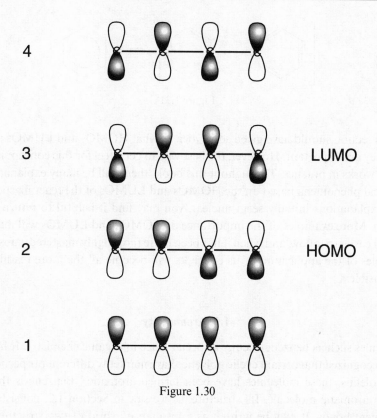

LUMO

HOMO

Figure 1.30

therefore will not take part in reactions as easily as the π system. In this molecule, the two π bonds are adjacent and therefore conjugated, so they cannot be considered in isolation. The four p orbitals give rise to four molecular orbitals. Remember that an interaction between two orbitals is bonding if they have the same sign (shown by the shading) and antibonding if their signs differ. The two molecular orbitals of lower energy are bonding and the higher two are antibonding. It should be obvious that the lowest is bonding and the highest is antibonding. We can see that orbital 2 is bonding because it has two bonding interactions and one antibonding, whilst orbital 3 has two antibonding and one bonding interaction and is therefore antibonding.

This diagram has been simplified by drawing all the atomic orbitals to the same size. In practice, the different atomic orbitals will contribute different amounts to the molecular orbitals, but that need not worry us for the time being; the signs of the orbitals are more important.

As there are four electrons in this system (one from each of the p orbitals), the two lower energy orbitals (i.e. the bonding orbitals) will be filled. We can see that orbital 2 is the HOMO and that orbital 3 is the LUMO. Any reactions of butadiene would be expected to involve one of these two orbitals, and indeed they do.

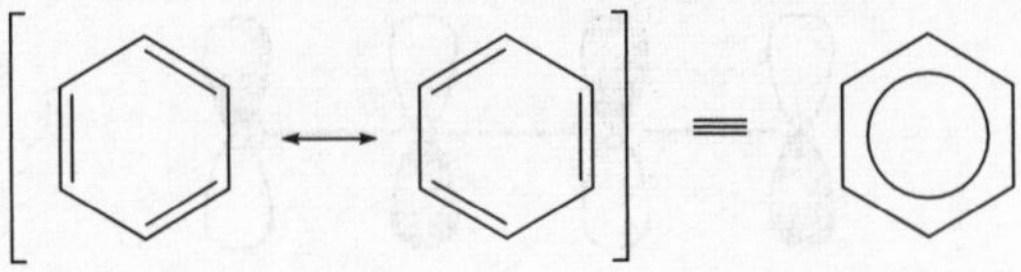

Figure 1.31

This section should have given some idea of what HOMOs and LUMOs are and why they are important. However, the best way to get a feel for this concept is to see how it works in practice. Throughout this book, there will be many explanations of chemical phenomena based on the HOMOs and LUMOs of the reacting species. If such explanations initially seem unclear, you may find it helpful to return to this section. More examples of the importance of HOMOs and LUMOs will make this section easier to follow, and when it has been more thoroughly mastered, subsequent examples of the application of its concepts will become all the more readily comprehensible.

1.5 Aromaticity

Molecules such as benzene occupy a special place in organic chemistry. It has long been recognized that certain cyclic polyenes have markedly different properties from other olefins; these molecules have been termed aromatic.[5] Benzene is the best-known aromatic molecule. Its structure, as we saw in Section 1.2, consists of six equivalent bonds. It may be written as a resonance hybrid of two structures with alternating single and double bonds, but is also commonly written with a circle in the middle as in Figure 1.31. This shows more clearly that all six bonds are equivalent.

Benzene does not have the properties that might be expected for a six-membered cyclic hydrocarbon with three double bonds. Its heat of hydrogenation (i.e. the energy released when the molecule is hydrogenated) is significantly less than three times that of the hydrogenation of one double bond, suggesting that the molecule has special stability. This stability may be seen in other ways; for example, benzene does not react with bromine. Olefins normally react rapidly with this reagent, so much so that decoloration of bromine water has been used as a diagnostic test for olefins.

What is it about benzene that gives it these special properties? To answer this, it is helpful to consider a molecular orbital diagram. Benzene has six p orbitals contributing to the bonding of the π system, so there must be six molecular orbitals. These are shown in Figure 1.32, in which all orbitals are drawn to the same size for simplicity. Three of the orbitals are bonding, and three antibonding. One of the bonding orbitals is lower in energy than the others, but the other two are of equal energy, or

---

[5] The term aromatic was given to these molecules historically because of their interesting odours. Nowadays, however, benzene is known to be a powerful carcinogen, so sniffing it is not a recommended practice!

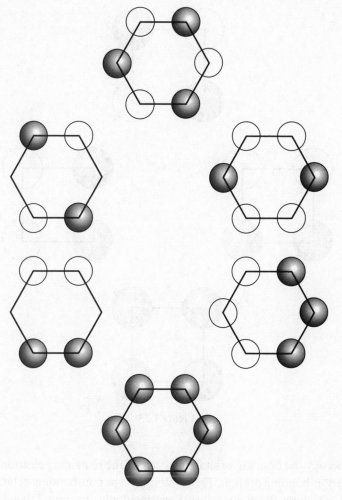

Figure 1.32

degenerate. Similarly, there are two degenerate antibonding orbitals. Note that all the bonding orbitals are filled. This is a common feature of aromatic molecules; when all the bonding orbitals of a molecule are filled, this gives it stability in the same way that atoms are more stable when their electronic structures resemble those of inert gases, when their valence shells are completely filled.

To see the importance of this, let us consider the molecular orbital diagram for 1,3-cyclobutadiene (Figure 1.33). This has one bonding orbital, two degenerate non-bonding orbitals[6] and one antibonding orbital. There are four electrons. Two

---

[6] They are termed non-bonding because they have two bonding and two antibonding interactions, i.e. equal bonding and antibonding interactions, which cancel each other out.

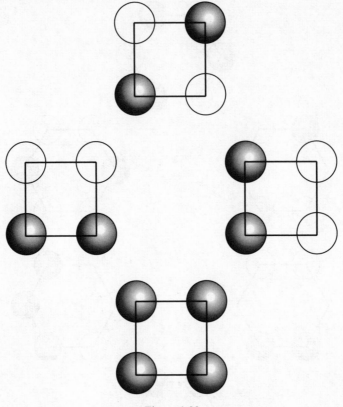

Figure 1.33

of these occupy the bonding orbital and each of the remaining electrons goes into one of the non-bonding orbitals. The result is that the non-bonding orbitals are only half filled, which results in instability. Experimentally, it is found that 1,3-cyclobutadiene is an extremely unstable molecule, and may be isolated only at very low temperatures. Ring strain (see Section 3.2.2) plays a part in this instability, but it does not account for all of it; the partially filled molecular orbital shell has a significant effect.

How are we to know which conjugated cyclic polyenes are stabilized by aromaticity (e.g. benzene) and which are not (e.g. cyclobutadiene)? Fortunately, there is a simple rule that allows us to predict which are which. This is known as the Hückel rule, and states that those cyclic conjugated polyenes with $(4n+2)$ π electrons are aromatic. We see that benzene, with its six π electrons ($4n+2$, $n=1$) is aromatic, whilst cyclobutadiene, with four π electrons, is not. Molecules such as cyclobutadiene, with $4n$ π electrons, are known as antiaromatic.

It is easy to see the basis for this rule. We can see from the molecular orbital diagrams that both molecules have one bonding orbital of lower energy than all the

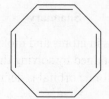

Figure 1.34

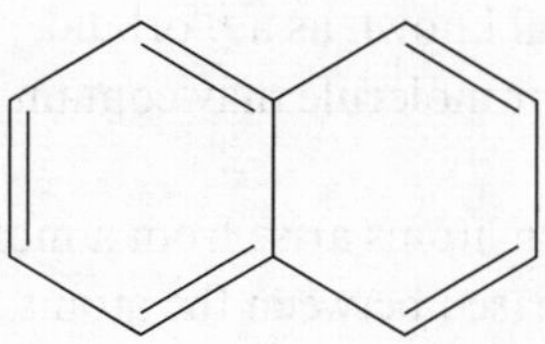

Figure 1.35

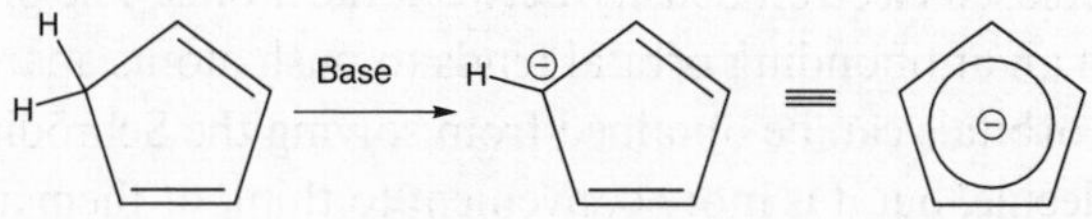

Figure 1.36

others, which may accommodate two electrons. Also, each molecule has two degenerate orbitals above this. For all cyclic systems, there will be one orbital of lowest energy, with degenerate pairs of orbitals above it. In order to fill complete shells of orbitals, therefore, there must be two electrons for the bottom orbital, and a further four for each pair of orbitals above this, or in other words, $(4n+2)$ electrons.

This rule allows us to predict that cyclooctatetraene, with its eight π electrons (Figure 1.34), is not aromatic. The properties of this molecule are indeed markedly different to those of benzene. The molecule is not planar, and has two different carbon–carbon bond lengths, corresponding to alternating double and single bonds. It reacts with bromine, and undergoes many other reactions typical of olefins.

The Hückel rule tells us that naphthalene is aromatic (Figure 1.35). It has 10 π electrons ($4n+2$ with $n=2$). It also explains the unusually high acidity of cyclopentadiene ($pK_a=16$). The resulting anion has six π electrons, and so is aromatic (Figure 1.36). This aromaticity confers special stability on the anion, which is not available to neutral cyclopentadiene. We shall see in Section 2.2.2 that this increases the acidity.

The stability of aromatic molecules and their other properties are important themes in organic chemistry and will be encountered again frequently in the rest of this book.

25

Summary

- The behaviour of electrons in atoms and molecules can be described with a wavefunction, which is obtained by solving the Schrödinger equation.
- The wavefunction of an atomic orbital is defined by four quantum numbers: n, l, m, and s.
- Atomic orbitals have different shapes according to the value of l: an orbital with $l=0$ is spherical and is called an s orbital, $l=1$ corresponds to a dumbbell-shaped orbital known as a p orbital.
- An orbital in an atom or molecule may contain no more than two electrons.
- Covalent bonds between atoms arise from a molecular orbital in which the electron density is increased between the atoms. When the region of electron density is along the internuclear axis, the bond is known as a σ bond.
- Another type of molecular orbital is called an antibonding orbital. In this, there is decreased electron density between the atoms. The presence of electrons in an antibonding orbital tends to push atoms apart.
- Molecular orbitals can be obtained from solving the Schrödinger equation for the molecule, but it is more convenient to think of them as combinations of atomic orbitals.
- It may also be convenient to consider molecular orbitals as combinations of hybrid orbitals, imaginary atomic orbitals which are made up of different amounts of different types of atomic orbitals.
- Double and triple bonds are made up of bonds with a sideways overlap of p orbitals, known as π bonds, in addition to σ bonds.
- Some molecules cannot be described with localized covalent bonds; in these, the electrons are said to be delocalized.
- Delocalized molecules can be drawn by writing two or more resonance structures.
- These resonance structures do not exist as discrete entities; a delocalized molecule is an average of its possible resonance forms.
- The movement of electrons can be shown with curly arrows.
- This movement can be imaginary, as in the relationship between resonance structures, or real, as in reactions.
- A curly arrow shows the movement of a pair of electrons.
- Single-headed arrows show the movement of single electrons; these are used in reactions of free radicals.
- Curly arrows begin at electron sources and end at electron sinks.
- The only exception to the above is in pericyclic reactions, when the electrons move in a circuit.
- The most important orbitals for chemical reactivity are frontier orbitals.
- The two types of frontier orbitals are known as HOMOs and LUMOs.

- In most reactions, the HOMO of one reactant will interact with the LUMO of another.
- HOMOs and LUMOs can be particularly useful when considering conjugated π systems.
- Cyclic polyenes with $(4n+2)$ π electrons are known as aromatic.
- Aromatic molecules are particularly stable.
- Cyclic polyenes with $4n$ π electrons are known as antiaromatic and are considerably less stable than aromatic molecules.

Problems

1. What is the state of hybridization (e.g. sp$_3$) of the asterisked carbon atom in each of the following structures?

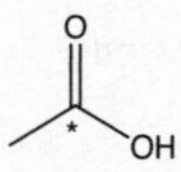

    a 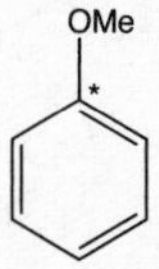   b    c 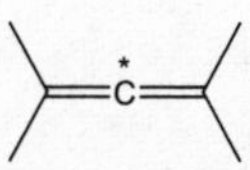   d

2. Which of the following sets of curly arrows show the imaginary movement of electrons in resonance structures, and which show real movement of electrons in reactions? Draw the alternative resonance structures/reaction products to which they lead.

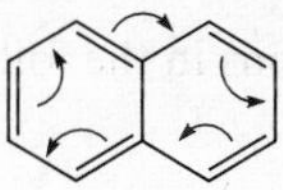

    a 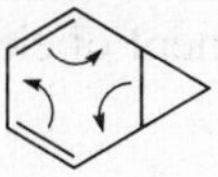   b 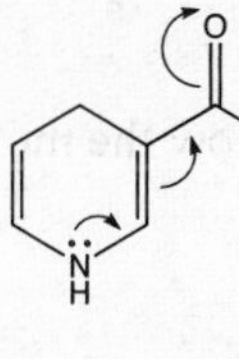   c 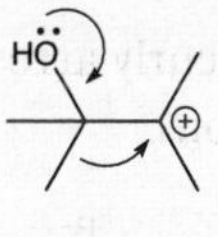   d

3. Use curly arrows to show how the following pairs of resonance structures relate to one another.

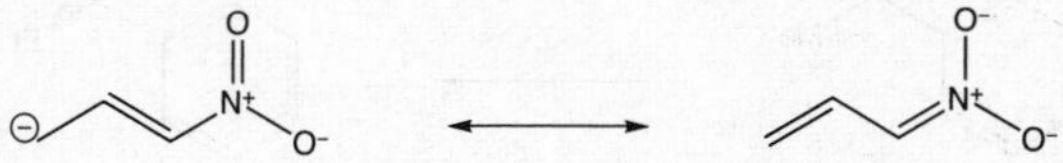

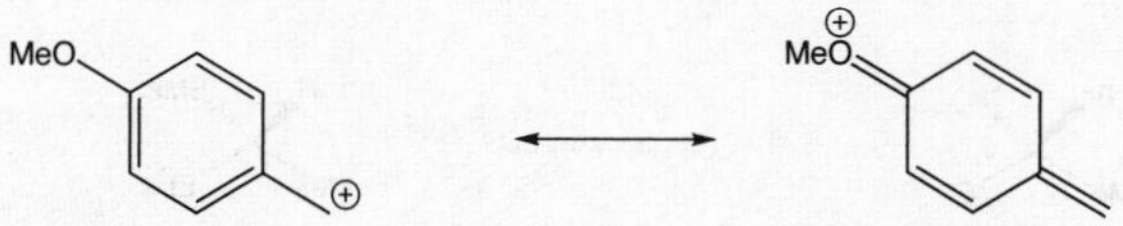

27

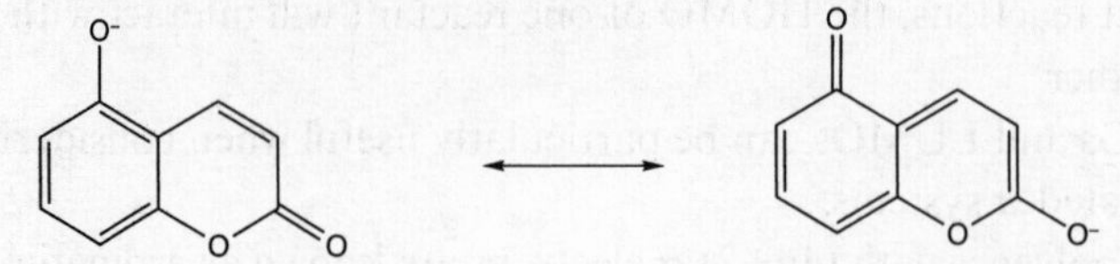

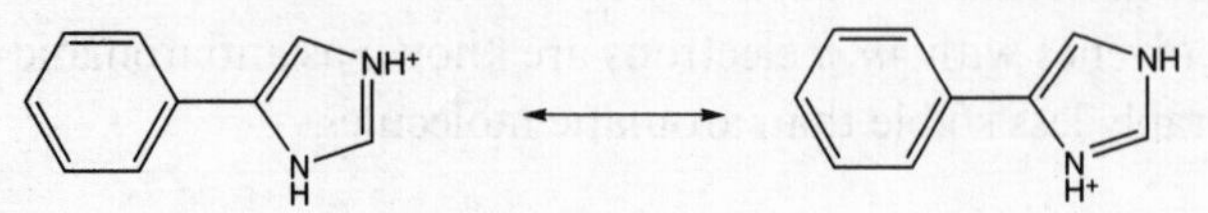

4. In each of the following structures, what is the most likely electron source?

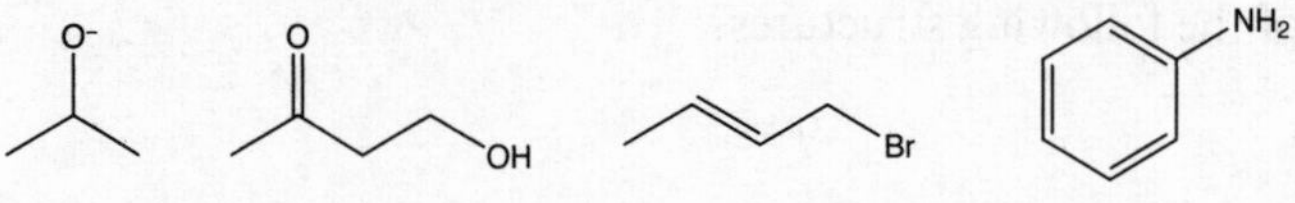

a   b   c   d

5. In each of the following structures, which atom is the most likely electron sink?

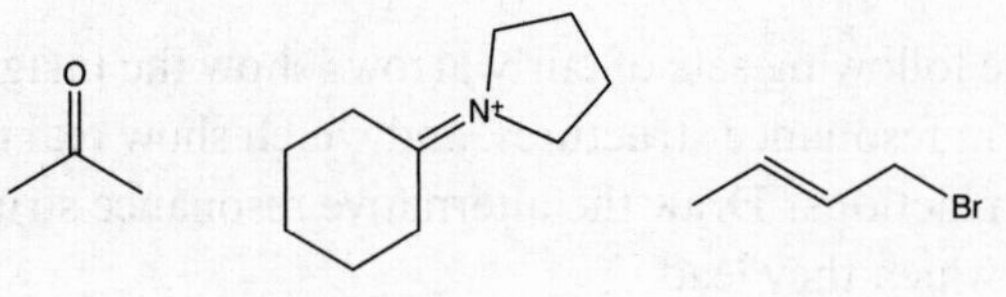

a   b   c

6. Draw curly arrows to show the movement of electrons in the following reactions.

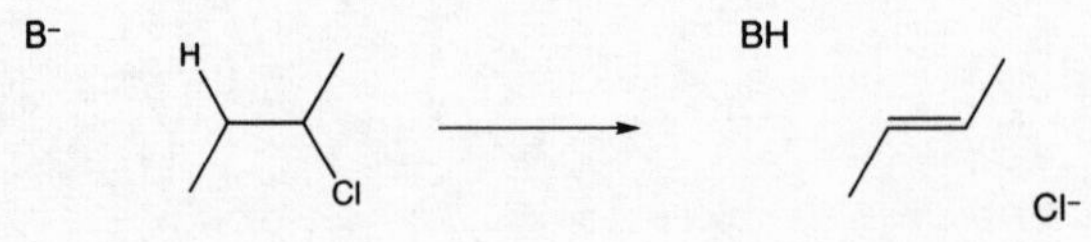

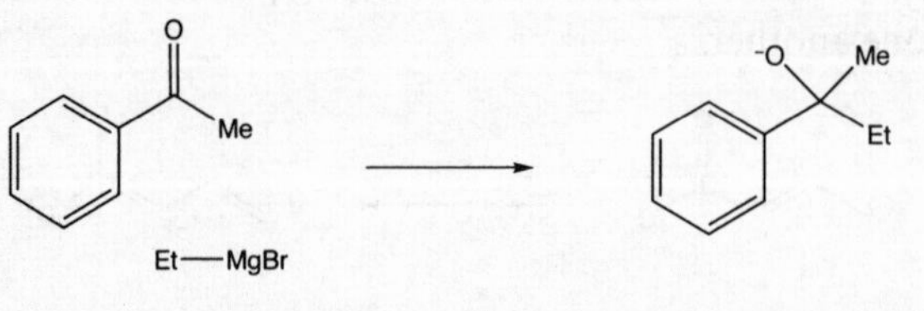

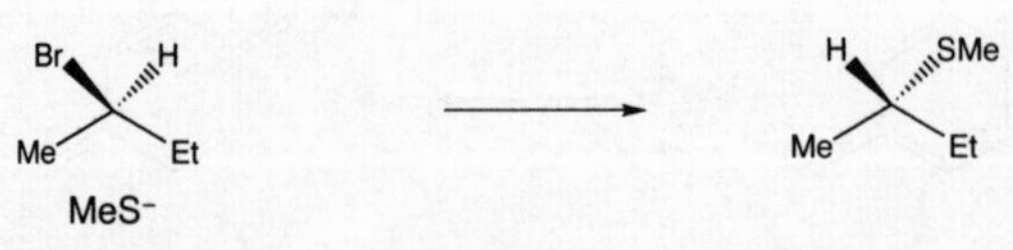

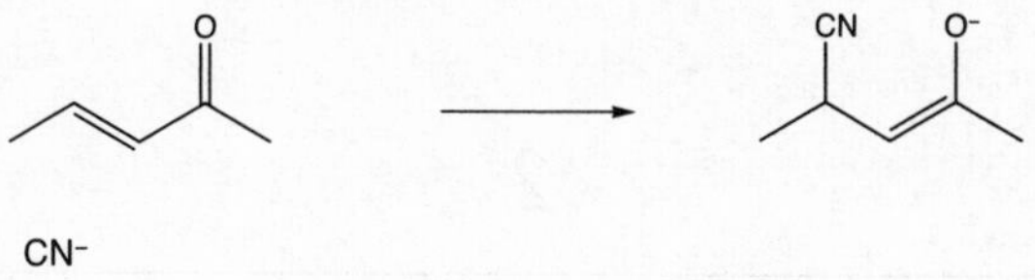

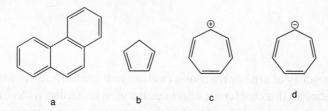

7. Draw a molecular orbital diagram for hexatriene (draw π orbitals only, and do not worry about the relative sizes of the coefficients of the orbitals). Which is the HOMO and which is the LUMO?

8. Which of the following are aromatic?

a   b   c   d

9. Cyclopropene is an extremely unstable molecule. However, it is possible to prepare one of its ions. Would you expect this to be the cation or the anion?

10. The molecule shown below has a significant dipole moment. Which end of the molecule would you expect would be positively charged, and which end would you expect to be negative?

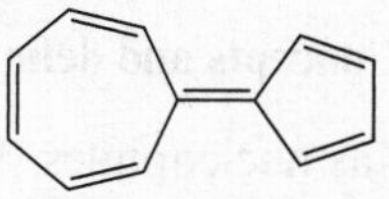

29

2

Ionic species

In the vast majority of organic reactions, electron-rich species interact with electron-deficient species. In this chapter we will concentrate on molecules with either a whole or a partial electric charge. We will look at the sorts of charged species we can expect to find in reactions, and the factors that govern their reactivity.

One of the most common ways for charge to be introduced into a molecule is by reactions with acids or bases, so a discussion of these will form a large part of the chapter. We will see what makes some acids and bases stronger than others, and we will see how they may catalyse reactions.

2.1 Concepts and definitions

Electron-rich species are known as nucleophiles. The name suggests that they love nuclei, and since nuclei are positively charged, this makes sense. An electron-rich species will certainly be attracted to positive charges. Nucleophiles may have either a negative charge or an available electron pair. We saw in Section 1.3 that curly arrows begin at electron sources; organic reactions are systems of electron flow, and the nucleophile is the source of the electrons.

The name given to an electron-deficient species is an electrophile. Again, the name is a clue to its nature; electrophiles love electrons, as they are deficient in them. Electrophiles are often positively charged, but need not be. One of the commonest electrophiles is the carbonyl group, in which the carbon is electron-deficient because the electronegative oxygen pulls the electrons in the bond towards itself.

A great many chemical reactions are simply the interaction of a nucleophile with an electrophile. This is a straightforward concept, but a tremendously important one. Often, a knowledge of which is the most nucleophilic species in a reaction mixture and which is the most electrophilic will enable us to predict much about the reaction.

Many species that are not very nucleophilic or electrophilic in themselves can be made so by treatment with acid or base. For example, a carbonyl group may become

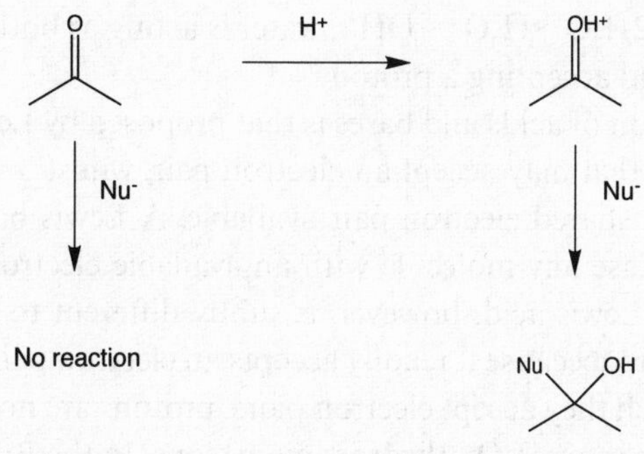

Figure 2.1

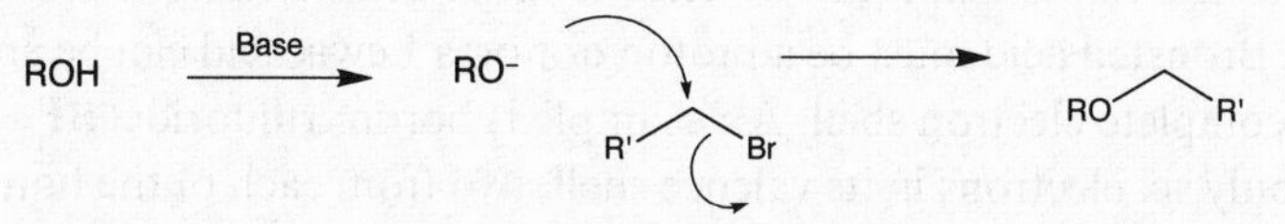

Figure 2.2

protonated under acidic conditions (Figure 2.1). The positive charge thus introduced to the group makes it more electron deficient, and hence more electrophilic. It will therefore be much more reactive towards nucleophiles than its uncharged equivalent.

Similarly, removal of a proton from functional groups will make them more nucleophilic. Alcohols are not very good nucleophiles in themselves, but treatment with base to form alkoxide anions makes them much better nucleophiles. This is the basis of the Williamson ether synthesis; an alcohol is treated with base and the resulting alkoxide reacts with an alkyl halide to form the ether (Figure 2.2). Normally, the alcohol will not react with the alkyl halide in the absence of base.

2.2 Acids and bases

2.2.1 Definitions of acids and bases

There are two definitions of acids and bases in common use, and they are used for slightly different purposes. The more frequently used of these is the Brønsted definition. This states simply that an acid is a species that can give up a proton (a proton donor), and a base is one that can receive a proton (a proton acceptor).

It is easy to see how this definition applies to simple acids and bases. If we consider the reaction between hydrochloric acid and the hydroxide ion, we can see that hydrochloric acid (the acid) gives up a proton, and the hydroxide ion (the base) accepts it. This is an example of a strong acid and a strong base; acids and bases need not be this strong, however, to comply with the Brønsted definition. In the self-

31

ionization of water, $2H_2O \rightarrow H_3O^+ + OH^-$, water is acting as both an acid and a base by both donating and accepting a proton.

The other definition of acids and bases is that proposed by Lewis. A Lewis acid is defined as a species that may accept an electron pair, whilst a Lewis base is something that has an unshared electron pair available. A Lewis base is the same as a Brønsted base, because any molecule with an available electron pair can use it to accept a proton. A Lewis acid, however, is subtly different to a Brønsted acid. A proton is a Lewis acid, because it readily accepts an electron pair. However, because of the ease with which they accept electron pairs, protons are not found free in solution; in water they exist as H_3O^+ (hydroxonium) ions. In the Brønsted definition, it is not the proton itself which is the acid, but the species that donates it.

Note that the Lewis definition of acids is more general than the Brønsted one. Whereas a Brønsted acid must be a proton donor, a Lewis acid can be any molecule with an incomplete electron shell. An example is boron trifluoride, BF_3. The boron atom has only six electrons in its valence shell, two from each of the boron–fluorine bonds. For a complete valence shell, it would need eight, and so it may accept two more electrons. Boron trifluoride in pure form is a gas, and so to make it more convenient to handle it is usually supplied as a complex with diethyl ether, in which a lone pair on the oxygen of ether is donated to the boron.

Although there are two different definitions of acids in common use, this seldom leads to confusion. Because the Brønsted definition is used more often, the term 'acid' is almost always taken to mean Brønsted acid. When reference is made to a Lewis acid, this is usually stated explicitly. Moreover, it is normally clear from the context which type of acid is being discussed.

Not all acids are equally strong, so it is useful to have a measure of the strength of an acid. By strength, we mean the ease with which the acid gives up its proton. An aqueous solution of a strong acid will have a lower pH than that of a weak acid (i.e. a higher H_3O^+ concentration). This is not to be confused with the effect of the concentration of an acid on the pH of its solution; whilst it is correct to say that hydrochloric acid is stronger than acetic acid, it is not appropriate to talk of a 1 M solution of hydrochloric acid being stronger than a 0.1 M solution of hydrochloric acid, even though the 1 M solution will have a lower pH than the 0.1 M solution.

The way in which the strength of a Brønsted acid is usually measured is based on the position of equilibrium for its reaction with water. (It is less meaningful to measure the strength of Lewis acids, as their strength will be different for different bases.) A general reaction for a Brønsted acid with water may be written:

$$HA + H_2O \rightarrow A^- + H_3O^+.$$

The equilibrium constant, K_a, is given by the formula

$$K_a = \frac{[A^-][H_3O^+]}{[HA]}$$

(the concentration of water does not come into the formula because K_a is defined in aqueous solution, so the water is present in large excess and its concentration is as near as makes no difference constant).

We could use K_a to define the strength of an acid, but in practice it is more convenient to use a measure called pK_a (analogous to pH), which is related to K_a by the formula

$$pK_a = -\log_{10} K_a.$$

This measure is commonly used. Stronger acids have lower values of pK_a than weaker acids (a lower value of pK_a means a higher value of K_a, and hence a higher equilibrium dissociation of the acid).

We have described an acid, HA, as being something that dissociates to H_3O^+ and A^-. In the reverse of this reaction, A^- gains a proton, and so it acts as a base. It is known as the conjugate base of the acid. For example, if the acid is HCl, then the conjugate base is Cl^-. Similarly, when a base becomes protonated, the resulting molecule is called the conjugate acid. The conjugate acid of ammonia (NH_3), for example, is the ammonium ion (NH_4^+).

We would normally describe HCl as an acid and ammonia as a base. However, there is no reason in principle why we should not describe Cl^- as a base, whose conjugate acid is HCl, and NH_4^+ as an acid, whose conjugate base is ammonia. The reason why this is unusual in practice is simply because Cl^- is a weak base and NH_4^+ is a weak acid. It is worth remembering, however, that acids and bases are not entirely separate entities, they are merely different forms of the same thing. In other words, an acid may become a base by losing its acidic proton, and a base may become an acid by gaining a proton. Or, to put it another way, an acid is a protonated base and a base is a deprotonated acid.

For this reason, the strengths of bases may also be described with the pK_a scale, simply by measuring the strength of the conjugate acid. A high pK_a value signifies a weak conjugate acid and so a strong base. Note that the pK_a value refers to the conjugate acid of the base, not the base itself. An alternative measure of base strength is pK_b, which is based on the reaction between the base and water to give hydroxide. This is now regarded as rather old-fashioned. The pK_a scale is preferable, because it is simpler to have one measure for both acids and bases.

The pK_a measure is defined for aqueous solution. This causes problems for acids significantly stronger than H_3O^+ or bases significantly stronger than OH^-. For example, if one tried to measure the pK_a values for the series of hydrohalic acids HCl, HBr, and HI, then no differences would be found in an aqueous solution, because they would all be almost completely dissociated. However, it may be shown from measuring their dissociation constants in other solvents that their acidities increase in the order HCl<HBr<HI. Any measure of pK_a values obtained for acids such as these, which are more acidic than H_3O^+, must be taken to be approximate. For the reason stated above, it is not possible to measure their pK_a values in aqueous

solution. However, pK_a values refer only to aqueous solutions, so any measures of dissociation constants taken in other solvents, no matter how carefully obtained, cannot give a true value of the pK_a.

The situation is similar for strong bases. An attempt to measure the base strength of the acetone enolate in water will result in almost complete conversion to acetone and hydroxide. It may be possible to obtain a precise measure of the dissociation constant in another solvent, but this will not give the pK_a. There are many pK_a values quoted for the conjugate acids of such strong bases, but their accuracy cannot be relied upon. They may often be a useful guide, however.

Not only are pK_a values impossible to measure for very strong acids and bases, but in aqueous solution they are also irrelevant. Any acid stronger than H_3O^+ will, when dissolved in water, simply give a solution of H_3O^+ ions. It is immaterial, there-fore, how strong the acid is; the effective strength of the acid will be simply that of H_3O^+. Similarly, strong bases dissolved in water will have no more effect than a solu-tion of hydroxide ions. This reduction in the power of strong acids and bases is called the levelling effect of water.

2.2.2 *Factors affecting acid strength*

The strength of an acid is a function of the stability of its conjugate base; the more stable the conjugate base, the stronger the acid. A number of factors are at work here and the situation can often be more complicated than it first appears. In addition to such simple effects as electronegativity, less predictable factors can also have an important role, such as the relative enthalpies and entropies of solvation of the acids and their conjugate bases. In practice, this subject is not as difficult to master as it might be, because there are a number of rules which can be used to predict the rel-ative strengths of different acids with a fair amount of success. A little care is needed here, however. There are principles that are often used, which, although they may frequently give the right answers, do so for entirely the wrong reasons. Perhaps the easiest approach if you are new to this subject is to concentrate on the trends that are observed, without worrying too much about the details of why they are so.

2.2.2.1 Nature of the atom donating the proton

One important factor affecting acid strength is the electronegativity of the element to which the acidic proton is bonded. Even this seemingly straightforward effect is more complex than we would like it to be, as electronegativity affects acid strength differently according to whether we are considering trends in going from left to right or top to bottom of the periodic table. In going across the periodic table, we find that electronegative elements can more easily bear a negative charge than less electro-negative ones, and so their attached protons will be more acidic. As an illustration of this, we can consider the first-row hydrides from CH_4 to HF. Their pK_a values are

Table 2.1. *pK$_a$s of first row hydrides*

Molecule	CH$_4$	NH$_3$	H$_2$O	HF
pK$_a$	48	36	15.7	3.2

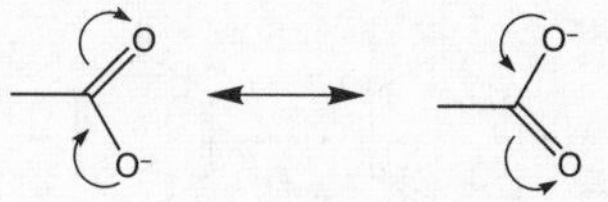

Figure 2.3

given in Table 2.1, and show a clear trend of increasing acidity in going from left to right across the periodic table (i.e. in order of increasing electronegativity). It is so difficult to remove a proton from methane that we would not normally think of this as an acid at all. In contrast, HF is sufficiently acidic for it to be known as hydrofluoric acid.

In going down the periodic table, the situation is reversed. The acidities of the hydrogen halides increase in the order HF<HCl<HBr<HI. Here, the electronegativity of the element is less important than its size. As the anions become larger, moving from F$^-$ through to I$^-$, the charge is more diffuse; this makes it more stable.

Although other factors are at work here, we nonetheless have two useful rules that are generally applicable. Acidity increases both to the right and to the bottom of the periodic table. Using these rules, we could successfully predict that it is easier to remove a proton from acetic acid (MeCOOH) than acetamide (MeCONH$_2$), and that H$_2$S is more acidic than water.

2.2.2.2 Delocalization

There are, of course, many factors other than the element to which the proton is attached that affect acidity, as we can see from the large variation in the strength of different acids in which the acidic proton is attached to the same element. Anions may be stabilized by delocalization; this can have a significant effect on the strength of their conjugate acids. For example, while ethanol has a pK$_a$ of 18, acetic acid, in which the acidic proton is also bound to oxygen, has a pK$_a$ of 4.8. This increase in acidity is due to the delocalization possible in the acetate anion, which spreads the charge over the two oxygens, thereby helping to stabilize it (Figure 2.3).

Delocalization is a very important effect, and we can reliably predict that an anion in which the charge is delocalized will have a stronger conjugate acid than a similar anion with a localized charge. A dramatic illustration of the effect of delocalization is that in the gas phase, toluene is a stronger acid than water! This is markedly different to the situation in aqueous solution, in which toluene is a very weak acid

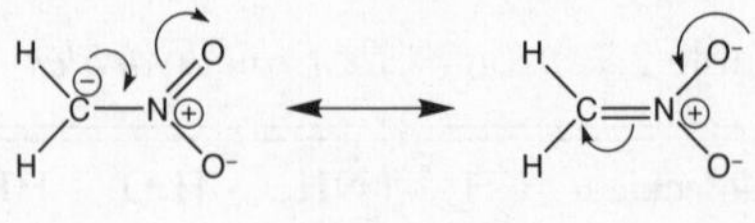

Figure 2.4

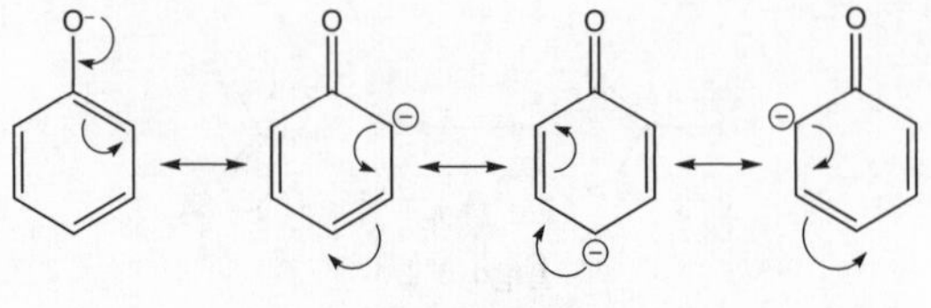

Figure 2.5

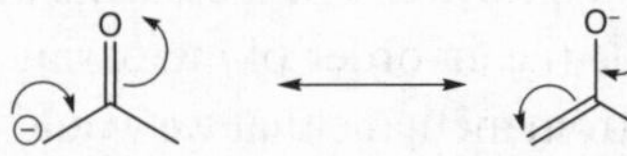

Figure 2.6

(pK_a=41). In water, the hydroxide anion is considerably stabilized by solvation, which is not possible to the same extent for the hydrophobic benzyl anion. When solvation effects are absent, toluene is revealed as the stronger acid, with the effect of delocalization more than compensating for the negative charge's residing on carbon, rather than on the more electronegative oxygen.

There are many examples of the importance of delocalization. Nitromethane is a moderately strong acid. This can be explained by the delocalization of the negative charge onto the oxygen atoms (Figure 2.4). Phenols are stronger acids than simple aliphatic alcohols, because of the delocalization of the negative charge into the aromatic ring (Figure 2.5). Protons attached to carbon atoms become acidic when next to a carbonyl group; for example, the pK_a of acetone is 20. Although this is by no means a strong acid, it is possible to remove a proton from acetone with a strong base such as sodium hydride, which is not possible for a simple hydrocarbon. The reason for this acidity is the delocalization of the resulting negative charge onto the oxygen atom (Figure 2.6). The subject of carbon acids will be dealt with more fully in Section 4.1.

2.2.2.3 Hybridization

The state of hybridization of the atom to which the hydrogen is attached plays a part in acidity. The general rule is that the more an orbital resembles a pure s orbital, the more acidic is the proton. An illustration of this is the acidity of acetylene, from which a proton may be removed by strong bases. Although it is usually not possible to remove protons from hydrocarbons in this way, the sp hybridization of the carbon

Table 2.2. *pK$_a$s of chlorinated acetic acids*

Molecule	CH$_3$COOH	CH$_2$ClCOOH	CHCl$_2$COOH	CCl$_3$COOH
pK$_a$	4.8	2.9	1.3	0.7

Table 2.3. *pK$_a$s of water and some alcohols*

Molecule	H$_2$O	MeOH	EtOH	^{i}PrOH	^{t}BuOH
pK$_a$	15.7	16	18	18	19

orbitals results in greater acidity. The reason for this is that s orbitals are held more tightly to the nucleus than p orbitals. A greater s character to the orbital therefore results in greater stabilization of a negative charge by interaction with the nucleus.

2.2.2.4 Field effects

Acidities are also affected by field effects. Acetic acid becomes progressively stronger as the hydrogens in its methyl group are substituted by chlorine (Table 2.2). As chlorine is electron-withdrawing, it helps to diminish the negative charge on the carboxylate group, spreading it over the rest of the molecule. Other electron-withdrawing groups have similar effects.

If we look at the series of increasing alkyl substitution in going from water to t-butanol (the pK$_a$s of which are shown in Table 2.3) it is tempting to think of alkyl groups as being electron-donating, and hence decreasing the acidity of adjacent groups. This is quite convincing at first, but the argument breaks down when the gas phase acidities are considered. In the gas phase, the order is reversed, t-butanol being the strongest acid and water the weakest.

Although alkyl groups are normally considered to be electron-donating, the effect is small. As we can see from the gas phase acidities, they help to stabilize the negative charge. This is because they contribute more bulk to the molecule, and so help to spread the negative charge, although they do not do this nearly as efficiently as true electron-withdrawing groups, or groups that can delocalize charge through a π system.

What is more important in aqueous solution is that the increasing hydrophobicity resulting from alkyl substitution makes solvation of the anion less efficient. Since the anions of the more highly substituted alcohols are less well solvated, they are not stabilized by solvation to the same extent as hydroxide, so the equilibrium concentration of the anion is lower.

This example should serve as a useful reminder that a simple theory that seems to account for a set of experimental observations is not necessarily correct.

37

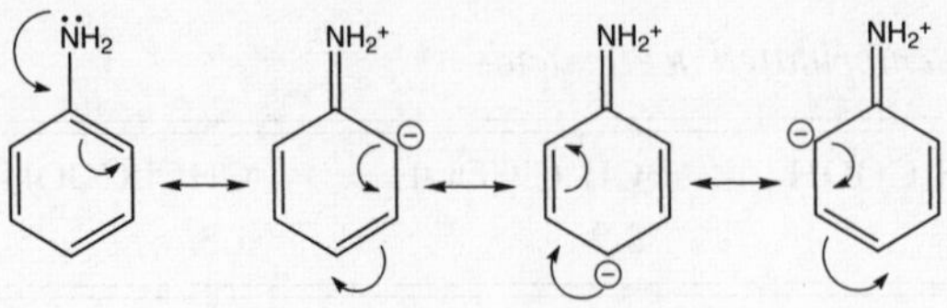

Figure 2.7

2.2.2.5 Special considerations for nitrogenous bases

Many of the factors that affect acid strength apply equally well to predicting the strength of bases, as any base which is the conjugate of a weak acid will be strong and vice versa. There are some other points, specific to nitrogenous bases (e.g. amines), which are also worth mentioning in this context.

We would expect that increasing alkyl substitution of amines would increase their basicity, as the inductive effect[1] of the extra alkyl groups should stabilize the positive charge present in the protonated form. However, in aqueous solution, we might also expect the more highly substituted amines to be less well solvated because of their increased hydrophobicity, which would mean that the cations would be less stabilized by solvation. These two effects act in opposite directions, and it is not obvious which will be dominant.

What is found experimentally is that as ammonia is substituted with methyl groups, the base strength increases up to dimethylamine, but that trimethylamine is weaker than expected, being intermediate in strength between ammonia and methylamine. In other words, the base strength increases in the order $NH_3 < NMe_3 < NH_2Me < NHMe_2$. Clearly, both the inductive effect of the methyl groups and the solvation of the cation play important parts. In the gas phase, the order is different, trimethylamine being the strongest base, as we would expect.

Aromatic amines (e.g. aniline) are a special case. They are much weaker than ammonia or aliphatic amines; the pK_a of the conjugate acid of aniline is 4.7, compared with 9.2 for the ammonium ion. This is sometimes explained in older textbooks as being due to the resonance stabilization of the nitrogen lone pair in aniline, which is not possible in the protonated form. At best, this is only part of the story, and at worst, it is downright misleading.

Here, as for many other reactions, measurements in the gas phase reveal the important nature of solvent effects. Aniline is actually a stronger base than ammonia in the gas phase. Although it is possible to write a neat series of curly arrows to show delocalization of the lone pair around the ring (Figure 2.7), but not to show stabilization of the positive charge, it does not mean that the ring cannot

---

[1] An inductive effect is an electronic effect (either electron-donating or electron-withdrawing) acting through σ bonds. Methyl groups have an electron-donating inductive effect. Inductive effects should not be confused with mesomeric effects, which act through conjugated π systems.

Table 2.4. *Classification of some acids and bases as hard, soft, or borderline.*

	Hard	Borderline	Soft
Acids	H^+, Li^+, Mg^{2+}, BF_3	Zn^{2+}, Cu^{2+}, Fe^{2+}, R_3C^+	Cu^+, Ag^+, Hg^{2+}, I^+
Bases	OH^-, F^-, H_2O, NH_3	Br^-, pyridine, N_3^-	RS^-, I^-, R_3P, olefins

stabilize this charge. An aromatic ring is highly polarizable, and can spread the positive charge over the molecule, even though it is not possible to draw this with conventional resonance structures. The reason for the low basicity of aniline in aqueous solution is that the hydrophobic nature of the ring makes it much less well solvated than ammonia, and so stabilization of its cation by solvation is much less efficient than for the ammonium ion.

The explanations presented in this section do not provide the means to predict the precise strength of any acid. However, they do show some useful trends, and perhaps more importantly, warn against the dangers of relying on simple rationalizations to explain phenomena that are affected by a number of interrelated factors. It should not come as any surprise to learn that the relative strengths of acids in water may be different in other solvents, and can even change with temperature.

2.2.3 Hard and soft acids and bases

Another property of acids and bases, besides their strength, is how hard or soft they are. This property can be important in considering the character of a base or of a Lewis acid. (It is not a useful concept to apply to Brønsted acids, as the hardness of a proton is the same no matter what its conjugate base.) It becomes more important when considering nucleophiles and electrophiles, as we shall see in Section 2.3.2.

Hardness and softness are not as precisely defined as the strength of acids and bases, and there is no convenient measurement in common use analogous to pK_a. Instead, it is useful to group acids and bases merely into three groups: hard, soft and borderline. A selection of acids and bases are classified into these groups in Table 2.4.

So what do we mean by hardness and softness in this context? In general, hard acids and bases are those with a tendency towards ionic bonding, and soft ones are more inclined to form covalent bonds. This leads to the rule that interactions of soft acids with soft bases (soft–soft interactions) and hard–hard interactions are more effective than hard–soft interactions. This applies both to kinetic and thermodynamic processes (see Chapter 3).

As an example of the effect of hardness and softness, let us consider the well-known reaction that takes place when hydrogen sulphide gas (H_2S) is passed through an aqueous solution of a mercury (II) salt, such as mercuric chloride, giving a black precipitate of mercuric sulphide:

$$HgCl_2 + H_2S \rightarrow HgS + 2HCl.$$

If we were to think of this reaction purely in terms of pK_a, we might find it surprising, as a weak acid (H_2S) has protonated a weak base (Cl^-), resulting in a much stronger acid (HCl). We can explain this, however, by considering the hard/soft properties of the species involved. The mercuric ion (Hg^{2+}) is a soft acid, and the sulphide ion (S^{2-}) is a soft base. They therefore have a much stronger affinity with each other than they do with hard species, such as the proton or the chloride ion.[2]

How are we to predict, then, whether a given species is hard or soft? There are a number of properties which are characteristic of the different types of acids and bases, which can be summarized as follows.

- Hard bases have high electronegativity and low polarizability. They are usually negatively charged, and are not easily oxidized.
- Soft bases have low electronegativity and high polarizability. They are easily oxidized, and do not necessarily have a negative charge.
- Hard acids are small and normally have a positive charge. They have low polarizability, and do not contain unshared electron pairs in their valence shells. Note that this description fits the proton, and this is indeed classified as a hard acid.
- Soft acids are large and do not necessarily have a positive charge. They are highly polarizable, and often have lone pairs in their valence shells.

It is instructive to consider hardness and softness in terms of frontier orbitals (i.e. HOMOs and LUMOs). When the idea of hardness and softness was first described, it was purely an empirical observation. The aim was to find a common way of describing a number of seemingly unrelated phenomena. In this respect, the theory was successful, as it allowed predictions to be made about what would react with what, but it was unable to explain these phenomena. The development of frontier orbital theory has allowed hardness and softness to be described more rationally. As a complement to the above definitions, we may define the four species in terms of their frontier orbitals.

- Hard bases have low-energy HOMOs.
- Soft bases have high-energy HOMOs.
- Hard acids have high-energy LUMOs.
- Soft acids have low-energy LUMOs.

---

[2] Strictly speaking, we should be more concerned with water than with the chloride ion, because the Brønsted acid formed in aqueous solution is H_3O^+, rather than HCl. However, this makes little difference to the argument, as water is also a hard base.

We saw in Section 1.4 that the interaction of a HOMO of one species with the LUMO of another is important in the formation of a covalent bond. If two orbitals are to interact, the interaction will be stronger when the two orbitals are close in energy. Thus a covalent bond will be most easily formed when a species with a high-energy HOMO interacts with one with a low-energy LUMO.

We are now in a position to see the rational explanation for the behaviour of hard and soft acids and bases. We have just seen that soft bases have high-energy HOMOs and soft acids have low-energy LUMOs. These are precisely the conditions necessary for the easy formation of a covalent bond, and this explains why soft acids and soft bases have an affinity with each other. We can see that the frontier orbital interaction of a soft base (high-energy HOMO) with a hard acid (high-energy LUMO) will be less effective, as will the interaction of a hard base (low-energy HOMO) with a soft acid (low-energy LUMO).

What about the interaction of a hard base with a hard acid? By the above argument, this should be the worst of all, because we now have a low-energy HOMO and a high-energy LUMO. We stated earlier that hard acids and bases have a tendency to form ionic bonds, and frontier orbital interactions are not important here. What is important is the electrostatic attraction between the two species. This depends on the charge on the reactants and their size, and hence their charge density. Hard acids and bases tend to be both highly charged and small, both of which factors contribute effectively to ionic interactions. Hard acids and hard bases can, therefore, interact effectively with each other, but the type of interaction is different to that of a soft acid with a soft base.

In summary, then, bases and Lewis acids may be divided into two classes: hard and soft. The best interactions are obtained between acids and bases of the same class. Hard acids and bases are those with high charge density and low polarizability, and they tend to participate in strong ionic interactions. Soft acids and bases are more highly polarizable, and interact with a more covalent character. This soft–soft interaction is facilitated by favourable frontier orbital interactions, in other words a high-energy HOMO and a low-energy LUMO.

2.2.4 Catalysis by acids and bases

Many everyday chemical reactions do not proceed at a satisfactory rate, if at all, in the absence of a catalyst. Acids and bases are by far the most common form of catalysts used in organic chemistry. In acid catalysis, the first step of the reaction mechanism will normally be protonation of the organic substrate; in base catalysis, deprotonation is the first step.

These forms of catalysis are so common that whenever one is faced with trying to guess at the mechanism of an unfamiliar reaction, it is always worth asking the question 'Is this reaction catalysed by acid or base?' (see Chapter 8). If the answer to this question is yes (the answer will usually be apparent from an examination of the

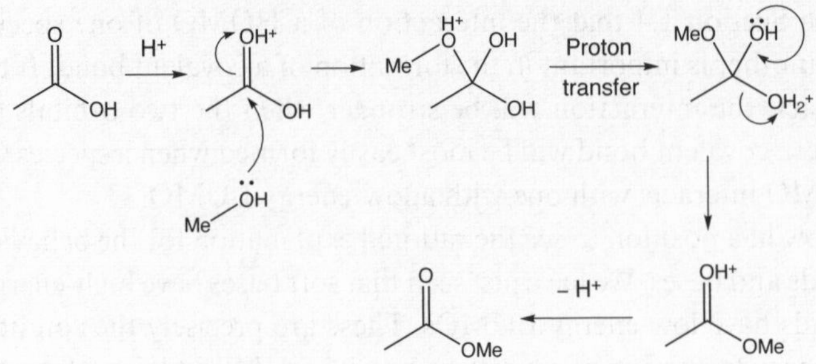

Figure 2.8

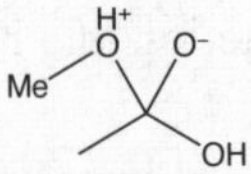

Figure 2.9

reaction conditions: is it necessary to add an acid or base to the reaction?) then this provides an excellent starting point for writing down a possible mechanism. If acid is present, the first step is likely to be a protonation, and if base is present, then the first step will almost certainly be a deprotonation. The subject of making educated guesses at unknown reaction mechanisms will be treated in depth in Chapter 8.

Let us consider an example of how acid catalysis can work in practice. If a simple carboxylic acid, such as acetic acid, is dissolved in methanol, there will be no reaction. If, however, a small amount of mineral acid such as HCl is added, acetic acid will react with methanol to produce methyl acetate:

$$MeCOOH + MeOH \rightarrow No \ reaction$$
$$MeCOOH + MeOH + HCl \ (trace) \rightarrow MeCOOMe + H_2O + HCl.$$

The mineral acid has the effect of protonating the carbonyl oxygen of the acetic acid. This introduces extra electron deficiency into the molecule, making it more susceptible to nucleophilic attack by methanol (Figure 2.8). After appropriate proton transfer steps, the resulting tetrahedral intermediate can lose water to give the ester (carbonyl addition–elimination reactions such as this will be discussed further in Section 6.3.5).

In the absence of the acid catalyst, the carbonyl group will remain neutral, and so will be less easily attacked by methanol. Furthermore, if methanol were to attack this neutral molecule, there would be charge separation in the resulting tetrahedral intermediate (Figure 2.9), which is energetically unfavourable.

We may distinguish between two types of acid catalysis. These are specific acid catalysis and general acid catalysis. Specific acid catalysis is so called because it is only the concentration of one acid which is important to the reaction, namely the protonated solvent. As measurements are often carried out in water, the specific acid

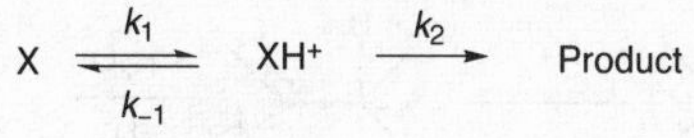

Figure 2.10

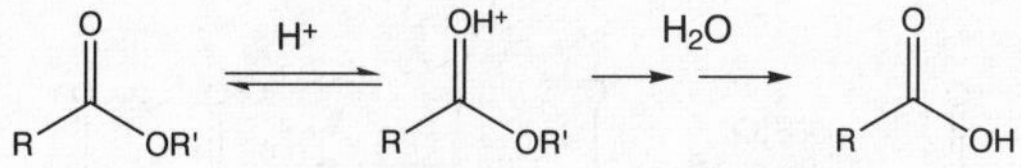

Figure 2.11

is usually H_3O^+. The rate of reaction is proportional to the concentration of the specific acid. The concentration and nature of other acidic species make no difference to the rate of the reaction, other than through the effect they have on the pH.

In general acid catalysis, the rate of reaction is increased by any acid, even when this does not have an effect on the pH. For example, the addition of phenol to water will have a negligible effect on the pH, because it is too weak an acid to be dissociated to any significant extent. Nonetheless, phenol is still an acid, and its acidity will help to increase the rate of a reaction which is general-acid-catalysed. Stronger acids will also increase the rate of general-acid-catalysed reactions, but this will often be difficult to distinguish in practice from specific acid catalysis, as stronger acids will also lower the pH.

If we know whether a reaction is specific- or general-acid-catalysed, this tells us something about the mechanism. We may write an acid-catalysed reaction as proceeding as shown in Figure 2.10. If k_1 and k_{-1} are both fast, establishing a rapid equilibrium between the starting material and its protonated form, and k_2 is rate-determining, then we have specific acid catalysis. The speed of protonation is unimportant here; what matters is the equilibrium concentration of the protonated substrate. This depends only on the concentration of the protonated solvent, which is the strongest acid present.

If k_2 is fast and k_1 is rate-determining, then the overall rate of reaction will depend on how fast the substrate can be protonated. Any acid can protonate the substrate, so the rate will be increased by an increase in the concentration of any acid; this is general acid catalysis.

An example of a specific-acid-catalysed reaction is the hydrolysis of a carboxylic ester (Figure 2.11). We see that an equilibrium is set up between the neutral ester and its protonated form, and the protonated form then reacts in a rate-determining step.

General acid catalysis is found in the hydration of double bonds to give alcohols (Figure 2.12). Initial protonation of the double bond is slow, and the resulting carbocation reacts rapidly with water to form the alcohol.

We may also define general and specific base catalysis in a similar way. The Claisen ester condensation (Figure 2.13) is specific-base-catalysed. The reaction is often carried out in ethanol, so the specific base is the ethoxide ion.

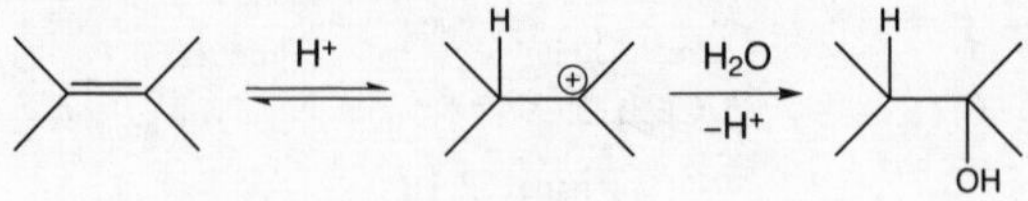

Figure 2.12

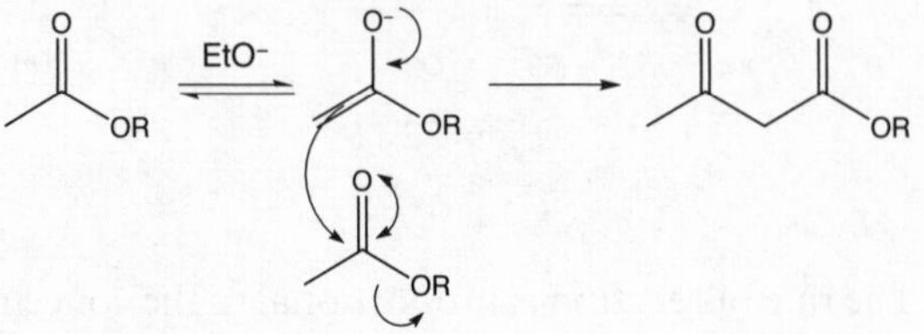

Figure 2.13

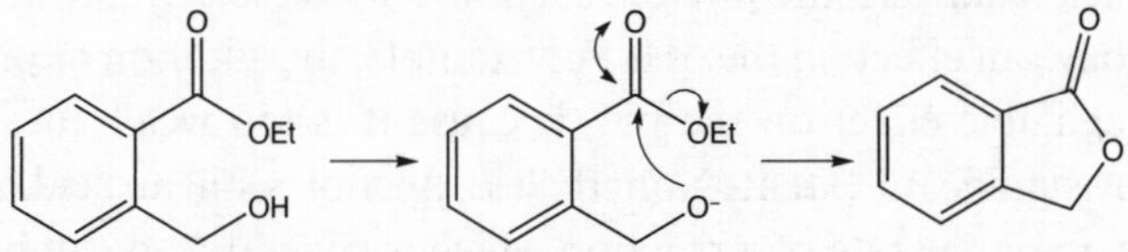

Figure 2.14

An example of general base catalysis is the intramolecular transesterification of ethyl 2-(hydroxymethyl)benzoate shown in Figure 2.14. Once the proton is removed from the hydroxyl group, attack on the adjacent carbonyl group follows rapidly, leading to the cyclization.

2.3 Nucleophiles

2.3.1 Relationship between nucleophilicity and basicity

A nucleophile is any species that has an unshared electron pair and can make use of this to react with an electron-deficient species, or electrophile. This is similar to the definition of a base, as bases also have unshared electron pairs. Nucleophiles have much in common with bases, and the preceding discussion of bases is also largely applicable to nucleophiles. Any species which is capable of acting as a base can also act as a nucleophile and vice versa.

The main distinction between basicity and nucleophilicity is that basicity is a thermodynamic property, while nucleophilicity is kinetic (see Chapter 3 for a more detailed explanation of these terms). In other words, basicity affects the position of equilibria and nucleophilicity affects the rate of reactions.

As a rule of thumb, strong bases are also strong nucleophiles, although this is by no means always so. For example, if we look at anions of the first row of the periodic table from CH_3^- to F^-, we find that the order of nucleophilicity is the same as basicity, i.e. $CH_3^- > NH_2^- > OH^- > F^-$.

Nucleophilicity will also tend to reflect base strength in a series of nucleophiles based on the same element. For example, the oxygen anions SO_4^{2-}, $MeCO_2^-$, PhO^-, and OH^- follow this trend, basicity and nucleophilicity both increasing in going from sulphate to hydroxide. However, hydroxide is also a stronger nucleophile than t-butoxide (Me_3CO^-), despite its lower base strength. This is for steric reasons; the greater bulk of the t-butoxide anion inhibits its reaction with electrophiles.

Nucleophilicity is much more sensitive to steric effects than base strength; steric hindrance is seldom an issue in proton transfer processes, because the proton is so small. The extent to which nucleophilicity is influenced by steric effects depends on the electrophile with which the nucleophile is reacting. Steric effects will be more important in reactions with more hindered electrophiles.

As we look at trends within columns of the periodic table, we find that nucleophilicity no longer parallels basicity. The nucleophilicities of the halogens usually follow the order $I^- > Br^- > Cl^- > F^-$, which is the reverse of the order of their base strengths. This can largely be attributed to the degree of solvation of the halide ions. The charge in F^- is tightly concentrated, which leads to strong solvation. The iodide ion has a much more diffuse charge, and consequently holds on to its solvent molecules much less strongly. It is therefore easier for iodide to approach an electrophile than it is for fluoride, because its solvent molecules are less in the way of the reaction. Evidence for this comes from studies of nucleophilicity in the gas phase, where solvent effects are absent and it is found that the order of nucleophilicities is reversed, matching the order of basicity.

Nucleophile strength generally increases on going down the periodic table. Thus phosphorous nucleophiles are stronger than their nitrogen analogues (despite their being weaker bases), and sulphur nucleophiles are stronger than those based on oxygen. Mercaptide anions (RS^-) are very powerful nucleophiles indeed, but only moderate bases, comparable in strength to amines.

2.3.2 Hard and soft nucleophiles

It is important to realise that although we have talked of nucleophilicity as an absolute property so far, arranging nucleophiles in order of their strength in a reaction with one electrophile may not give the correct order when we consider their reaction with a different electrophile. The above rules are useful generalizations, but we must remember that they are generalizations and there may be reactions in which nucleophilicities differ from those expected from the arguments presented here. The properties of hardness and softness apply to nucleophiles and electrophiles just as they do to acids and bases, and we find that soft nucleophiles react best with soft electrophiles, and hard nucleophiles react best with hard electrophiles.

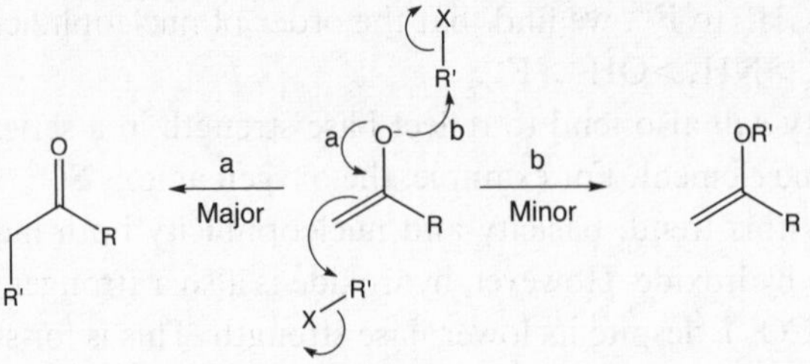

Figure 2.15

It is useful to consider the frontier orbitals of nucleophiles. By definition, a nucleophile reacts by means of its HOMO. The higher the energy of the HOMO of a nucleophile, the more effective this frontier orbital interaction will be. Soft nucleophiles have high-energy HOMOs; the situation here is exactly analogous to that described earlier for bases. When a soft nucleophile interacts with a soft electrophile, the frontier orbital interaction will be effective, and the reaction is likely to be efficient.

It is not necessary to have a high-energy HOMO to be an efficient nucleophile. Species with low-energy HOMOs can be effective nucleophiles if they have a sufficient concentration of charge; they will then be classed as hard nucleophiles. They will tend to react better with hard electrophiles. For example, amines are hard nucleophiles, and they react extremely rapidly with acid chlorides, which are hard electrophiles. This is not a perfect illustration of the hard/soft principle, as acid chlorides are in any case extremely reactive. However, the reaction of acid chlorides with thiols (soft nucleophiles) is less vigorous in comparison with their reaction with amines than we would expect from the relative reactivities of amines and thiols with a soft electrophile, such as an alkyl halide.

Returning to our consideration of the nucleophilic properties of the halogens, we should note that the halide ions become not only more nucleophilic, but also softer on going down the group. Iodide is the softest nucleophile, so we might expect it to react best with soft electrophiles. This is another way of saying that the outer electrons of the iodide ion are more polarizable, and so iodide is better able than other halide ions to take part in reactions in which frontier orbital interactions are important.

2.3.3 *Ambient nucleophiles*

Some species do not have their nucleophilicity localized on one atom, but have two sites at which they may react. These are called ambident nucleophiles. A frequently encountered example is the enolate anion, which normally reacts at the carbon atom, but can sometimes react on oxygen. A common reaction of enolates is alkylation with alkyl halides. This takes place mostly on carbon, although O-alkylation is sometimes found as a minor product (Figure 2.15).

46

This behaviour can be explained by considering whether the species are hard or soft. Alkyl halides are soft electrophiles, and so we would expect reactions with them to take place at the softer part of the nucleophile. Recall that hard nucleophiles are characterized by high charge density, and soft nucleophiles by the readiness with which their HOMOs can interact with the LUMOs of electrophiles. An enolate is normally written with its negative charge on the oxygen, representing the greater concentration of charge on this atom. We would expect the oxygen to be harder than the carbon for this reason. Also, molecular orbital calculations show that the HOMO of the enolate anion has a larger coefficient on the carbon atom than on the oxygen. The carbon, then, is the soft nucleophilic centre. This fits with the observed behaviour of the reaction with alkyl halides.

In practice, enolates are usually made as their lithium salts. The lithium cation is usually tightly associated with the enolate anion. Lithium is a hard electrophile, and so we would expect it to be associated with the oxygen atom, as indeed it is.

Another ambient nucleophile is the cyanide ion, which may react either on carbon or nitrogen. Again, the carbon atom is the softer centre, as it is more polarizable. The nitrogen is harder, as it is more electronegative and hence bears more of the negative charge. Cyanide normally reacts with alkyl halides on its carbon atom, as would be expected because alkyl halides are soft electrophiles. However, if silver cyanide is used, this promotes an S_N1 reaction (see Section 6.3.3), as the silver complexes to the halide ion, forming (or at least partially forming) a carbocation. A carbocation is a harder electrophile than an alkyl halide, and reacts at the nitrogen of the cyanide ion. Thus the reaction of silver cyanide with alkyl halides gives isonitriles.

2.3.4 Alpha effect

When a nucleophile is joined to an atom with a lone pair, its nucleophilicity is found to be much greater than expected. This is known as the alpha effect. One molecule that displays this effect is hydrazine, NH_2NH_2, which is considerably more nucleophilic than ammonia, although it is less basic. This can be explained by considering the frontier orbitals.

The two lone pairs on the adjacent nitrogens of hydrazine interact with each other, as shown by the molecular orbital diagram of Figure 2.16. This is similar to the π bond in ethylene (Section 1.1.4), but there are now four electrons instead of two. Two of these must therefore occupy the higher energy orbital, which is antibonding in nature. There is no net bonding between the two lone pairs, because both the bonding and antibonding orbitals are filled, so despite this interaction, hydrazine is still drawn with a single bond between the two nitrogens.

The effect of this interaction is to raise the energy of two of the lone pair electrons, relative to their energy in the absence of the other lone pair. In other words, the energy of the HOMO has been raised. This is what causes the increased

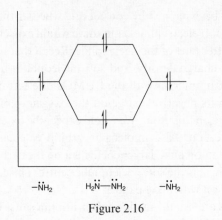

Figure 2.16

nucleophilicity of hydrazine, particularly in reactions with soft electrophiles. The basicity is not increased (it is actually decreased, because of the electron-withdrawing effect of the second nitrogen), because the proton is such a hard electrophile that it is insensitive to frontier orbital effects. However, most organic electrophiles are sufficiently soft for the effect to be important.

There are other molecules that show this effect, such as hydroxylamine (NH_2OH) and hydrogen peroxide (HOOH). The explanation is the same for all of them, although the relative change in nucleophilicities and basicities might be slightly different.

2.3.5 Solvent effects

The strength of a given nucleophile is not constant, but depends to a large extent on the solvent. We have already seen that the order of nucleophile strength amongst the halogens is reversed in the gas phase relative to their strengths in solution. This is an extreme example, but changing from one solvent to another can also have a profound effect.

Nucleophiles are at their weakest when they are strongly solvated, as it is more difficult for them to approach closely to an electrophile. The solvent molecules get in the way. Charged nucleophiles, such as halide ions, are most strongly solvated in hydroxylic solvents, particularly water, as they form hydrogen bonds. They are considerably less well solvated in aprotic solvents such as acetonitrile (MeCN), dimethylformamide (DMF, $HCONMe_2$) and dimethyl sulphoxide (DMSO, Me_2SO). The rates of some reactions in such solvents are much faster than in hydroxylic solvents if the rate of reaction is dependent on the strength of the nucleophile.

It is as well to remember the limitations of changing to solvents which are less able to solvate nucleophiles. Bromide ion would be very poorly solvated in a hydrocarbon solvent such as hexane, and it might be expected that it would be a very effective

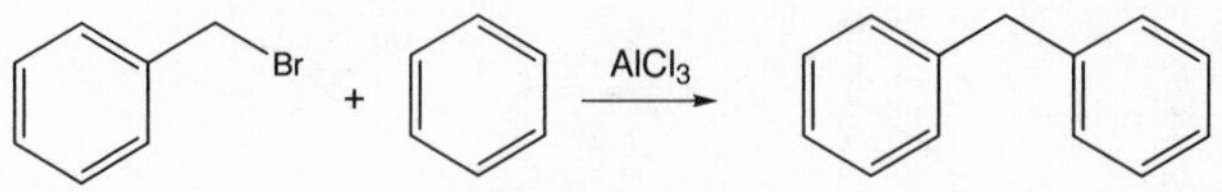

Figure 2.17

nucleophile. However, bromide salts do not dissolve in hexane, so this would not be true in practice.

2.4 Electrophiles

2.4.1 Relationship of electrophiles to nucleophiles

An electrophile is any species that will react with a nucleophile. The most important organic electrophiles are carbonyl compounds. They take part in a great many reactions, which are based on the electron-deficient character of the carbonyl carbon. They therefore react with nucleophiles. Carbon atoms with leaving groups such as halides attached also often react as electrophiles. More electrophiles than nucleophiles are carbon-based compounds. Many nucleophiles are either inorganic, such as bromide ion, or centred on a heteroatom within an organic molecule, such as ethoxide ion (a heteroatom is any atom other than carbon or hydrogen). The most common forms of carbon nucleophiles are those in which a carbon bears a negative charge, for example enolate ions or organometallic compounds. Olefins and aromatic compounds may also be carbon nucleophiles.

It is conventional (although possibly somewhat artificial) to class reactions as either nucleophilic or electrophilic. In the former the substrate is attacked by a nucleophile, and in the latter by an electrophile. An example of a nucleophilic reaction is the hydrolysis of ethyl bromide to give ethanol, in which the substrate (ethyl bromide) is attacked by a nucleophile (hydroxide ion). The bromination of ethylene is an electrophilic reaction, as the substrate (ethylene) is attacked by an electrophile (bromine).

These examples appear clear-cut, but the distinction between these two modes of reaction (nucleophilic and electrophilic) is somewhat artificial. How do we decide which of the reactants is the substrate? In these reactions there was one organic molecule and one inorganic reagent, so we classed the organic component as the substrate. Both examples, however, are simply reactions between a nucleophile and an electrophile. It is not always this easy to say which is the substrate. For example, which is the substrate in the Friedel–Crafts alkylation (see Section 6.3.6) of benzene with benzyl bromide (Figure 2.17)? Is this an electrophilic reaction of benzene, with benzyl bromide as the electrophile, or a nucleophilic reaction of benzyl bromide, with benzene as the nucleophile? Both are perfectly valid ways of looking at the reaction, and we should not get bogged down with an arbitrary classification system.

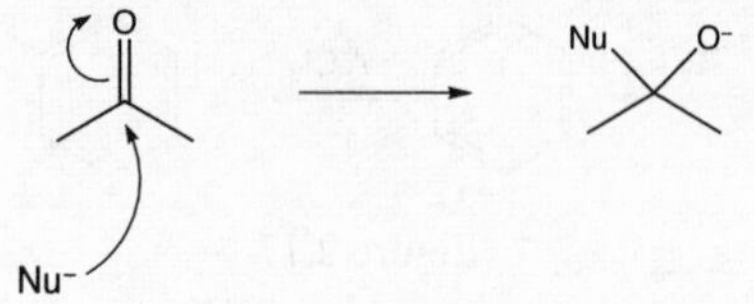

Figure 2.18

Much of what was said above about nucleophiles also applies to electrophiles. However, whereas we may usefully compare nucleophilicity with base strength, it is not worthwhile to try to correlate the reactivity of electrophiles with their strengths as Lewis acids. This is simply because there is no convenient measure of Lewis acidity in common use, analogous to the pK_a scale for Brønsted acids. Moreover, many carbon electrophiles (alkyl halides for example) do not act as Lewis acids.

In addition to being electron-deficient, an electrophile must also have a vacant orbital. Of course, all atoms and molecules have vacant orbitals, but what we really mean here is that an electrophile must have a vacant orbital that is not too high in energy, and can therefore take part in reactions. This may be an empty orbital that would normally be filled, such as in an alkyl cation, or it may be an antibonding orbital. This vacant orbital is the LUMO of the electrophile. When a carbonyl group acts as an electrophile, for example, the electron pair from the nucleophile enters the antibonding π orbital, changing the carbon–oxygen bond from a double bond to a single bond. We use curly arrows to depict this as shown in Figure 2.18.

2.4.2 Hard and soft electrophiles

The terms hard and soft apply to electrophiles just as they do to nucleophiles. A soft electrophile is one with a low-energy LUMO, and a hard electrophile is one with a high charge density. Alkyl halides are soft electrophiles, because they have no charge, and have a carbon–halogen antibonding orbital which is low in energy. The reaction of an alkyl halide with a soft nucleophile, such as a thiolate ion, will therefore be faster than with a hard nucleophile such as hydroxide. The order of hardness of different alkyl halides parallels that of the halide ions, so alkyl iodides are the softest electrophiles of the alkyl halides.

Carbonyl compounds are harder than alkyl halides, because of the high partial positive charge on the carbon caused by the inductive effect of the carbonyl oxygen. The LUMO of carbonyl compounds, however, is still relatively low in energy, so the carbonyl group still has some soft character. Carbocations are harder still, as they have a full positive charge. The effect of the hardness and softness of different electrophiles on their reaction with cyanide ion was discussed in Section 2.3.3.

Although most of the important electrophiles in organic chemistry are carbon-based, there are also examples of inorganic electrophiles, of which the halogens form

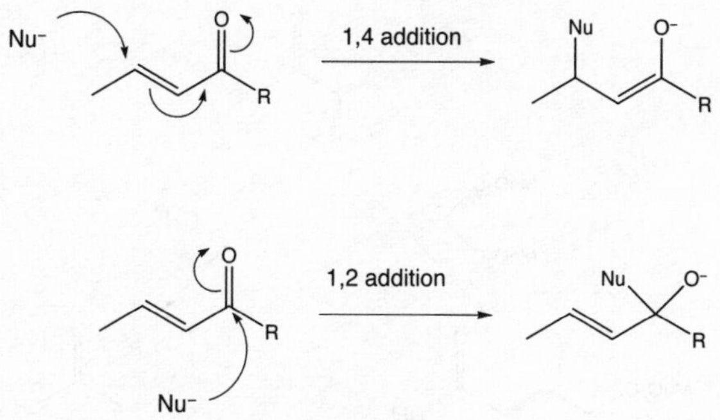

Figure 2.19

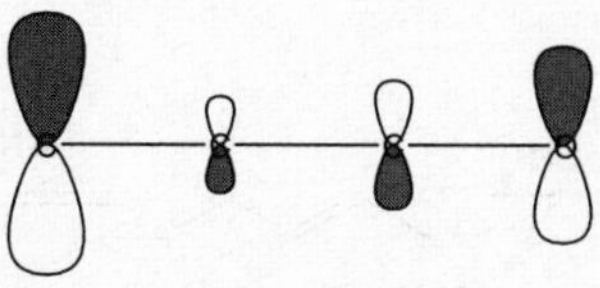

Figure 2.20

an important class. They have a low-energy LUMO (the antibonding orbital between the two halogen atoms) and so are soft. This explains their ready reaction with olefins, which are soft nucleophiles.

2.4.3 Ambident electrophiles

Just as some nucleophiles may react on different sites, so may some electrophiles. The most important class of ambident electrophiles are α,β-unsaturated carbonyl compounds, which may react with a nucleophile either at the carbonyl group (1,2 addition) or at the double bond (1,4 addition, conjugate addition, or Michael addition) as shown in Figure 2.19.

In general, the carbonyl carbon is the preferred site of attack by hard nucleophiles, and the β-carbon is preferred by soft ones. This is because the carbonyl carbon is more electron-deficient, while the LUMO normally has a larger coefficient at the β-carbon (Figure 2.20). An example of this is provided by ethyl acrylate, which undergoes ester hydrolysis on reaction with hydroxide (a hard nucleophile) but gives Michael addition with the anion of diethyl malonate (a soft nucleophile) (Figure 2.21).

It is as well to be wary of such straightforward explanations, however. Cyanide ion reacts with α,β-unsaturated ketones at the β-position, but this is because it gives the thermodynamically favoured product. Direct attack at the carbonyl group also occurs, but this is a reversible process (Figure 2.22).

It is more likely that 1,2 addition will take place if the carbonyl group is more

51

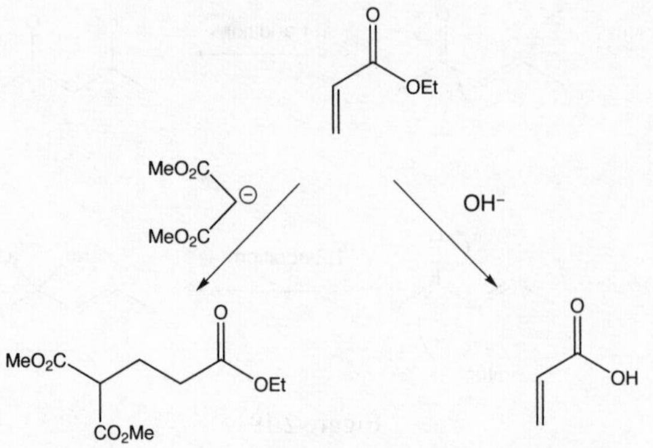

Figure 2.21

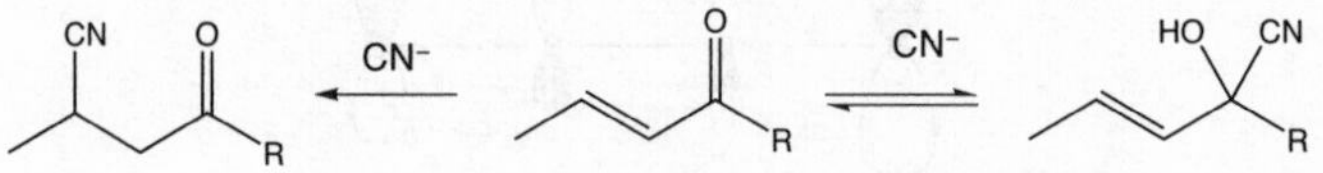

Figure 2.22

reactive. Thus acid chlorides and aldehydes undergo 1,2 addition more readily than less reactive carbonyl compounds, such as esters and amides, which are more likely to participate in Michael additions, other things being equal. As we have seen, however, much depends on the nature of the nucleophile.

Steric factors also play a part here; if the carbonyl group is sterically hindered, Michael addition will be favoured, and conversely a hindered olefin will favour direct addition.

2.5 Leaving groups

2.5.1 Definitions of leaving groups

A great many organic reactions involve what is known as a leaving group. This is exactly what it says it is: a group that leaves the rest of the molecule in some fragmentation process. For example, bromide ion is the leaving group in the hydrolysis of ethyl bromide. There will be a leaving group in any substitution or elimination reaction, so the concept of leaving groups is obviously very important in organic chemistry. Knowledge of the character of leaving groups is essential for a proper understanding of these reactions.

Leaving groups may be of two types, nucleofugal and electrofugal. These names are reminiscent of nucleophiles and electrophiles, and they are indeed analogous to them. A nucleofugal group takes a pair of electrons with it when it leaves, that pair of electrons having formerly been the bond between the leaving group and the rest

$$Nu{\overset{\frown}{}}E \longrightarrow Nu^- \quad + \quad E^+$$

Figure 2.23

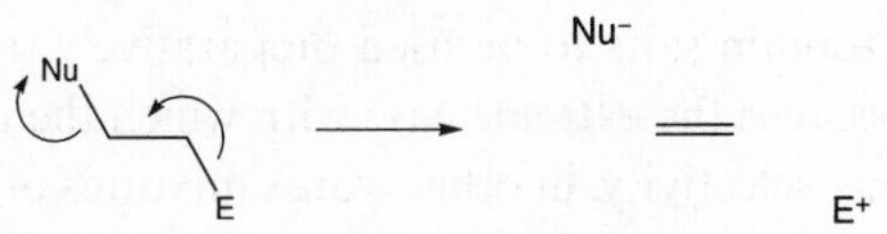

Figure 2.24

of the molecule. An electrofugal leaving group leaves without its electron pair. A nucleofugal leaving group, once it has left, can therefore act as a nucleophile in, for example, the reverse reaction, while an electrofugal leaving group can act as an electrophile after leaving. We almost always refer to nucleofugal groups when we talk of leaving groups. If an electrofugal leaving group is being described, this will normally be stated explicitly.

The process of leaving is shown schematically in Figure 2.23. We have a similar problem to the one we had earlier when we considered nucleophiles and electrophiles, namely that it is not always entirely clear which group is the leaving group, and what constitutes the rest of the molecule. Again, the distinction is arbitrary. It is worth noting that whenever we have a nucleofugal leaving group, the rest of the molecule that it leaves can be considered to be an electrofugal leaving group.

The simple bond fission shown in Figure 2.23 is one possible way in which a molecule may fragment; this is the first step of an E1 or an S_N1 reaction (see Sections 6.2.3 and 6.3.3). There is also a leaving group in the S_N2 reaction. In an E2 reaction, an olefin is formed by loss of both a nucleofugal and an electrofugal group from the starting material (Figure 2.24). Leaving groups are also important in addition–elimination reactions, the most important of which are carbonyl substitutions (see Section 6.3.5).

2.5.2 Nucleofugal leaving groups

The leaving group ability of a nucleofugal leaving group is closely connected to its base strength, the strongest bases being the poorest leaving groups. The most commonly encountered leaving groups are the halide ions, which are the conjugate bases of strong acids. The order of leaving group ability is $I^- > Br^- > Cl^- > F^-$, which is the reverse of their basicities.

Sulphonate esters contain still better leaving groups. The sulphonate ions that leave are the conjugate bases of strong acids. The most common examples are tosylates (p-toluenesulphonates, $p\text{-}MeC_6H_4SO_3R$), although brosylates ($p\text{-}BrC_6H_4SO_3R$) and mesylates ($MeSO_3R$) are also sometimes encountered. The leaving group in triflate esters (trifluoromethanesulphonate esters, CF_3SO_3R) is the conjugate base of a very strong acid indeed, and so is an even better leaving group than the other sulphonate esters.

The best leaving group of all is nitrogen, which can leave from diazonium salts. Although substitution reactions of aromatic diazonium salts are well known, it is rare for aliphatic diazonium salts to be used preparatively (i.e. in the synthesis of other compounds), because the extreme ease with which the nitrogen leaves makes it difficult to obtain any selectivity, in other words mixtures of many different products are usually formed.[3]

Other leaving groups exist, but their leaving group ability depends on their base strength. Carboxylate anions may act as leaving groups, as their conjugate acids are still moderately strong. Phenolate anions can also be leaving groups, although this is less common; alkoxide ions are poorer still. This is reflected in the relative efficacy of anhydrides and phenyl and alkyl esters as acylating agents.

For an amide to be cleaved by a base-catalysed process, NH_2^- would need to act as a leaving group. This anion is a very strong base, and so a very poor acid. This is why amides are usually stable under basic conditions. Carbanions (see Section 4.1) are even poorer leaving groups, and do not leave unless the carbanion is stabilized in some way.

It is possible to turn some groups into far better leaving groups under acidic conditions (see Chapters 5 and 6). For example, amides may be hydrolysed in strong acid. The acidic conditions lead to protonation of the nitrogen, so that the leaving group is neutral ammonia. Similarly, the hydroxyl group of alcohols may be protonated, so that the leaving group is water, rather than hydroxide. By this means, alcohols may take part in nucleophilic substitution reactions or eliminations.

2.5.3 Electrofugal leaving groups

These are a much smaller class than nucleofugal leaving groups. The most important is the proton. This leaves in elimination reactions, as well as the nucleofugal group. In any reaction in which an organic substrate is treated with a base to form a carbanion, the proton can be thought of as an electrofugal leaving group, although the term is seldom used this way in practice.

Another common electrofugal leaving group is CO_2, which leaves in decarboxylation reactions. This leaves an anion, so the reaction is only possible if the resulting molecule has some way of stabilizing the negative charge. The most common way in which this occurs is if some part of the molecule is protonated first. Examples are shown in Figure 2.25.

Metals may also be electrofugal leaving groups. In the reaction of Grignard reagents, magnesium acts in this manner, as shown in Figure 2.26 (Grignard

---

[3] There are exceptions to this, the most important of which is the reaction of diazonium salts derived from α-amino acids, discussed in Chapters 5 and 6.

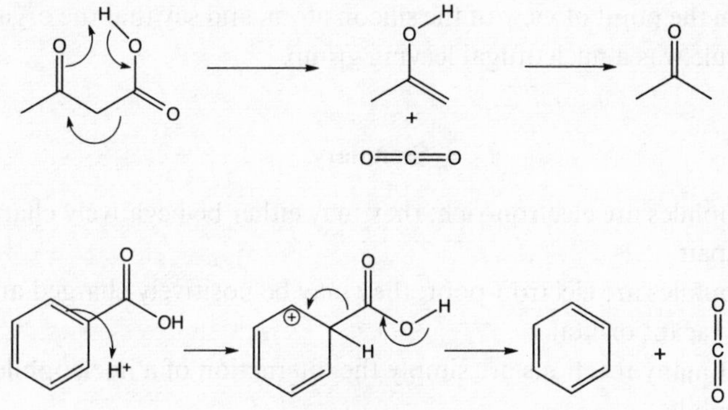

Figure 2.25

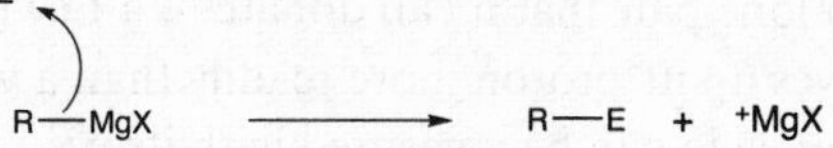

Figure 2.26

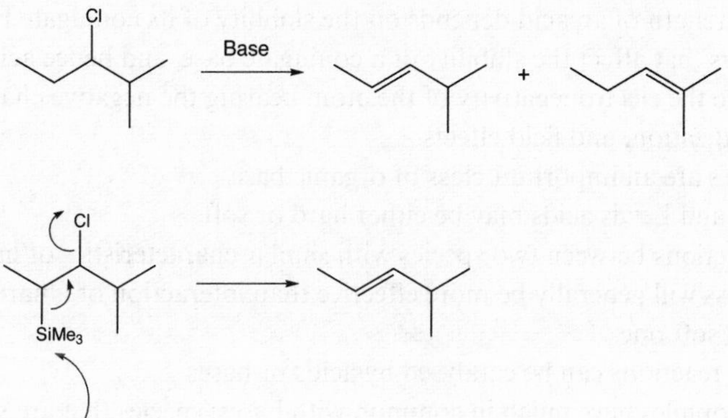

Figure 2.27

reagents are a source of nucleophilic carbon, see Section 4.1). Silicon can act in this way, and is sometimes used to direct elimination reactions. Base-catalysed elimination (See Section 6.2.2) of the chloride in Figure 2.27 leads to a mixture of products. However, when a trimethylsilyl group is present, the double bond may be introduced specifically on treatment with fluoride.[4] Note that, although we would normally think of the silicon as being an electrofugal leaving group, we could consider this

---

[4] Fluoride ions have a great affinity with silicon, and are often used for displacement at silicon in this way.

reaction from the point of view of the silicon atom, and say that the organic portion of the molecule was a nucleofugal leaving group.

Summary

- Nucleophiles are electron-rich; they may either be negatively charged or have a lone pair.
- Electrophiles are electron-poor; they may be positively charged and must have a vacant orbital.
- A great many reactions are simply the interaction of a nucleophile with an electrophile.
- A Brønsted acid is a proton donor; a Brønsted base is a proton acceptor.
- A Lewis acid has an empty orbital that can accommodate a lone pair; a Lewis base has a lone pair that it can donate to a Lewis acid.
- A strong acid gives up its proton more readily than a weak acid.
- The strength of an acid can be measured with its pK_a; the lower the value of the pK_a, the stronger the acid.
- The conjugate base of a strong acid is a weak base, and vice versa.
- The strength of an acid depends on the stability of its conjugate base.
- Factors that affect the stability of a conjugate base, and hence acid strength, include the electronegativity of the atom bearing the negative charge, delocalization, and field effects.
- Amines are an important class of organic base.
- Bases and Lewis acids may be either hard or soft.
- Interactions between two species with similar characteristics of hardness or softness will generally be more effective than interaction of a hard species with a soft one.
- Many reactions can be catalysed by acids or bases.
- Nucleophiles have much in common with bases; species that are strong bases are often, but not always, also good nucleophiles.
- The main distinction between basicity and nucleophilicity is that basicity is a thermodynamic effect and nucleophilicity is kinetic.
- Nucleophiles can be hard or soft; soft nucleophiles react best with soft electrophiles and hard nucleophiles react best with hard electrophiles.
- Nucleophilicity can be enhanced by an adjacent lone pair; this is known as the alpha effect.
- The solvent can have an important effect on nucleophilicity.
- Electrophiles must have a vacant orbital to accept electrons from nucleophiles.
- The reactivity of ambient nucleophiles and electrophiles can often be explained by consideration of hardness and softness and of frontier orbitals.

- Leaving groups may be nucleofugal or electrofugal; nucleofugal groups leave with a pair of electrons, electrofugal groups leave without it.
- The leaving group ability of nucleofugal leaving groups is closely related to their base strength; strong bases are poor leaving groups.
- The most important electrofugal leaving groups are the proton, CO_2, and metals.

Problems

1. Which of the following species are nucleophiles and which are electrophiles?

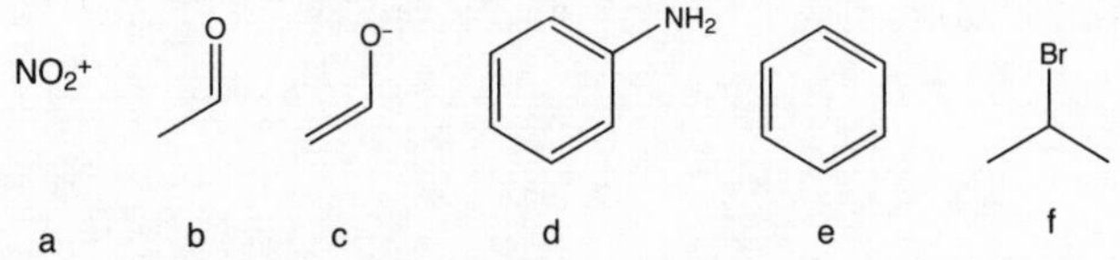

NO_2^+

a   b   c   d   e   f

2. Arrange the following in order of acid strength, showing the most acidic proton.

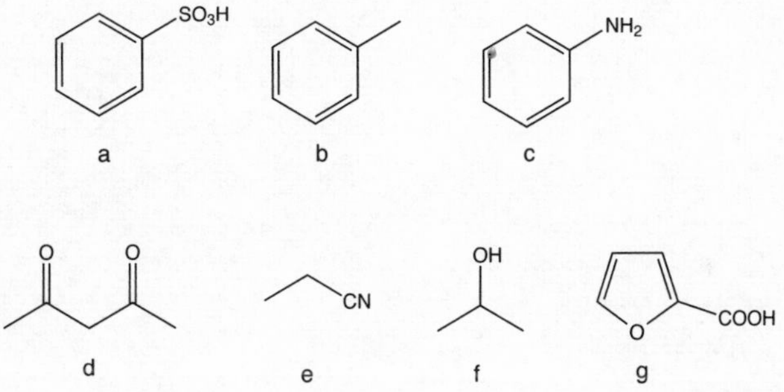

a   b   c

d   e   f   g

3. Arrange the following in order of their rate of reaction with ethyl iodide (an S_N2 displacement). Why is this not the same order as their base strength?

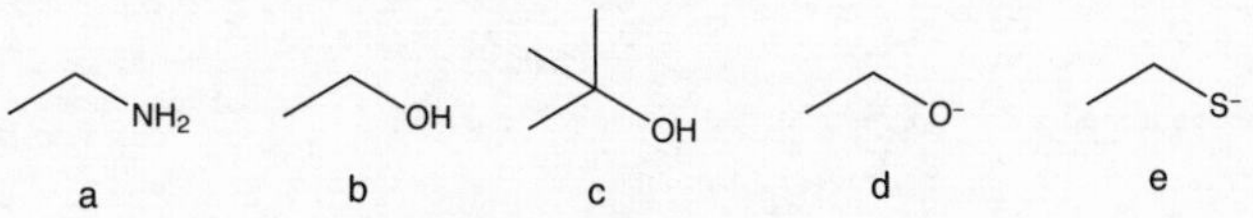

a   b   c   d   e

4. Which is more basic of the hydroxide (OH^-) and hydroperoxide (HOO^-) anions, and which is more nucleophilic? Why?

5. Which would you expect to give the greater proportion of O-alkylation on reaction with the sodium salt of ethyl acetoacetate, propyl iodide, or propyl bromide? Why?

6. What would you expect would be the product of the reaction between the α,β-unsaturated carbonyl compound below and a thiol?

7. Arrange the following in order of leaving group ability.

Br⁻   OH⁻

a   b   c   d   e   f

3

Why reactions happen

If a chemical reaction takes place, there is always a reason why. In this chapter, we will look at the thermodynamics and kinetics of reactions, which tell us why some reactions happen and some do not. We will see some of the general features that can make reactions favourable, and look at why some reactions may be favoured over others if there are alternative pathways.

3.1 Thermodynamics and kinetics compared

3.1.1 Free energy

It is pertinent to ask 'why do reactions happen?' The answer is that a chemical reaction will take place if it leads to a decrease in free energy. Free energy changes can be thought of as the thermodynamic driving forces behind all reactions. This is true not only of chemical reactions, of course. A decrease in energy is the driving force behind all physical processes. For example, apples fall out of trees because they move to a position of lower gravitational energy in this way.

The driving force for a reaction is a very important concept. No chemical reaction can happen without it. If we are not sure whether a reaction is feasible or not, one of the first things we should do is to look for a plausible driving force.

Free energy is made up of two components, enthalpy and entropy, which we will describe below. The free energy (G) of a substance is related to its enthalpy (H) and its entropy (S) by the formula:

$$G = H - TS$$

where T is the absolute temperature. We are not normally interested, however, in the absolute free energy of a substance, but rather in the change in free energy (ΔG) that occurs in a reaction. This is related analogously to the change in enthalpy (ΔH) and the change in entropy (ΔS) by the more useful equation:

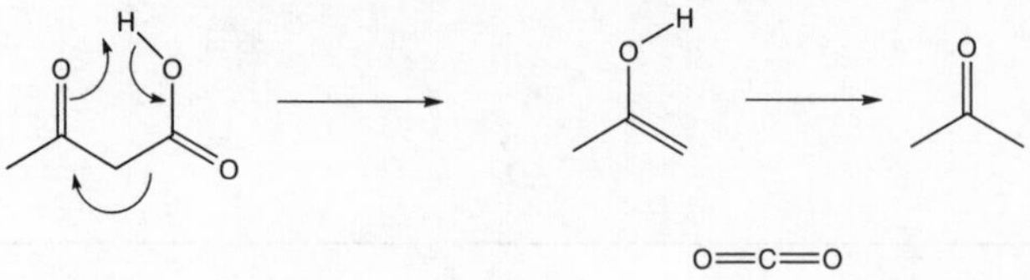

Figure 3.1

$$\Delta G = \Delta H - T\Delta S$$

It is important not to get confused by the signs of the various terms. A reaction is favourable if ΔG is negative. This means that the system will have lost energy, which is what happens in any spontaneous thermodynamic process. An everyday analogy is the flow of water, which will spontaneously flow downhill (i.e. to a position of lower energy), but will only flow uphill if energy is supplied externally.

Either a negative ΔH or a positive ΔS will contribute to a negative ΔG. Clearly, if ΔH is negative and ΔS is positive, ΔG will be negative and the reaction will be favourable. Conversely, a positive ΔH and a negative ΔS will result in a positive ΔG and an unfavourable reaction. If ΔH and ΔS are of the same sign, the reaction may or may not be favourable, depending on the relative sizes of the terms and on the temperature. Note that the effect of entropy becomes more important at higher temperatures.

The enthalpy change of a reaction is also known as the heat of reaction. A reaction with negative ΔH gives out heat and is said to be exothermic. Reactions with positive ΔH absorb heat from their surroundings; these are endothermic. ΔH is mostly determined by the bond strengths of the reactants and products. The formation of strong bonds in the product at the expense of weaker bonds in the reactants will lead to an exothermic reaction; this is a common driving force for reactions. Other factors can also influence ΔH, such as ring strain and resonance energy – these are discussed further below.

Entropy is a measure of the amount of disorder in a system. This is best explained with the aid of some examples. A common way in which entropy might increase is if the total number of molecules increases. An example is the decarboxylation of acetoacetic acid to give acetone and carbon dioxide, a reaction that takes place readily (Figure 3.1). This increases the amount of disorder because the acetone and carbon dioxide molecules can now move independently of each other, which is less ordered than when they are attached to each other within the acetoacetic acid molecule. Entropy is also increased by the liberation of gases, as the entropy of a gas is usually much higher than that of a liquid or a solid.

3.1.2 Equilibrium constants

All chemical reactions are in principle reversible. Some are readily reversible, whereas other reactions are so favourable that the reverse reaction happens to such

Table 3.1. *Relationship between free energy and equilibrium constant*

ΔG (kJ/mol)	K
0	1
−5	8
−10	60
−20	3000
−30	200 000

a small extent that we may safely ignore it. We can quantify the extent to which a reaction is reversible with the equilibrium constant, K. For a reaction

$$A+B \rightleftharpoons C+D$$

the equilibrium constant is defined by the formula

$$K = \frac{[C][D]}{[A][B]}$$

where [A] etc. are the concentrations of the components after the reaction has reached equilibrium. If K is large, this means that the forward reaction is favourable, and if K is less than 1, then the reverse reaction is favourable.

There is a simple relationship between the free energy change for a reaction, ΔG, and its equilibrium constant, K, which is

$$\Delta G = -RT \ln K$$

where R is the gas constant and $\ln K$ is the natural logarithm of K. If ΔG is zero, then $K=1$, which means that reactants and products will be present in equal concentrations at equilibrium. Positive values of ΔG will result in $K<1$, or a greater concentration of reactants than products. Negative ΔG will mean values of $K>1$, or more products than reactants; this is what we normally think of as a favourable reaction. Note that the logarithmic nature of the relationship means that K rises very quickly with increasing ΔG. Some approximate values of ΔG and their associated equilibrium constants are shown in Table 3.1 (note that the ΔG values are negative, showing that we are looking at a thermodynamically favourable reaction).

3.1.3 Kinetics

Although a negative ΔG is a necessary condition for a reaction to take place, it is not sufficient. If we are to observe a reaction, it must also take place at a measurable rate. Many reactions with large negative ΔG are also fast, although by no means all of them.

61

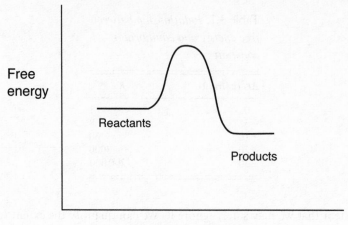

Reaction coordinate

Figure 3.2

Some reactions may be very favourable thermodynamically (i.e. have large negative ΔG), but take place so slowly that we might be forgiven for thinking that they do not happen at all. For example, it would be a thermodynamically favourable process to turn all the organic material in the human body into carbon dioxide, water, and nitrogen, although few of us live in daily fear of spontaneous combustion. Similarly, the conversion of diamond to graphite is thermodynamically favourable at atmospheric pressure, but there is such a large kinetic barrier to its occurrence that there is not much danger that your expensive diamond jewellery will turn into a worthless pencil lead.

If a reaction is fast, then it is said to be kinetically favourable. The kinetics of a reaction is a term relating to the rate of the reaction.

3.1.4 Reaction profiles

It is useful to plot the absolute free energy of a reaction against the course of the reaction. This gives a graph called a reaction profile (Figure 3.2). The energy of the reaction components is plotted on the vertical axis, and the reaction coordinate is plotted along the horizontal axis. The reaction coordinate is a measure of the extent of the reaction, which is related to the time course of the reaction, although reaction coordinate and time are not quite equivalent.

We plot the free energy of the reactants on the left of the graph, and that of the products on the right. The difference between these two free energies is ΔG. The next question to ask is what is the shape of the line joining the reactants and products. The simplest answer would be a straight line, but this does not happen in practice. The precise shape of the line varies according to the nature of the reaction, but in general it rises to a maximum somewhere between the two ends .

The kinetics of a reaction are dependent on this maximum. The higher it is, the

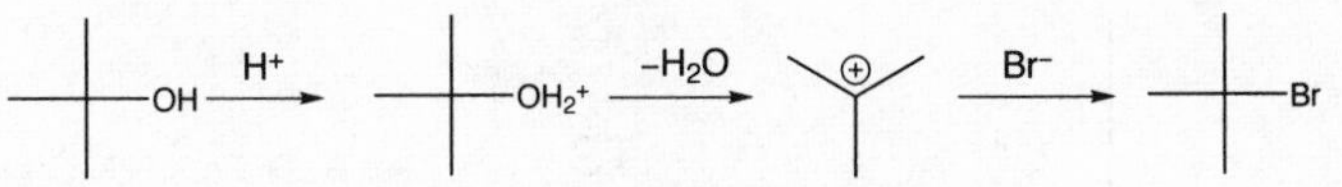

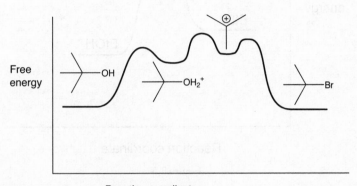

Figure 3.3

Free energy / Reaction coordinate

Figure 3.4

slower the reaction will be. The difference in energy between the starting material and the energy maximum is known as the activation energy. If the activation energy is small, it will not be difficult for the molecules to overcome this barrier, and the reaction will proceed quickly. If, on the other hand, the activation energy is large, it will be more difficult for the reactants to find this amount of energy, and the reaction will be correspondingly slower.

3.1.5 Intermediates and transition states

When discussing the pathways of reactions, we often draw the structures of entities between the two ends of the reaction. These fall mostly into two categories: intermediates and transition states. It is important not to confuse the two.

An intermediate is a genuine chemical entity which has a finite existence, even though this may be very short. An example is in the S_N1 substitution of t-butyl alcohol with HBr to give t-butyl bromide (Figure 3.3; see Section 6.3.3 for a fuller discussion of this reaction). The reaction proceeds through protonation of the hydroxyl oxygen and cleavage of the C−O bond. In this reaction, both the protonated alcohol and the carbocation are intermediates.

The reaction profile for this is shown in Figure 3.4. There are two local energy minima in this profile; these are the intermediates. This is the characterizing feature of an intermediate: intermediates correspond to local energy minima, although they are often considerably higher in energy than the starting materials or products, as the carbocation is in this example.

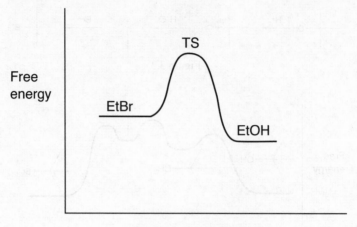

Reaction coordinate

Figure 3.5

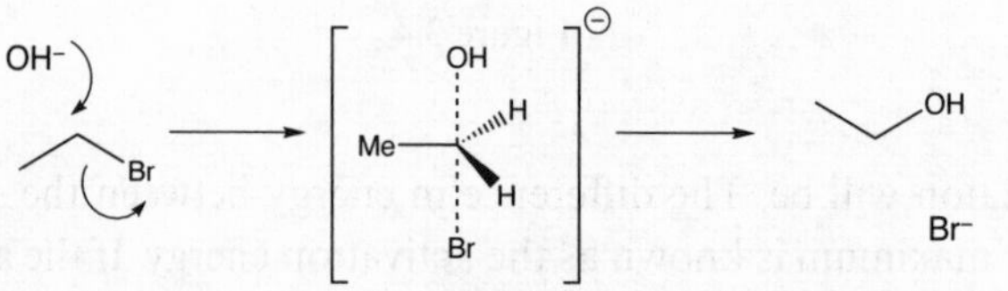

Figure 3.6

The S_N2 substitution of ethyl bromide by hydroxide ion has no intermediate. Its reaction profile is shown in Figure 3.5. The maximum of the curve corresponds to the transition state for the reaction, which is commonly depicted as shown in Figure 3.6 (this reaction is also discussed further in Section 6.3.2). A transition state has no finite lifetime, but is a state which, as the name suggests, is passed through transiently during the course of the reaction.

Although we cannot observe transition states directly, we can draw inferences about them from high-energy intermediates. The Hammond postulate states that the transition state between two chemical species is closer in structure to the species to which it is closer in energy. For high-energy intermediates, the transition state will therefore be much more similar to the intermediate than to the starting material. Normally, if a reaction has a high-energy intermediate, any change that reduces the energy of this intermediate is likely also to reduce the energy of the transition state leading to it, and hence increase the rate of the reaction.

Note that all reactions have transition states – reactions with intermediates must have transition states leading to the intermediates – but not all reactions have intermediates.

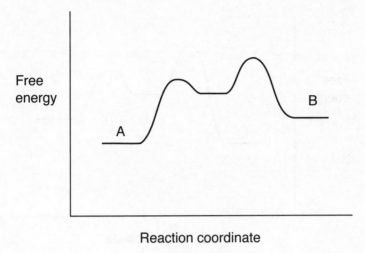

Figure 3.7

3.1.6 *Thermodynamic versus kinetic control*

Not all reactions have only one possible product. Many reactions have a number of different products, some of which may be more favourable than others. Consider the reaction profile in Figure 3.7. The middle of the curve corresponds to starting materials, and two reactions A and B are shown to either side. It is obvious that the major reaction here will be reaction A, as it is more favourable both thermodynamically (lower final energy) and kinetically (lower energy transition state). Reaction B will also take place to a greater or lesser extent, depending on the difference between these energy levels, but A will always predominate.

Now consider the reaction profile in Figure 3.8. Reaction C leads to a product lower in energy than reaction D, in other words it is thermodynamically more favourable. On the other hand, the transition state leading to reaction D is lower in energy, so reaction D is the more kinetically favourable reaction. It is not such a simple matter here to predict which reaction will predominate. In general, the thermodynamic product will be favoured when an equilibrium can be established; this will happen under harsher conditions, such as higher temperatures. If a reaction takes place under mild conditions and is irreversible, then the kinetic product will be formed preferentially.

An example of a reaction under thermodynamic control is the acid-catalysed E1 dehydration (see Section 6.2.3) of t-amyl alcohol (Figure 3.9). The intermediate carbocation may lose a proton from the methylene group or one of the methyl groups adjacent to the cationic centre. The protons on the methyl groups are more acidic for both steric and electronic reasons, and so the removal of these is kinetically favoured. However, removal of one of the methylene protons is favoured because this gives rise to a more stable product with a more substituted

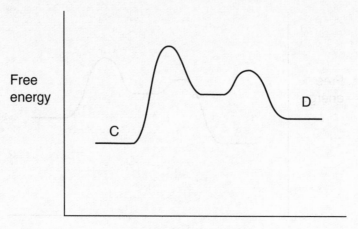

Reaction coordinate

Figure 3.8

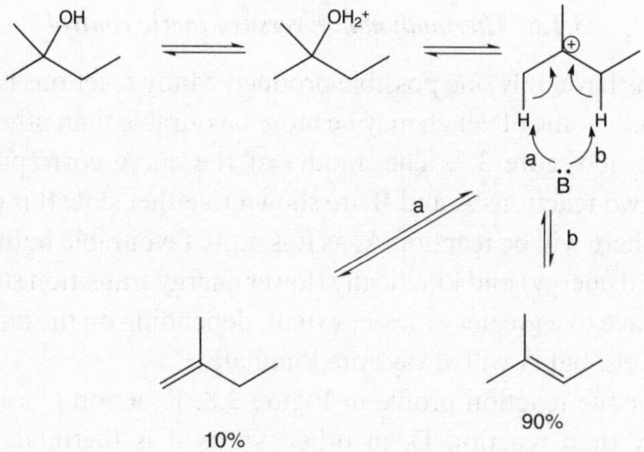

Figure 3.9

double bond.[1] In practice, the reaction gives about 90% of the thermodynamic product.

This can be explained by the reversibility of the reaction. The olefin, once formed, can protonate again to regenerate the carbocation. This allows the establishment of equilibrium, resulting in the preference for the thermodynamic product.

The kinetic/thermodynamic dichotomy also arises in the protonation of an extended enolate (Figure 3.10). Protonation at the α-position (reaction a) is kinet-

---

[1] The greater stability of more substituted double bonds is explained by hyperconjugation, as discussed in Chapter 1. The double bond can hyperconjugate with any adjacent alkyl group, and so the more alkyl groups that are adjacent, i.e. the more highly substituted the double bond, the greater the number of hyperconjugated resonance forms that can be written, and hence the greater the stability of the double bond.

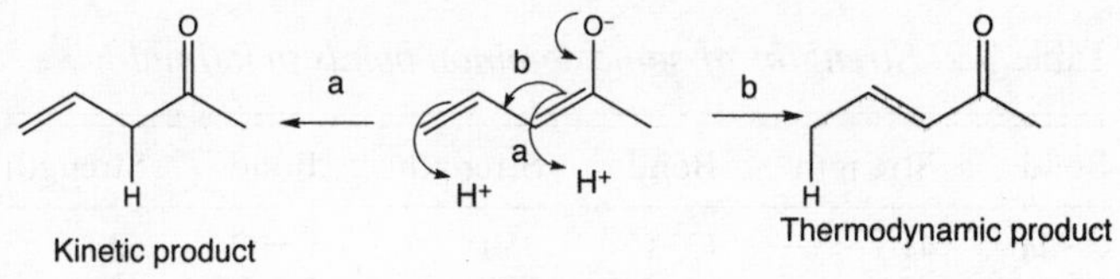

Kinetic product

Thermodynamic product

Figure 3.10

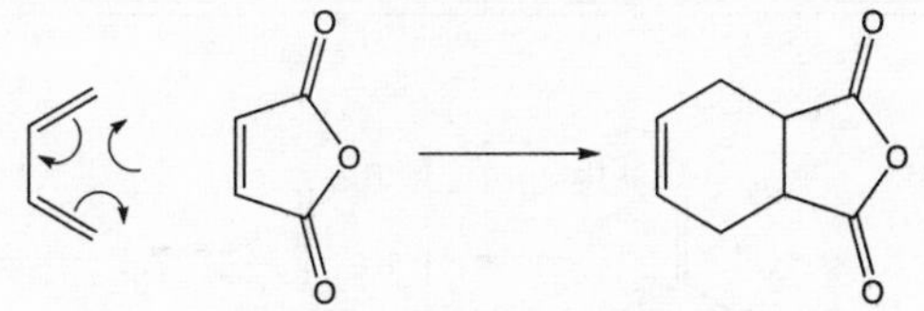

Figure 3.11

ically favoured, because it carries a greater negative charge, but protonation at the γ-position (reaction b) leads to a product in which the double bond is conjugated with the carbonyl group, which is more stable. If the anion is protonated at low temperatures under conditions at which extremes of pH are avoided (e.g. rapid quenching with a large excess of aqueous ammonium chloride) the kinetic product will be dominant. If higher temperatures are used, or strongly acidic or basic conditions allow equilibration of the product, then the more conjugated (thermodynamic) product will be formed instead.

3.2 Thermodynamic effects

3.2.1 Bond strengths

One of the most important factors affecting thermodynamic stability is bond strength. If strong bonds are formed, this helps to make a reaction thermodynamically favourable. The factors that determine the strength of a bond are complicated and we shall not worry too much about them here. There are, however, a few simple rules worth bearing in mind.

Although double bonds are stronger than single bonds, they are not twice as strong (see Section 1.1.4). A reaction that forms single bonds at the expense of the π component of double bonds is therefore likely to be favourable. An example is the Diels–Alder reaction between maleic anhydride and butadiene (Figure 3.11), which goes in almost quantitative yield. Two single bonds are formed in this reaction, whereas there are three double bonds in the starting materials and only one in the product. Thus although the total number of bonds remains the same, the formation of σ bonds from π bonds in this way provides the thermodynamic driving force for the reaction.

Table 3.2. *Strengths of some common bonds in kJ/mol*

Bond	Strength	Bond	Strength	Bond	Strength
C−H	410	C−C	350	C=C	680
N−H	390	N−N	160	N=N	420
O−H	460	O−O	140	C=O	500
H−H	440	C−N	340	C−O	380

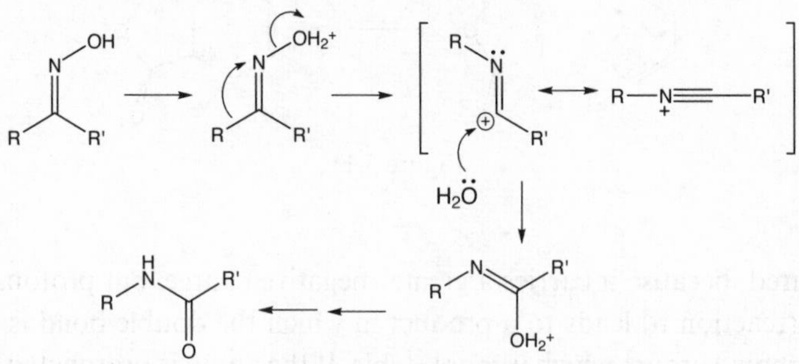

Figure 3.12

Clearly, bonds between some combinations of atoms are stronger than those between others. Some approximate strengths for bonds between common atoms are shown in Table 3.2. It is not a sensible exercise to try to commit these values to memory, but the general trends are worth noting. One important feature is that bonds between two heteroatoms are seldom as strong as the bond between the same heteroatom and carbon.

An example of a reaction in which the driving force comes from breaking a bond between two heteroatoms and joining these to carbon is the Beckmann rearrangement (Figure 3.12). In this reaction, one bond is broken between nitrogen and oxygen, and another between two carbons. The two bonds formed in their place (one C−O and one C−N) are stronger, and this provides a driving force for the reaction.[2]

The geometry of this reaction is also noteworthy. The group *trans* to the hydroxyl group (i.e. the group on the opposite side of the double bond from the hydroxyl group) migrates. This is true irrespective of the nature of the two alkyl groups, in other words it is not the relative migratory aptitude of the two groups which is important, but their geometry. The reason for this can be seen from a consideration of the orbitals. The alkyl group that migrates must enter the antibonding orbital of the N−O bond to break it. This orbital has a large lobe on the side of the nitrogen away from the oxygen, in other words adjacent to the alkyl group *trans* to the oxygen

[2] As well as this change, the product has a C=O bond and a C−N bond, whereas the starting material has a C=N bond and an N−O bond. This also helps to make the reaction thermodynamically favourable, as C=O bonds are stronger than C=N bonds.

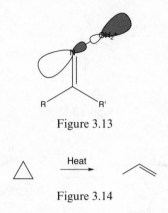

Figure 3.13

Heat

Figure 3.14

(Figure 3.13). It is this proximity to the antibonding orbital that allows this group to migrate.

3.2.2 *Ring strain*

When atoms are joined together in a ring, we have what is known as a cyclic molecule. If the ring contains, for example, six atoms, it is called a six-membered ring. Different sized rings have different stabilities, and small rings (three- or four-membered) are particularly unstable. The instability of small rings is known as ring strain.

Ring strain arises because it is not possible for the orbitals of the atoms to overlap at their optimum angles. The orbitals in saturated carbon atoms have their lowest energy when the angle between them is 109°, as in methane (see Section 1.1.3). In a three-membered ring, the atoms in the ring form a triangle, and so the angles between them are 60° (they may deviate slightly from this if the three atoms are not all the same, but even so this will be a good approximation). Because the orbitals cannot overlap at their preferred angle, the bonds are weaker than they would otherwise be, and it is this that causes ring strain.

An example of the practical consequences of ring strain is the reactivity of cyclopropane. Although alkanes are normally unreactive molecules, cyclopropane undergoes a variety of reactions because of the instability due to its ring strain. When heated, it rearranges to give propene (Figure 3.14). Note that a π bond has formed in place of a σ bond, in contrast to the rule described above. This illustrates the powerful effect of ring strain, and also serves as a useful reminder that all rules in chemistry can seldom be relied upon consistently, since there are frequently many complex factors that determine the course of a chemical reaction.

The ether bonds of epoxides (three-membered cyclic ethers) can be cleaved far more easily than those in open-chain ethers, again because of ring strain. Epoxides are easily opened by a variety of nucleophiles, in contrast to normal ethers, which are normally inert to nucleophilic cleavage. Protonated epoxides are opened very readily indeed, and will react with mild nucleophiles such as halide ions or even water (Figure 3.15).

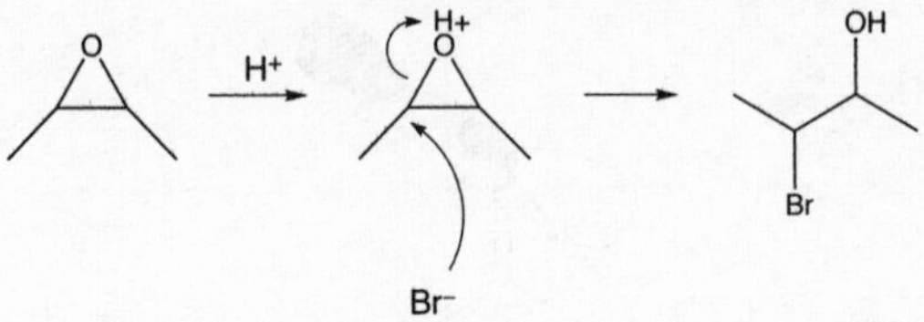

Figure 3.15

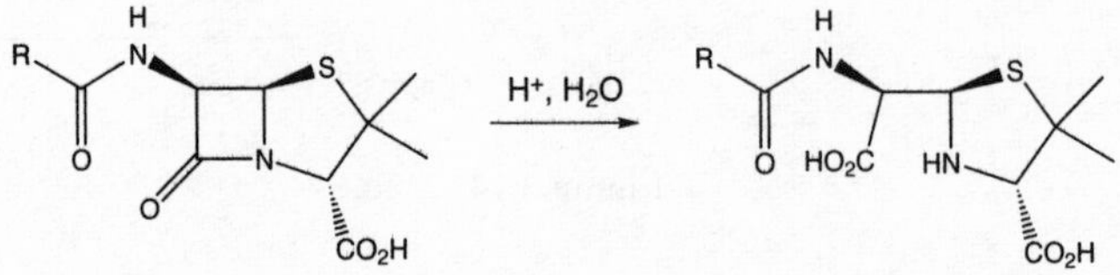

Figure 3.16

Four-membered rings are less reactive than three-membered ones, but still show noticeable effects of ring strain. The β-lactam group (the basis of the penicillin and cephalosporin antibiotics) can by hydrolysed by mild acid (Figure 3.16), whereas amides are normally stable except under harsh conditions. Five-, six-, and seven-membered rings are not strained, and usually have similar reactivity patterns to their open-chain counterparts.

Rings with between eight and eleven members are again unstable, but for different reasons. Because of the way the molecules are folded, the hydrogen atoms attached to the ring on one side will tend to interfere with those on the other side. The easiest way to see this effect is to try to make a model of such compounds. Figure 3.17 shows this interaction in cyclononane.

The consequence of this instability is that these compounds are difficult to prepare, and are therefore rarely encountered. This contrasts with the smaller strained rings, which can be prepared much more easily. This is for kinetic reasons; atoms that are close together will more frequently come close enough for bonding than atoms that are further apart, so cyclization is kinetically more favourable in smaller rings than in larger ones. To form large rings by cyclization of open-chain compounds, the chain must be folded in a particular way for the two ends to find each other.

Rings with twelve or more members are just as stable as open-chain compounds, but can be difficult to prepare for kinetic reasons.

3.2.3 Aromatic stability

We saw in Section 1.5 that aromatic compounds have a particular stability. This can affect the outcome of chemical reactions, because those that form aromatic compounds are likely to be more favourable than similar reactions that do not.

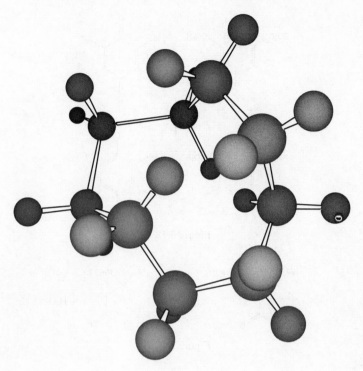

Figure 3.17

An example of the ease with which aromatic compounds are formed is provided by the Hantzsch pyridine synthesis. The reaction has two stages, an initial cyclization to the dihydropyridine (this is the subject of one of the problems for Chapter 8) and the subsequent oxidation (Figure 3.18). This oxidation can be done under mild conditions with various reagents, a common one being the quinone chloranil. It is the formation of the aromatic pyridine from the non-aromatic dihydropyridine which makes this reaction so easy; oxidations removing hydrogens from carbon atoms in this way normally require much harsher conditions.

3.2.4 Entropies

The effect of entropy on the thermodynamics of a reaction is important and must not be forgotten. One place in which entropy commonly plays a major part is in cyclization reactions.

Consider the important biological molecule mevalonic acid, a precursor in the synthesis of steroids, such as cholesterol and sex hormones. It is difficult to isolate this molecule as its free acid, as it spontaneously cyclizes to form the six-membered lactone (Figure 3.19). Although the equilibrium constant for the reaction between

71

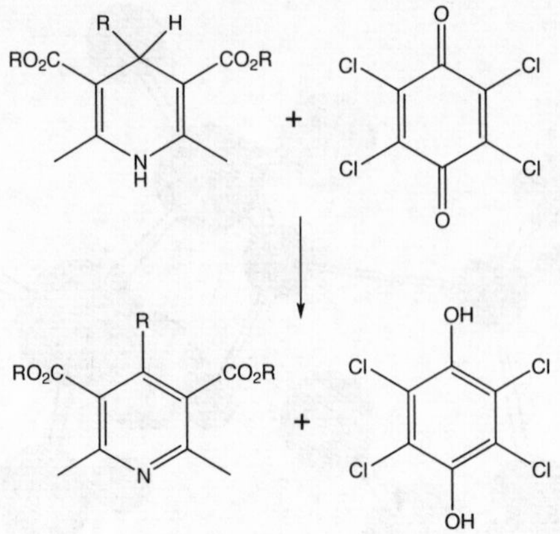

Figure 3.18

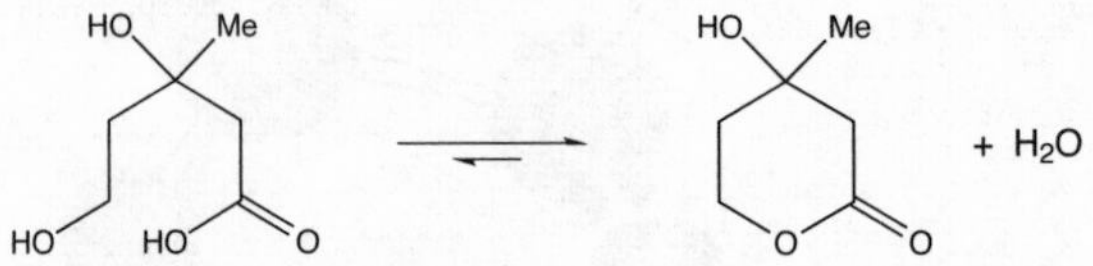

Figure 3.19

an acid and an alcohol to form an ester is normally close to 1, the equilibrium for this molecule lies greatly in favour of the ester.

The reason for this ready cyclization is entropy. In both this esterification reaction and intermolecular esterifications, the strength of the bonds formed is similar to the strength of the bonds broken, so the enthalpy change of the reaction is small. However, in the normal esterification reaction, there are two molecules of starting material and two product molecules, whereas cyclization of mevalonic acid produces two molecules of product from just one of starting material. Since one of the most important contributions to entropy is the number of molecules, the cyclization has a clear entropic advantage because the number of molecules increases.

Cyclization reactions in general will have an entropic advantage over their intermolecular counterparts. A reaction such as esterification between two molecules joins two molecules together, and thus has an inherent entropic disadvantage (although here this is offset by liberation of a molecule of water). This disadvantage is not present in the cyclization reaction, because the two groups that react are already part of the same molecule at the start of the reaction.[3]

---

[3] There is in fact a small entropy disadvantage in cyclization reactions, as the rotational freedom of the molecule is restricted by the cyclization. This effect, however, is small compared with the effect of the change in the number of molecules.

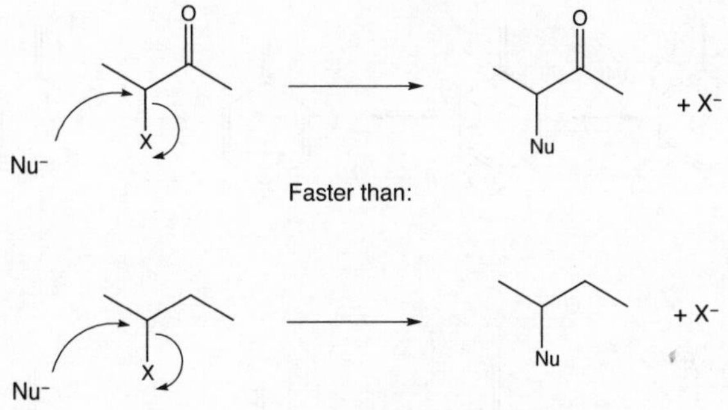

Figure 3.20

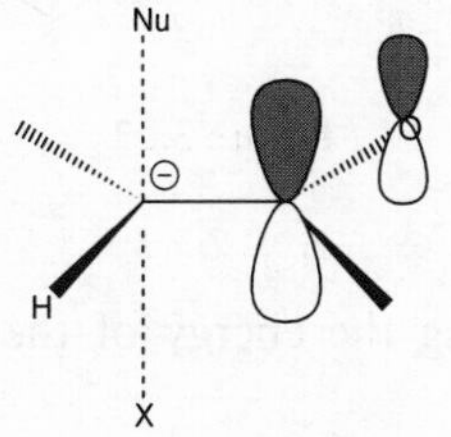

Figure 3.21

3.3 Kinetic effects

3.3.1 Electronic effects

If a reaction is to take place rapidly, there must be an efficient interaction between the orbitals of the reactants. We saw in Section 2.3.2 how reactions will be faster when soft nucleophiles react with soft electrophiles, and hard nucleophiles react with hard electrophiles. To recap, soft–soft reactions are fast because of the overlap between the orbitals, and hard–hard reactions are fast because the opposite charges on the two species attract each other.

Consideration of the reactivity of hard and soft nucleophiles and electrophiles using the principles described in Chapter 2 will tell us much about the kinetics we can expect for a given reaction. However, we should be wary of looking at each molecule just as a member of a large category, and must remember that each molecule is different. A number of factors can affect the reactivity of any given molecule, and so influence its rate of reaction.

For example, S_N2 displacements of α-halo ketones take place considerably faster than similar reactions in simple aliphatic alkyl halides (Figure 3.20). The reason for this is that the carbonyl group adjacent to the reaction centre is able to stabilize negative charges, and a negative charge builds up in the transition state for this reaction (Figure 3.21). This stabilization lowers the energy of the transition state, and as dis-

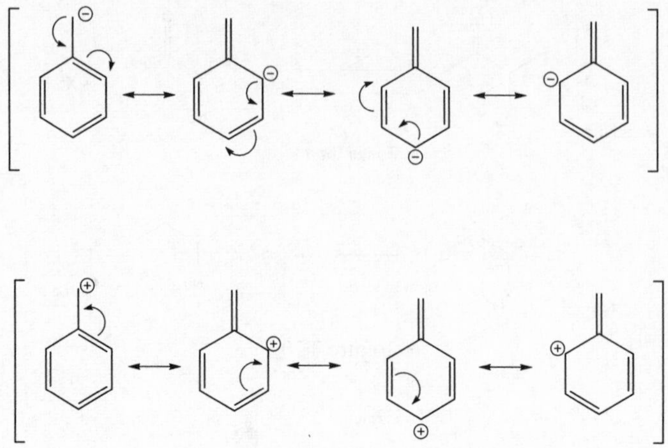

Figure 3.22

cussed in Section 3.1, lowering the energy of the transition state of a reaction increases its rate.

Similarly, S_N1 displacement of these compounds is very slow, because the intermediate has a positive charge, and is therefore destabilized by the adjacent carbonyl group. Since this intermediate is of high energy, the energy of the transition state leading to it is close to the energy of the intermediate (see Section 3.1.5). Therefore, raising the energy of the intermediate raises the energy of the transition state and thus slows the reaction.

Reactions in benzylic positions (i.e. immediately adjacent to aromatic rings) are usually faster than those in simple alkyl groups whether there is a build-up of positive or of negative charge in the transition state, since either charge can be stabilized by delocalization around the ring (Figure 3.22). Reactions with negatively charged transition states will be faster still if the ring has electron-withdrawing substituents (e.g. nitro groups, halides), and those with positively charged transition states will be favoured by electron-donating substituents (e.g. alkoxy groups, alkyl groups).

Hammett plots

Organic chemistry is primarily a qualitative science. It is seldom meaningful to try to describe it with numbers. However, one quantitative relationship that has proved useful in organic chemistry is the Hammett equation, which applies to reactions taking place close to aromatic rings.

One reason why organic chemistry is normally so hard to quantify is that structural differences between one molecule and another, even apparently trivial ones, can change the reactivity in a variety of intricately interacting ways, both electronic and steric. The Hammett equation addresses the variation in reactions taking place close to aromatic rings when substituents *meta* or *para* to the reaction centre are varied. The substituents are thus too far away to have any steric effect on the course of the reaction, and do not change the conformation of the rigid aromatic ring. The only effects the substituents have on the reaction are therefore due to their electronic properties.

The Hammett equation is

$$\log_{10}(k_X/k_H) = \rho\sigma_X$$

where k_X and k_H are the rates of the reaction or equilibrium constants with the ring substituted with X and the unsubstituted ring respectively, ρ is a parameter of the reaction (see below), and σ_X is a measure of the electronic effect of the substituent.

The definition of σ_X is based on the acidity of the substituted benzoic acid and is given by

$$\sigma = \log_{10}(K_X/K_H)$$

where K_X is the ionization constant of the substituted benzoic acid and K_H is the ionization constant of benzoic acid itself. An alternative way of writing this is

$$\sigma_X = pK_{a(H)} - pK_{a(X)}$$

(remember that $pK_a = -\log_{10}K$).

Substitution of benzoic acid with electron-withdrawing substituents increases its acidity because they help to stabilize the negative charge in the conjugate anion. For these substituents, the ionization constant of the acid is greater than that of benzoic acid, so K_X/K_H is greater than 1, and σ_X is positive. Similarly, electron-donating substituents have a negative σ_X.

We thus have a quantitative measure of the extent to which a particular substituent is electron-donating or electron-withdrawing. This will vary according to whether the substituent is in the *meta* or *para* position, and may even change its sign. For example, the σ values of the methoxy (MeO) group are about +0.1 in the *meta* position, and about −0.3 in the *para* position. The reason for this difference is that the overall electronic effect of the methoxy group and many other substituents has two separate components, an inductive effect and a resonance (or mesomeric) effect. The first effect is due to the electronegativity of the oxygen, is electron-withdrawing, and acts primarily through the σ framework of the molecule. The second effect is due to the availability of an oxygen lone pair to donate into the π system. The inductive effect varies with distance, and so is weaker at the *para* position, whereas the π effect is due to the distribution of the π wavefunction, and is stronger or weaker on alternate carbon atoms. This can be shown by drawing resonance structures (Figure 3.23). When the methoxy group is at the *para* position, therefore, its inductive electron-withdrawing effect is weaker and its mesomeric electron-donating effect is stronger.

The parameter ρ is a measure of the electronic requirements of the reaction. It may be positive or negative. We can see from the Hammett equation that ρ must be positive for reactions which are favoured by electron-withdrawing substituents.

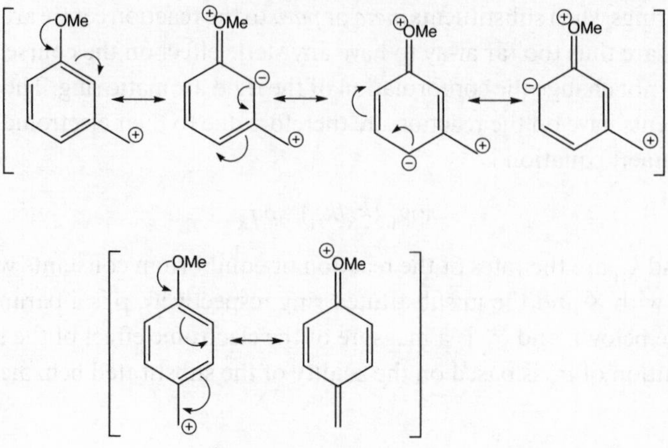

Figure 3.23

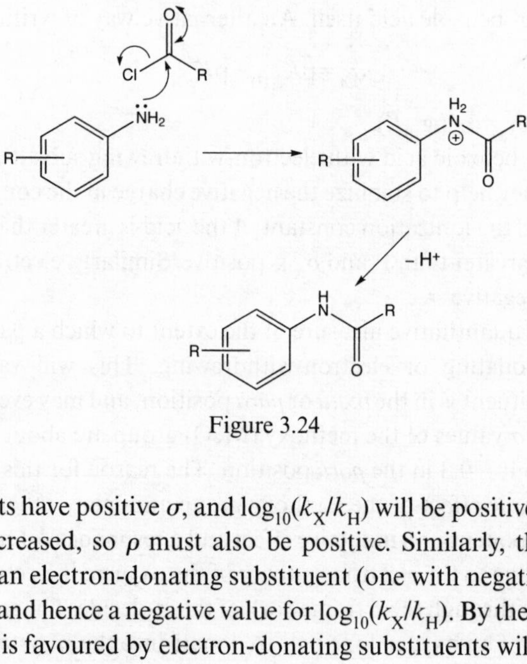

Figure 3.24

These substituents have positive σ, and $\log_{10}(k_X/k_H)$ will be positive because the rate of reaction is increased, so ρ must also be positive. Similarly, the same reaction carried out with an electron-donating substituent (one with negative σ) will lead to a decreased rate, and hence a negative value for $\log_{10}(k_X/k_H)$. By the same arguments, a reaction which is favoured by electron-donating substituents will have negative ρ.

Since the extent to which reactions are favoured by substituents with different electronic properties depends on the build-up of charge in the transition state, we may see ρ as a measure of this charge. For example, a reaction with a large negative value of ρ (about -2.7) is the acylation of aryl amines (Figure 3.24). In the transition state the nitrogen lone pair is partially bonded to the carbonyl carbon, so the nitrogen atom has a partial positive charge. This is stabilized by electron-donating substituents, resulting in an increase in the rate of reaction.

A reaction with a large positive ρ (about $+2.5$) is the base-catalysed hydrolysis of

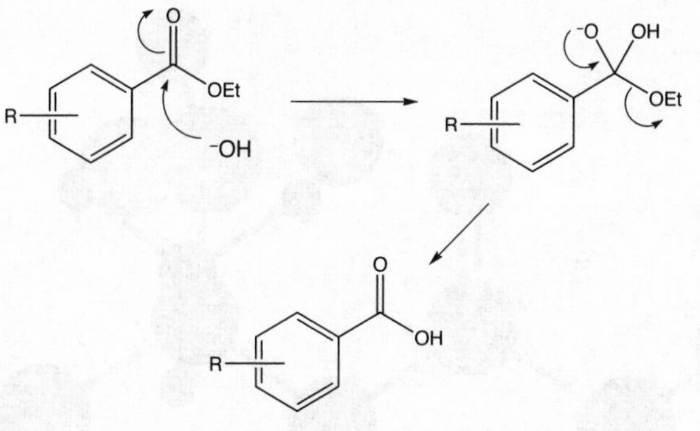

Figure 3.25

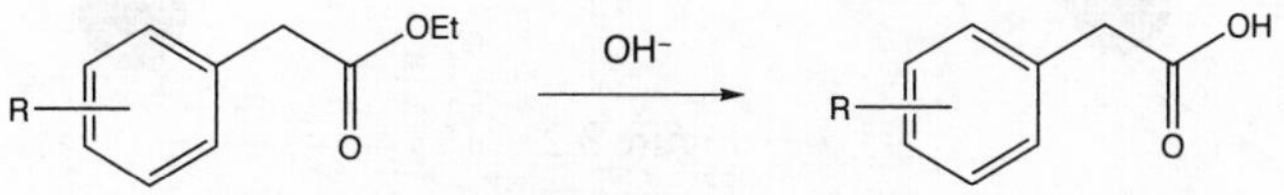

Figure 3.26

ethyl benzoates (Figure 3.25). Here there is a build-up of negative charge in the transition state, so the reaction is accelerated by electron-withdrawing substituents.

The magnitude of ρ is also significant; it indicates the extent to which the rate of reaction depends on the electronic properties of the substituents. For example, the hydrolysis of ethyl arylacetates (Figure 3.26) would be expected to be less sensitive to the effects of the substituents than the hydrolysis of ethyl benzoates, and this is reflected in the smaller ρ ($+0.82$).

Any parameter whose meaning depends on its sign in this way can be a source of confusion. To recap, electron-withdrawing substituents have positive σ and promote reactions with positive ρ. To help remember this, electron-withdrawing substituents tend to cause a *positive* charge and have *positive* σ.

It will come as no surprise to learn that plots of $\log_{10}(k_X/k_H)$ against σ for various substituents (known as Hammett plots) do not always give precisely straight lines, but the relationship is often reasonably good. Hammett plots can give useful information about the mechanism of reactions; this is discussed further in Section 7.5.

3.3.2 Steric effects

Although electronic effects are undoubtedly important for the rates of reactions, a reaction with favourable electronics will not necessarily take place quickly. Steric effects also play a major part.

Let us consider base-catalysed ester hydrolysis. While methyl esters are usually hydrolysed readily, t-butyl esters are extremely difficult to hydrolyse under basic conditions. Although there are electronic differences between a methyl and a t-butyl

77

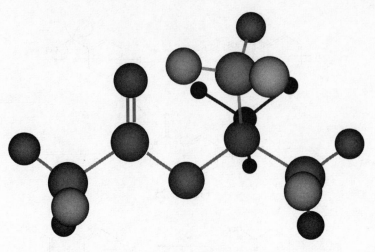

Figure 3.27

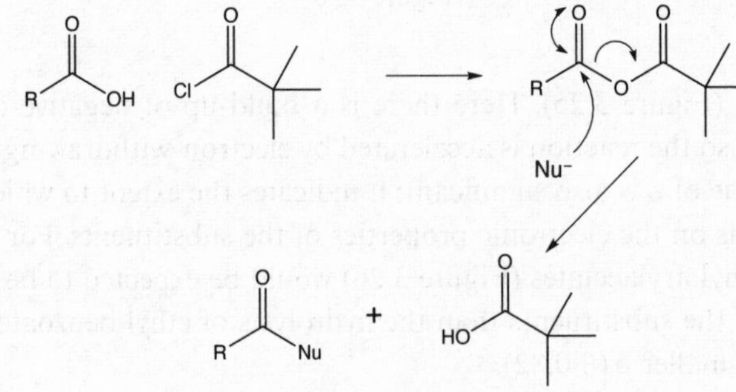

Figure 3.28

group, these groups are far enough removed from the site of reaction (the carbonyl carbon) that their effect on the reaction is small. The reason for this difference in reaction rates is the steric bulk of the t-butyl group. It is so large that it makes it difficult for an incoming nucleophile to approach the carbonyl group closely enough to react (Figure 3.27).

Ester hydrolysis can also be slowed by steric hindrance on the acid side of the molecule. Esters of pivalic acid are particularly difficult to hydrolyse. This low reactivity is exploited in a reaction sometimes used to make derivatives of carboxylic acids, in which a mixed anhydride is made using pivaloyl chloride (Figure 3.28). When this anhydride is treated with a nucleophile, the reaction takes place at the less crowded carbonyl group, and good yields can be obtained by this method. Figure 3.29 shows a three-dimensional representation of the mixed anhydride of acetic and pivalic acids.

Steric effects are common in many different types of reaction, and larger molecules frequently react more slowly than smaller ones in similar reactions. Another

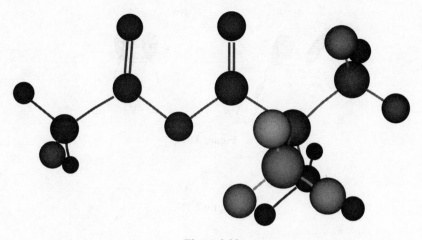

Figure 3.29

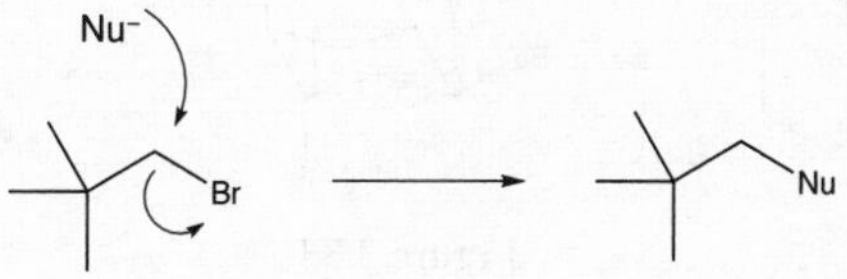

Figure 3.30

example is the S_N2 displacement of neopentyl bromide (Figure 3.30). When the nucleophile is the ethoxide ion, this displacement takes place at less than one millionth of the rate of the displacement of methyl bromide. Again, electronic effects contribute to the difference in rate, but their effect is dwarfed by the steric hindrance due to the bulky alkyl group.

3.3.3 Orbital overlap

One final requirement for rapid reactions is the alignment of the participating orbitals to enable them interact with each other. An example will make this clear.

To eliminate HBr by a base-catalysed E2 mechanism (see Section 6.2.2), the C−H and C−Br bonds must adopt the conformation shown in Figure 3.31, so that when the electrons in the C−H bond are released as the proton is removed, they may enter the C−Br antibonding orbital, causing the bond to break. In this conformation, the C−H and C−Br bonds are described as being antiperiplanar to each other (i.e. they are in the same plane, but on opposite sides of the bond from which they are being eliminated). The related term synperiplanar describes a conformation in which two bonds are in the same plane and on the same side of the bond (Figure 3.32). The bromocyclohexane shown in Figure 3.33 undergoes elimination rapidly, since in its most stable conformation the bromine and the hydrogen are antiperiplanar to each

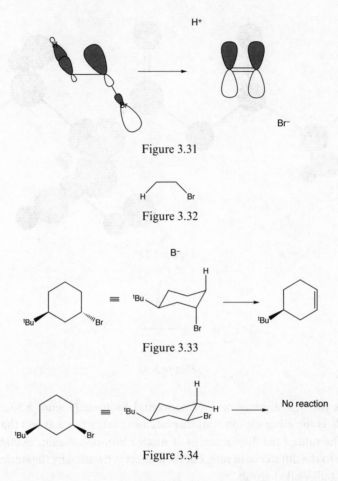

Figure 3.31

Figure 3.32

Figure 3.33

Figure 3.34

other (the bulky alkyl group must be equatorial, and therefore keeps the ring in the conformation shown). However, its isomer shown in Figure 3.34 cannot reach this conformation, so the bromide is not eliminated from this compound when treated with base. In the first isomer, the hydrogen and bromine are said to be *trans*-diaxial to each other.

Although orbital overlap effects can be at their most dramatic in cyclic systems, they are by no means limited to them. There are many reactions which are slower than might be expected because it is not possible for orbitals to overlap correctly. These effects are known as stereoelectronic effects, and we will meet them again when we look at E2 eliminations in Section 6.2.2.

Summary

- Reactions will take place only if they are thermodynamically favourable.

- A thermodynamically favourable reaction will result in a decrease in free energy.
- The equilibrium constant for a reaction is related to the free energy change.
- A reaction must also be kinetically favourable if we are to observe it, that is it must take place at a measurable rate.
- We can plot the energy of a reaction against its course in a reaction profile.
- Reactions with high activation energies are slower than those with low activation energies.
- An intermediate is a genuine chemical entity, and corresponds to a local energy minimum in a reaction profile.
- A transition state corresponds to an energy maximum and does not have a finite existence.
- Reactions may be under either thermodynamic or kinetic control.
- Reactions under thermodynamic control give the most stable product.
- Reactions under kinetic control give the product that is formed most easily.
- Factors affecting whether a reaction is thermodynamically favourable include bond strengths, ring strain, aromaticity, and entropy.
- Cyclization reactions are usually particularly favourable for entropic reasons.
- Factors affecting the kinetics of a reaction include electronic effects, steric effects, and, particularly in cyclic systems, orbital overlap.
- We can look at the electronic effects of substituents in aromatic systems with Hammett plots.

Problems

1. If, in the equilibrium $A+B \rightleftharpoons C+D$, the free energy of $C+D$ is lower than that of $A+B$, on which side does the equilibrium lie?
2. In the reaction profile below, which of the labels A, B, C, and D correspond to the kinetic product, the thermodynamic product, an intermediate, and a transition state?

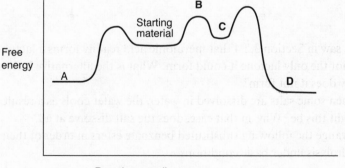

3. Which of the following reactions are shown with their intermediates, and which are shown with transition states?

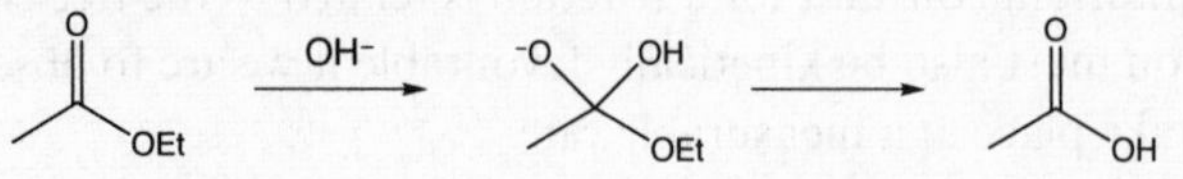

a

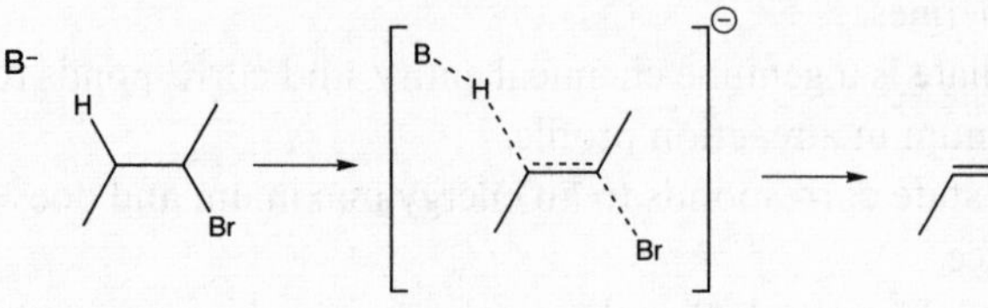

b

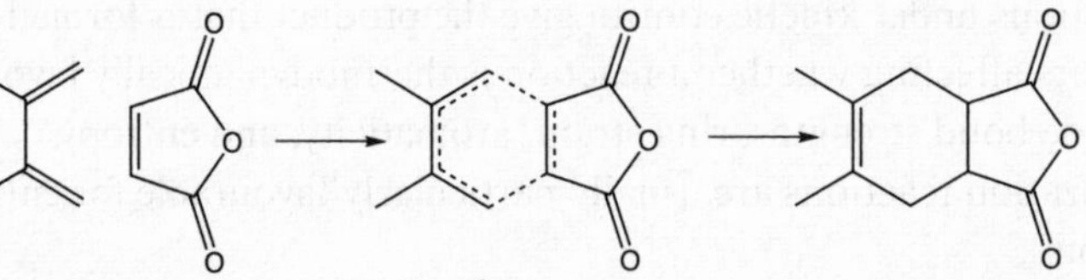

c

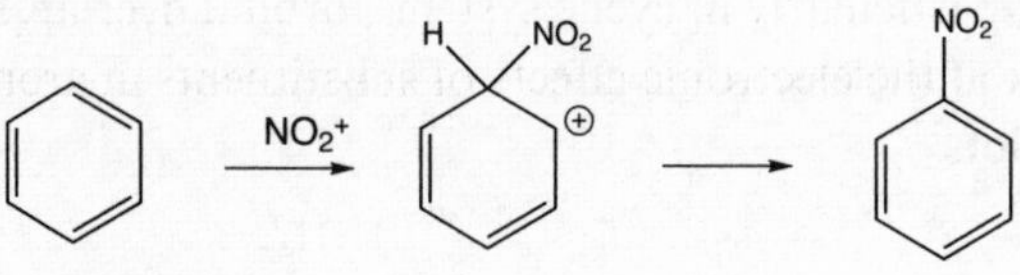

d

4. The elimination reaction shown below gives predominantly the terminal olefin. Is this reaction under kinetic or thermodynamic control? Why?

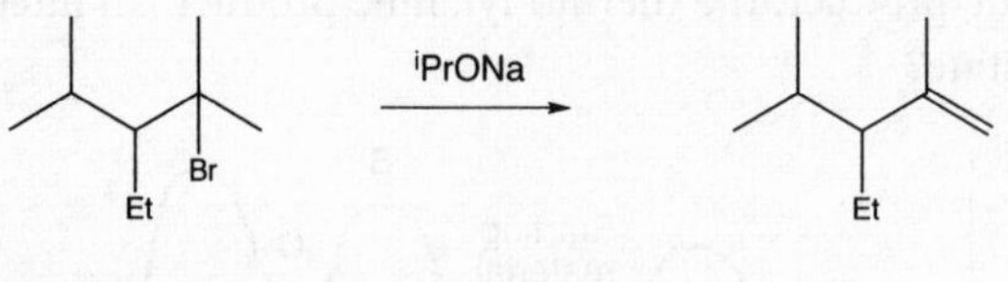

5. We saw in Section 3.2.4 that mevalonic acid readily forms a lactone, but this is not the only lactone it could form. What is the alternative lactone, and why does it not form?

6. When some salts are dissolved in water, the water cools as a result. Why might this be? Why, in that case, does the salt dissolve at all?

7. Arrange the following substituted benzoate esters in order of their rate of hydrolysis under basic conditions.

8. Which of these bromoalkenes will give an alkyne and which will give an allene on elimination with base?

4

Reactive Carbon Species

In any organic reaction, carbon plays a part. Although, as we shall see in Chapter 5, atoms other than carbon can have profound effects on the reactivity of organic molecules, reactions taking place at carbon atoms themselves are of great importance. In this chapter we shall look at reactive intermediates based on carbon. We shall see what types of intermediates carbon can form, and we shall look at the factors affecting their stability and at their common patterns of reactivity.

4.1 Carbanions

4.1.1 Structure and stability of carbanions

Firstly, we should be clear about what we mean by the term 'carbanion'. A carbanion is a molecule in which a negative charge resides on a carbon atom. This sounds like a straightforward definition, but in practice two factors can complicate it.

The first of these is that any anion must exist with a positive counter-ion; for carbanions this is almost always a metal ion. The complication here is that the carbanion and its counter-ion might not be independent of each other, but instead have a degree of covalent bonding (Figure 4.1). Although this covalent bonding may be negligible, as in the potassium salt of the triphenylmethyl anion, it can sometimes be particularly pronounced, as in organomercury compounds. Grignard reagents (organomagnesium compounds) are a common type of carbanion-like molecule in which the carbon–metal bond has significant covalent character.

This is not the place for a detailed discussion of the complex subject of bonding in organometallic compounds, and in this chapter we will refer to many compounds as carbanions, even if there is significant covalent bonding between the carbon and the metal. This is justified because in practice many of these compounds react largely as if they were pure carbanions. Nonetheless, we should not forget that this is a simplification.

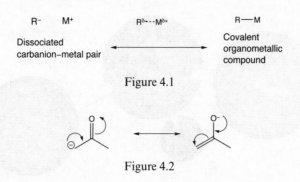

Figure 4.1

Figure 4.2

The second complication is that many carbanions are stabilized by delocalization of the negative charge onto heteroatoms, as in enolate anions (Figure 4.2). We could argue that such molecules are not carbanions at all, because most of the negative charge resides on the oxygen, and that they should therefore be described as oxyanions. This is largely a matter of semantics. What is important is that a carbon atom bears at least some of the negative charge. More to the point, the carbon atom is frequently the site at which the molecule reacts. We will therefore cheerfully describe all such molecules as carbanions.

The structure of carbanions will vary according to the extent of covalent bonding and the nature of any resonance stabilization of the negative charge. In the extreme in which bonding to the metal is entirely covalent, the structure will be just that of a normal covalent carbon compound, not that of a carbanion. As the bonding becomes less covalent, the carbon–metal bond gets longer, although the configuration of the carbon atom remains the same. When the bonding with the metal is purely ionic the negative charge occupies an orbital of its own, in the same way as the electrons of a lone pair on an atom such as nitrogen have their own orbital. Thus an isolated methyl anion[1] is sp_3 hybridized and has an approximately tetrahedral conformation, as shown in Figure 4.3. This is not a true tetrahedron because the lone pair orbital is not exactly equivalent to the three C−H bonds, just as ammonia is not a perfect tetrahedron.

If the negative charge can be delocalized, the carbon that bears the negative charge will normally change from sp_3 to sp_2 hybridization. An example is the allyl anion (Figure 4.4). In this, the lone pair occupies a p orbital which is able to interact with the neighbouring double bond. In fact, the two terminal carbon atoms are equivalent in the allyl anion because of this delocalization. The negative charge resides in an orbital which is spread over the molecule, rather than on any one atom. This orbital is the HOMO (see Section 1.4) of the molecule.

Enolate anions are similar. The negative charge is spread throughout a

---

[1] This is primarily a theoretical example – isolated methyl anions are not easy to prepare, although they have been detected spectroscopically.

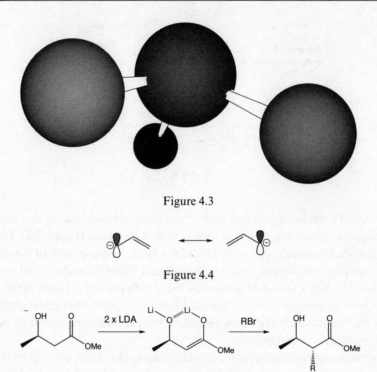

Figure 4.3

Figure 4.4

Figure 4.5

delocalized HOMO, some of the charge residing on carbon and some on oxygen. Because the anion now contains a heteroatom, the distribution of charge is no longer symmetrical, the oxygen being more negatively charged than the carbon. The associated cation (often lithium, as enolates are often prepared as their lithium salts) is usually bound more or less strongly to the oxygen atom. This can have an important effect on the reactivity. The metal ion can chelate to other negatively charged species (or those with lone pairs), and may thereby affect aspects of the reaction such as stereochemistry. For example, when the β-hydroxy ester shown in Figure 4.5 is treated with two equivalents of lithium diisopropylamide (LDA), it forms the dianion as shown, the lithium chelating the two oxygens. The molecule is therefore held in a cyclic conformation, and the methyl group above the ring inhibits the attack of an electrophile from this face, so the enolate is alkylated below the ring as shown.

Many of the factors affecting the stability of carbanions were discussed in Section 2.2.2. By far the most important of these is the delocalization of the negative charge. Delocalization may either be into a carbon π system, as in the benzyl anion, or, which is usually more favourable, onto an electronegative heteroatom, as in enolate anions and the anion derived from nitromethane.

Other important factors affecting carbanion stability are as follows:

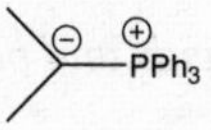

Figure 4.6

Figure 4.7

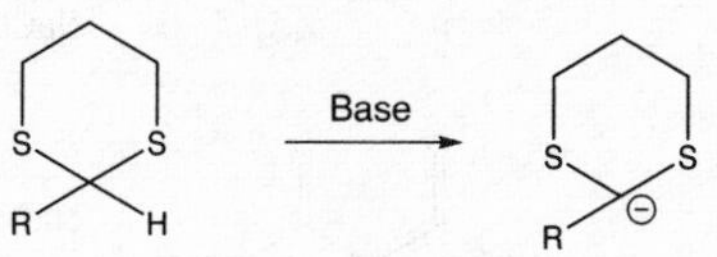

Figure 4.8

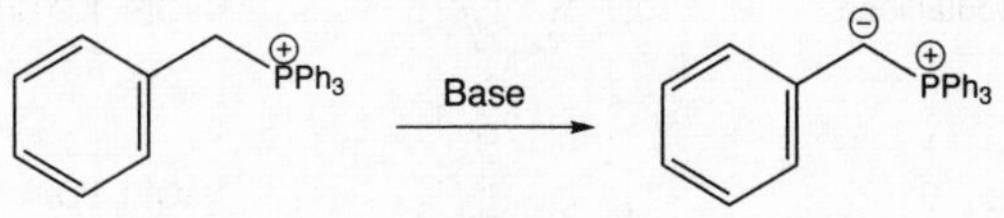

Figure 4.9

- Increasing alkyl substitution decreases the stability of the carbanion, thus tertiary carbanions are less stable than secondary, which are themselves less stable than primary.
- The hybridization of the carbon affects the stability, the most stable carbanions being sp hybridized and the least stable sp_3.
- Any electron-withdrawing group sufficiently close to the carbanion will help to stabilize it. Many electron-withdrawing groups (e.g. nitro groups) also act by resonance stabilization, but a common example of a pure electron-withdrawing group is the positively charged phosphorus atom in a phosphorus ylide (an ylide is a molecule with positive and negative charges on adjacent atoms), as shown in Figure 4.6.
- If deprotonation of the molecule creates an aromatic system, this will help to stabilize the carbanion. Hence the cyclopentadienyl anion (Figure 4.7) is considerably more stable than might be expected.
- Sulphur helps to stabilize an adjacent negative charge. The proton of thioacetals may be removed relatively easily (Figure 4.8).

These effects may of course act in combination. Thus the proton of the benzyl phosphonium salt in Figure 4.9 may be removed more easily than those of toluene or aliphatic phosphonium salts, in which only one of the stabilizing influences is at work.

Table 4.1. *Some carbon acids and bases used for their deportation*

Compound		Typical base used for deprotonation
β-diketo compounds	(structure: R–CO–CH₂–CO–R)	NaCO₃, NaOEt
Nitroalkanes	(structure: R–CH₂–NO₂)	NaCO₃, NaOEt
Malonate esters	(structure: RO–CO–CH₂–CO–OR)	NaOEt
Cyclopentadiene	(structure: cyclopentadiene ring)	NaOEt, KO^{t}Bu
Ketones	(structure: R–CH₂–CO–R)	NaOEt, LDA
Triphenylmethane	Ph₃C—H	BuLi

4.1.2 Formation of carbanions

The simplest and most common way of forming carbanions is treatment of the corresponding alkanes with a base. All carbanions may in fact be thought of as the conjugate bases of carbon acids, although these are often such weak acids that they would not normally be thought of as acids at all. Most carbanions are such strong bases that they would be completely protonated by water or other hydroxylic solvents, and so must be prepared in aprotic solvents (solvents with no acidic hydrogens) under anhydrous conditions. A common solvent for such reactions is tetrahydrofuran (THF). Table 4.1 shows some types of molecules from which carbanions may be prepared, and the bases commonly used for this.

Carbanions may be prepared in this way under equilibrating or irreversible conditions. For example, the treatment of ketones with LDA in THF forms the enolate salt of the ketone irreversibly. When ketones are treated with sodium ethoxide in ethanol, however, the enolate is formed reversibly (Figure 4.10). The position of the equilibrium will depend on the acidity of the ketone. For a simple ketone such as acetone, the equilibrium will lie mostly on the side of the neutral ketone, and there

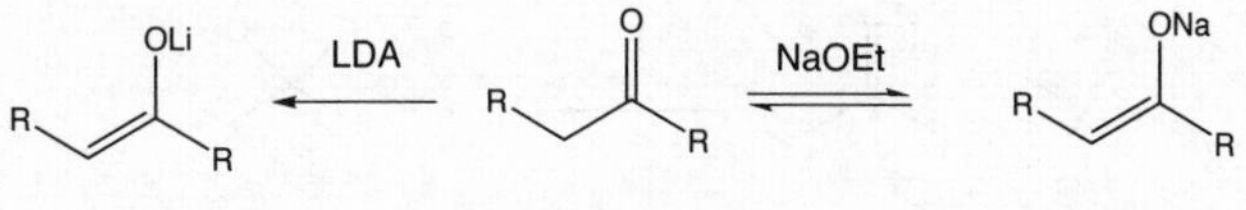

Figure 4.10

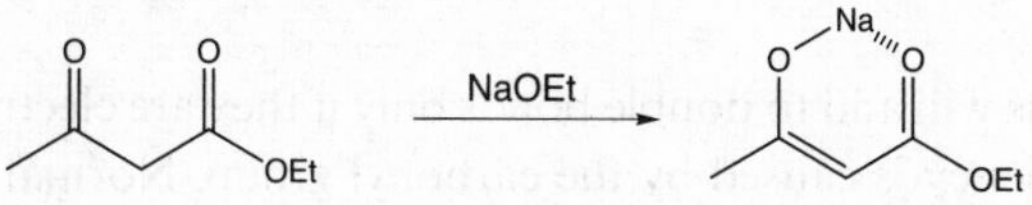

Figure 4.11

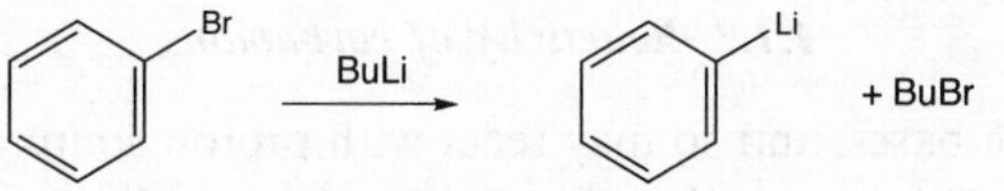

Figure 4.12

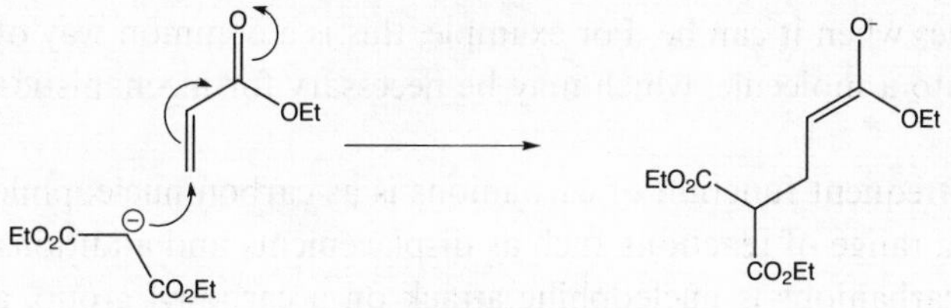

Figure 4.13

will be only a small concentration of the enolate (although this enolate is reactive and can play an important part in reactions). In contrast, a molecule such as ethyl acetoacetate, in which the negative charge is stabilized by delocalization over two carbonyl groups (Figure 4.11), will be present mostly as the enolate under these conditions.

Alkyl or aryl halides may be treated with certain metals to produce organometallic compounds. The most common example of this is treatment with magnesium to produce Grignard reagents. Alkyl lithium compounds are also commonly prepared in this way.

A similar reaction is halogen–metal exchange, in which an organic halide is treated with an organometallic compound. This is an equilibrium process, and if it is to be successful preparatively the two alkyl groups used must have very different reactivities. The most common use of this in practice is the treatment of relatively susceptible halides (frequently aryl) with butyllithium (BuLi) (Figure 4.12).

Carbanions may also be formed by addition of nucleophiles to multiple bonds. For example, the enolate derived from diethyl malonate may add to the double bond of ethyl acrylate to produce a new enolate (Figure 4.13). Of course, nucleophiles

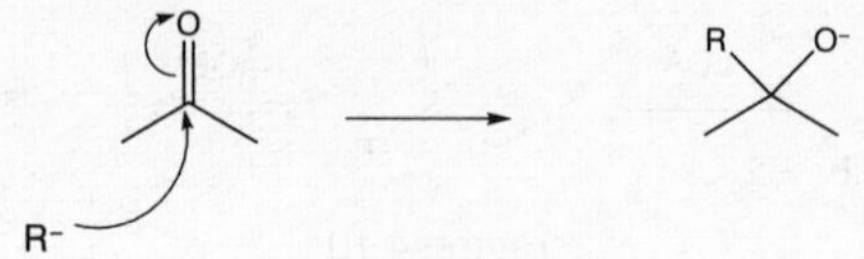

Figure 4.14

such as carbanions will add to double bonds only if they are electron-deficient; here, the electron deficiency is caused by the carbonyl group. Normal olefins react with electrophiles, not nucleophiles.

4.1.3 Reactivity of carbanions

All carbanions are bases, and so may react with proton sources to give the corresponding alkanes. This, as mentioned in Section 4.1.2, is the reason why carbanions must often be prepared under strictly anhydrous conditions. As the carbanions are normally prepared from these alkanes, this is seldom a useful reaction, although there are times when it can be. For example, this is a common way of introducing deuterium into a molecule, which may be necessary for mechanistic or biological studies.

The most frequent function of carbanions is as carbon nucleophiles. They thus take part in a range of reactions such as displacements and additions. A common reaction of carbanions is nucleophilic attack on a carbonyl group, a generalized form of which is shown in Figure 4.14. The oxyanion formed may become protonated on acid work-up[2] to give an alcohol, or, depending on the nature of the R group, it may react further. Reactions of carbonyl groups are discussed in more detail in Sections 5.1.3 and 6.1.2.

One such example, much loved by those setting undergraduate exams, is the reaction of Grignard reagents with esters (Figure 4.15). The intermediate oxyanion breaks down by loss of an alkoxide anion to form a ketone. This can then react further with another molecule of Grignard reagent to produce the tertiary alcohol. This reaction clearly requires two equivalents of Grignard reagent for one equivalent of the ester, but it will not stop at the ketone stage even if only one equivalent is used. This is because the ketone is more reactive than the ester (the lone pair of the extra oxygen in the ester is a good π donor, and so makes the ester carbonyl group less electrophilic), so the ketone will compete effectively with the ester for the

---

[2] Work-up is a term used to describe operations done when the main reaction is complete and before the product can be isolated. It may be relatively simple, as it is here, where acid is added to the reaction to protonate the oxyanion, or it may have more complex effects. Whilst some form of work-up procedure is used in almost all reactions, it is often ignored in descriptions of reactions, especially if it is merely a pH adjustment. You should therefore not be too surprised to see protons come and go seemingly by magic in some reaction schemes, although it is good practice always to write the steps that cause these transformations explicitly.

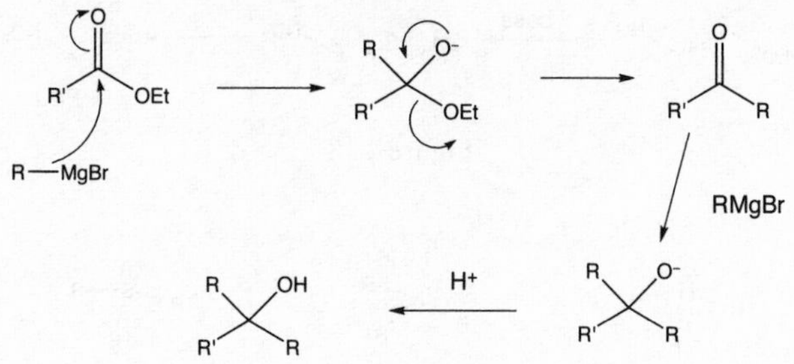

Figure 4.15

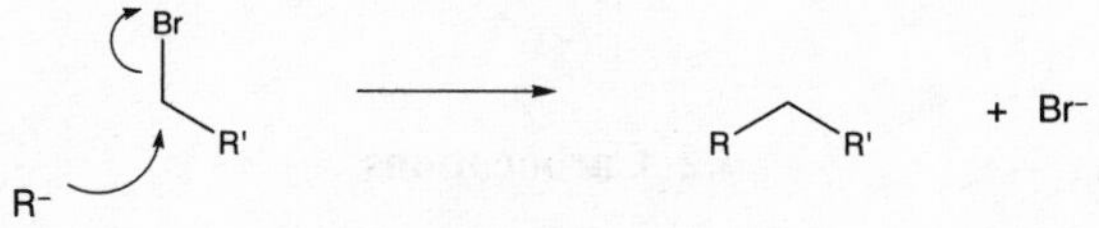

Figure 4.16

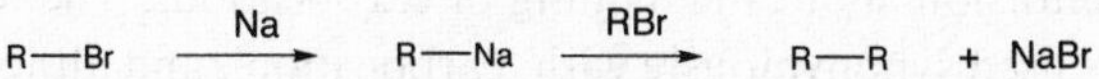

Figure 4.17

available Grignard reagent. The result is that half of the ester is converted to the tertiary alcohol and half remains unreacted. The reaction is of course an efficient way of making tertiary alcohols, if two of the alkyl groups are identical.

Although many carbanions will react in the expected manner with alkyl halides in substitution reactions (Figure 4.16), Grignard reagents are a notable exception. This should not come as a great surprise, as Grignard reagents are prepared from alkyl halides. If they did react readily in this manner, then they would react with the as yet unreacted alkyl halide under the conditions of their formation, and could not be made. This does indeed happen when sodium reacts with alkyl halides, and is the basis of the Wurtz coupling (Figure 4.17). This reaction tends to produce a large number of side products and is seldom used preparatively. The readiness of the alkyl halides to undergo Wurtz coupling in the presence of sodium is the reason why the more esoteric metal magnesium is so commonly used to prepare organometallic compounds from alkyl halides.

Carbanions may also undergo elimination reactions. This is known as the E1cB mechanism (see Section 6.2.4), and whilst it is not the most common mechanism for eliminations, there are some that proceed in this way. One example is the reaction of the nitro compound shown in Figure 4.18, which is discussed further in Section 6.2.4.

In addition to β-eliminations, carbanions may also undergo α-eliminations, which form carbenes rather than olefins. This is discussed in more detail in Section 4.4.2.

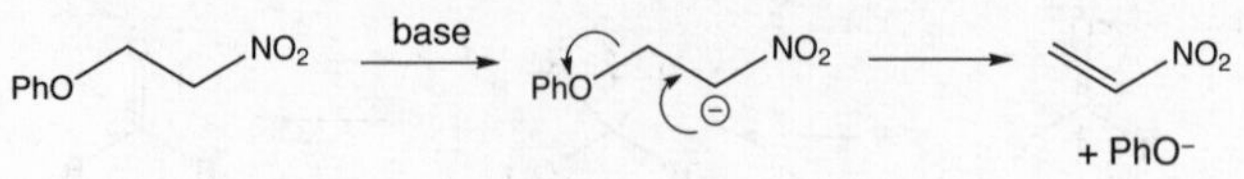

Figure 4.18

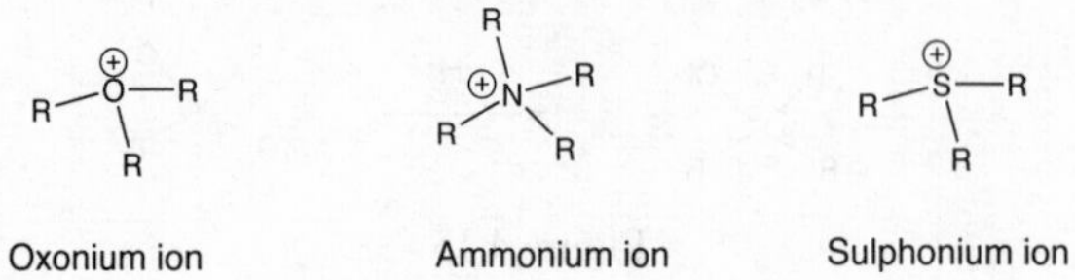

Oxonium ion   Ammonium ion   Sulphonium ion

Figure 4.19

4.2 Carbocations

4.2.1 Structure and stability of carbocations

There is some confusion about the naming of carbocations. The term 'carbonium ion' is sometimes used synonymously with 'carbocation', and although this has the justification of usage, it is frowned upon by many chemists. The suffix '-onium' is used for other atoms to describe cations which are formed by sharing what is usually a lone pair, in other words hypervalent (having a higher than usual valency) cations. Examples are oxonium, ammonium, and sulphonium ions (Figure 4.19). Most carbocations, in contrast, are formed by removing a substituent and its electron pair from the carbon, leading to a hypovalent (having less than its usual valency) cation. The inconsistency in calling these species carbonium ions is obvious, but this seldom leads to confusion, because hypervalent carbocations are extremely rare.[3] Nonetheless, the term is best avoided. The term 'carbenium ion' is sometimes used to describe hypovalent carbocations, although it is not heard often. 'Carbocation' will be used here. Although this is a general term which encompasses both hypervalent and hypovalent carbon, this will not lead to confusion as we shall not be considering any hypervalent carbocations.

Most carbocations have only three substituents and are therefore usually sp_2 hybridized. A typical structure is that of the methyl cation, which was discussed in Section 1.1.3. There is a vacant p orbital perpendicular to the plane of the molecule; this is the LUMO. In all reactions of carbocations there is an interaction between this LUMO and some source of electrons, usually the HOMO of another molecule.

Although carbocations may be tightly associated with their counter-ions in close ion pairs, the continuum between covalent and ionic bonding does not muddy the water here to anything like the same extent as it does in the description of carbanions. A carbocation is, in general, a purely ionic species.

---

[3] The most commonly encountered of such ions is the CH_5^+ ion, which is produced in chemical ionization mass spectrometry. This acts as a strong acid and protonates the molecules in the sample.

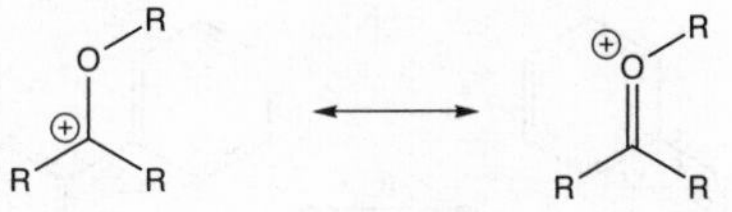

Figure 4.20

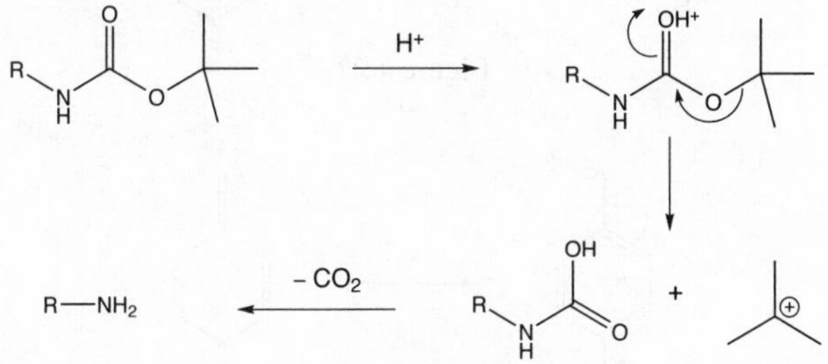

Figure 4.21

However, the positive charge on carbocations is frequently stabilized by hetero-atoms, and this gives us the same problem as we encountered in defining carbanions, namely whether the species should be called a carbocation at all. In general, if a carbocation is stabilized by interaction with the lone pair of an adjacent heteroatom, it is usually not called a carbocation. For example, the ion shown in Figure 4.20 would normally be called an oxonium ion.

The factors affecting carbocation stability are similar to those affecting carban-ion stability, although of course many of them act in the opposite direction. Simple alkyl carbocations are more stable with increasing alkyl substitution; in other words, tertiary carbocations are the most stable and primary carbocations the least so. The stability of the t-butyl cation is illustrated by the properties of the t-butoxycarbonyl (BOC) group, which is used as a protecting group for amines, primarily in peptide synthesis. When this is treated with acid and becomes protonated, the t-butyl cation spontaneously cleaves off, and the resulting carbamic acid decarboxylates to give the original amine (Figure 4.21). The reaction normally takes place under mild condi-tions, typically within minutes at room temperature.

The stability conferred by alkyl substitution is due mainly to two factors. One of these is hyperconjugation, which was discussed in Section 1.2. The other is the inductive effect of the alkyl groups. Alkyl groups are slightly electron-donating, and thus help to neutralize and so stabilize the positive charge.

An adjacent heteroatom with a lone pair (for example oxygen or nitrogen) can have a considerable stabilizing influence on a carbocation, although as described above, it is not always clear that such species should be called carbocations at all.

Like carbanions, carbocations can be stabilized by resonance. Thus the allyl cation is more stable than normal primary cations, and the benzyl cation even more so. The trityl cation is particularly stable (Figure 4.22). The stability of such cations

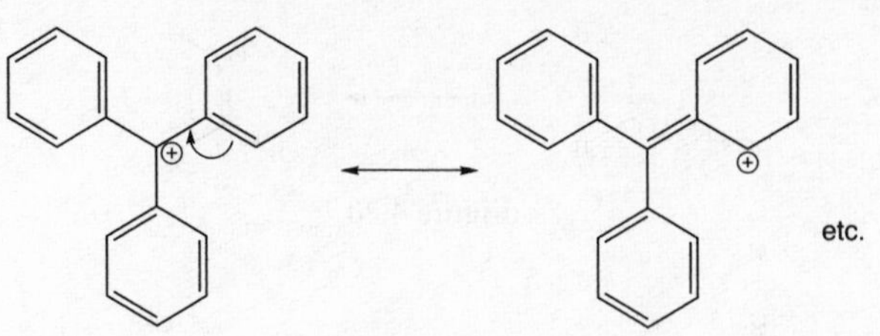

Figure 4.22

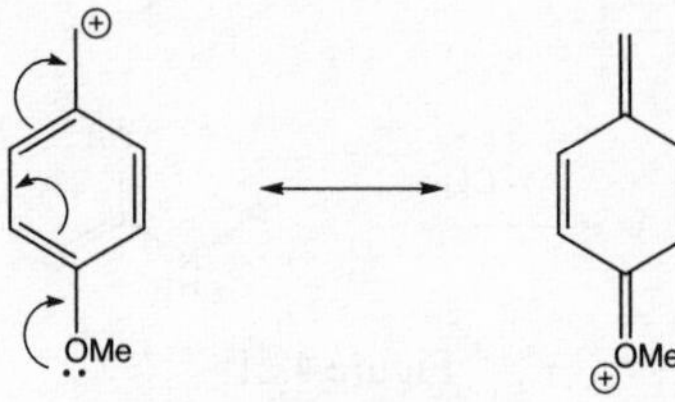

Figure 4.23

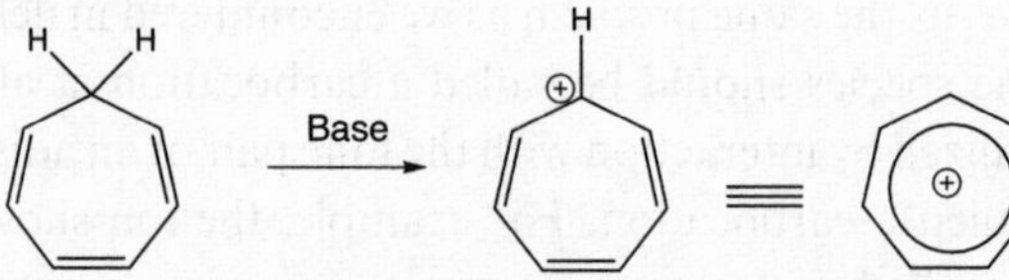

Figure 4.24

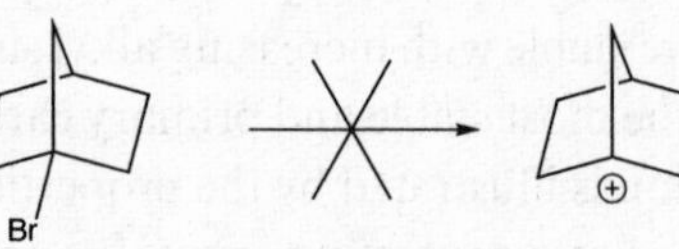

Figure 4.25

can be further increased if electron-donating substituents feed into the π system. For example, the p-methoxy benzyl cation (Figure 4.23) is considerably more stable than the unsubstituted benzyl cation .

As with carbanions, formation of an aromatic species helps to stabilize a carbocation. Probably the most common example of this is the tropylium cation (Figure 4.24). This is so stable that tropylium bromide does not exist as a covalent compound, but is completely dissociated as an ionic compound, both in the solid and in acetonitrile solution.

Finally, it is important to remember that carbocations are normally planar, and if a molecule cannot adopt a planar configuration, this will have the effect of destabilizing a carbocation. For example, it is not possible to form a carbocation from norbornyl bromide, since the bridgehead carbon cannot become planar (Figure 4.25).

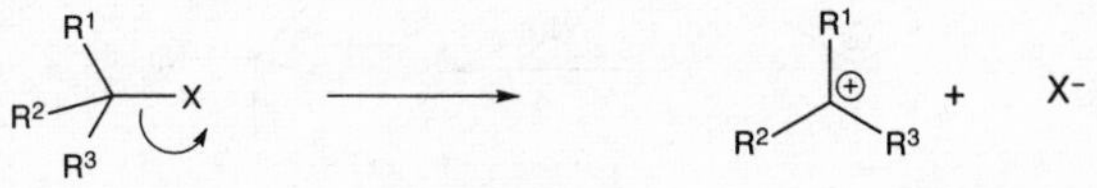

Figure 4.26

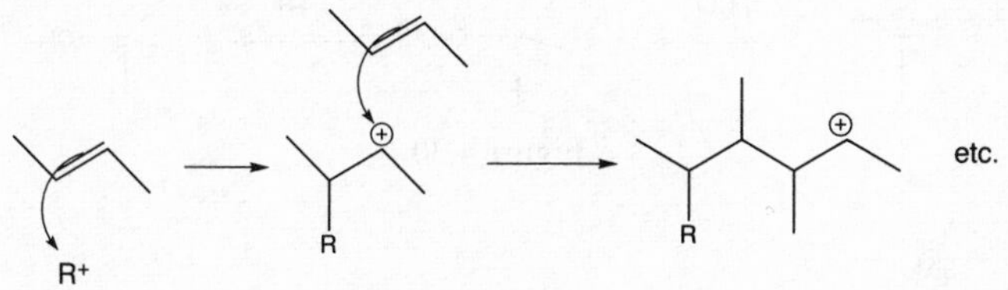

Figure 4.27

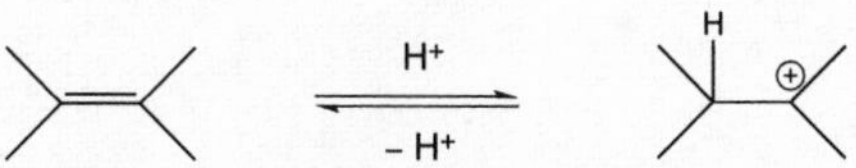

Figure 4.28

4.2.2 Formation of carbocations

There are two main ways in which carbocations can form. One (Figure 4.26) is the cleavage of a bond between carbon and another atom, which leaves with its lone pair (a nucleofugal leaving group, see Section 2.5). This reaction is favoured if the resulting carbocation is stable (see above), and if the nucleofugal leaving group is a good leaving group (factors affecting this were discussed in Section 2.5). Compounds are also more likely to dissociate in this way in polar solvents, as this can help to stabilize the charges formed.

The second way is for a positively charged species to add to a carbon–carbon double bond. The positively charged species is most often, but not always, a proton. It may be a carbocation, in which case a chain reaction can take place, leading to polymerization (Figure 4.27).

When the electrophile is a proton, this reaction is readily reversible. The formation of a carbocation from an olefin (Figure 4.28) is the reverse of the second step in an E1 elimination reaction (see Section 6.2.3).

4.2.3 Reactivity of carbocations

There are three major reaction pathways open to carbocations, two of which are simply the reverse of the two ways of formation described above. These two pathways should not come as a surprise if we are familiar with the properties of electrophiles in general (see Section 2.4), and remember that carbocations are just a type of electrophile.

95

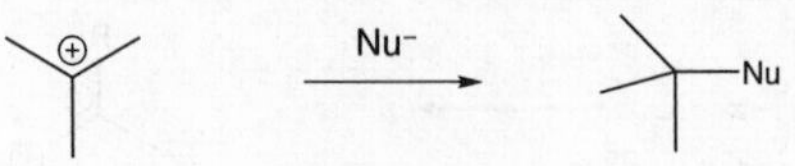

Figure 4.29

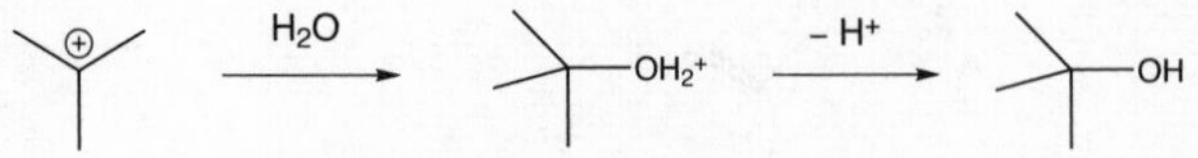

Figure 4.30

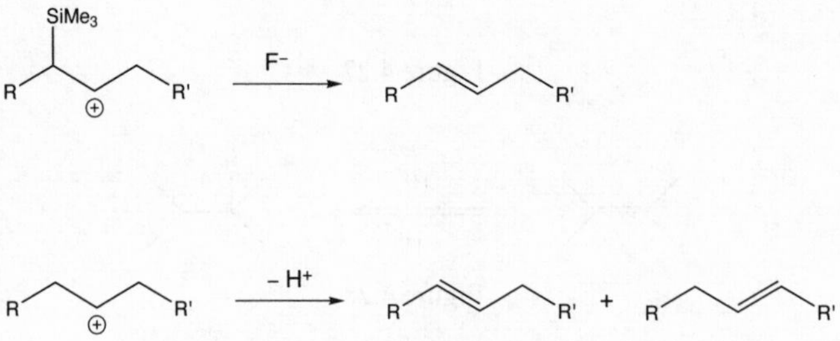

Figure 4.31

They may combine with a nucleophile, which will give a neutral species if the nucleophile is negatively charged (Figure 4.29). This will not necessarily always be so; for example, water is a good nucleophile and readily combines with carbocations to form protonated alcohols. These may then simply lose a proton to form the neutral alcohol (Figure 4.30).

Carbocations may lose a proton or other electrofugal leaving group from the position adjacent to the positively charged carbon, resulting in an olefin (the reverse of their formation from protonation of double bonds, Figure 4.28). It is almost always a proton which is lost, but other groups may take on this role. For example, the trimethylsilyl (TMS) group is able to leave in this way (Figure 4.31). It usually leaves in preference to a proton, and so the formation of a carbocation adjacent to a TMS group can lead to better definition of the position of the double bond in the resulting olefin (see also Section 2.5.3). This contrasts with the elimination of a proton from a carbocation, where the position of the double bond may vary if there is more than one α proton, and a mixture of products often results.

The third way in which carbocations commonly react is by rearrangement. This takes place so that a more stable carbocation is formed, such as a tertiary instead of a secondary carbocation. For example, the neopentyl cation rapidly rearranges as shown in Figure 4.32 so that a tertiary cation replaces a primary one.

This ready rearrangement can sometimes be a nuisance in reactions involving carbocations, and can lead to unexpected products. It is one of the major limitations

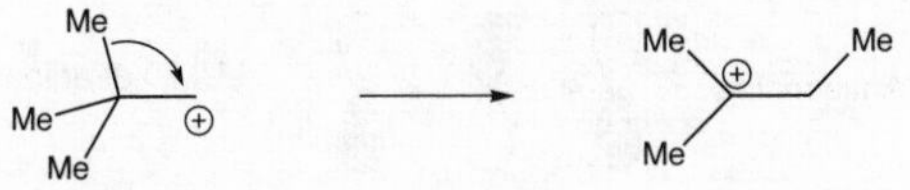

Figure 4.32

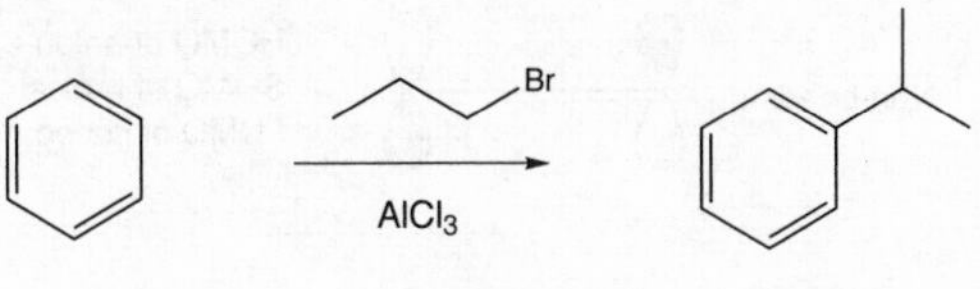

Figure 4.33

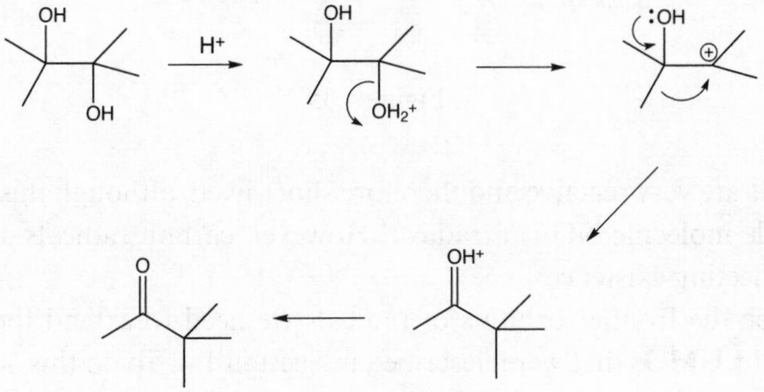

Figure 4.34

of the Friedel–Crafts alkylation of aromatic compounds (see Section 6.3.6). For example, attempted preparation of n-propylbenzene by this route, treating benzene with n-propyl bromide in the presence of a Lewis acid catalyst, produces iso-propylbenzene (Figure 4.33).

Nonetheless, rearrangement of carbocations can sometimes lead to predictable and useful reactions. One such is the pinacol rearrangement, in which a 1,2 diol is treated with acid (Figure 4.34). Protonation of one of the hydroxyl groups allows it to leave, giving a carbocation. Although this is tertiary, rearrangement gives an even more stable cation, because the positive charge on the carbon is stabilized by the oxygen lone pair.

4.3 Radicals

4.3.1 Structure and stability of radicals

Free radicals, normally referred to simply as radicals, are species containing an unpaired electron (denoted by a dot adjacent to the relevant atom). They are normally electrically neutral, but can sometimes be positively or negatively charged.

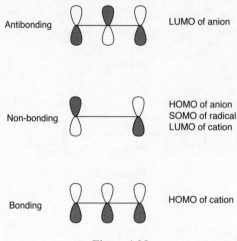

Antibonding   LUMO of anion

Non-bonding   HOMO of anion
SOMO of radical
LUMO of cation

Bonding   HOMO of cation

Figure 4.35

Most radicals are very reactive and therefore short-lived, although this is not always so. The stable molecule NO is a radical. However, carbon radicals almost always have only a fleeting existence.

To describe the frontier orbitals of radicals we need to extend the concepts of HOMOs and LUMOs that were described in Section 1.4. To do this, let us consider various forms of the allyl system. The structure is much the same whether we are considering the allyl cation, anion, or radical, and the molecular orbital diagrams are also similar. These are shown (π orbitals only) in Figure 4.35. The three atomic p orbitals combine to produce a bonding orbital, a non-bonding orbital, and an antibonding orbital.

The only qualitative difference between the three species is the number of electrons in the non-bonding orbital: none in the cation, one in the radical, and two in the anion.[4] Thus in the cation the bonding orbital is the HOMO and the non-bonding orbital the LUMO, and in the anion the non-bonding orbital is the HOMO and the antibonding orbital the LUMO. But what about the radical? Here the non-bonding orbital is only half occupied, so it is not entirely clear whether it should be called a HOMO or a LUMO. In fact, there is a special term for it: a SOMO, or Singly Occupied Molecular Orbital.

It is the SOMO that takes part in the reactions of radicals; this concept will be used below in Section 4.3.3 to describe many features of the reactivity of radicals. We saw in Section 2.3.2 that frontier orbital interactions are more important in soft nucleophiles and electrophiles, and less important in hard species. Most radicals are extremely soft, as they are uncharged, and so these frontier orbital interactions are important in explaining the reactivity of radicals.

---

[4] Note that this orbital is the frontier orbital for each species. We can see that the coefficient at the central carbon atom is zero in this orbital, which is why allyl systems always react at one of their ends.

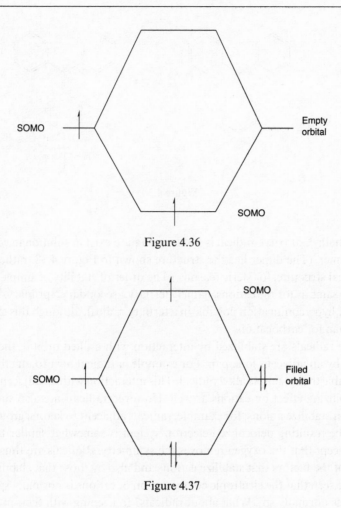

Figure 4.36

Figure 4.37

The SOMO is also important in explaining the stability of radicals. An interaction between a singly occupied orbital and either an empty or a full orbital leads to a decrease in energy, and so the interaction has a stabilizing effect. If the singly occupied orbital interacts with an empty orbital, it is clear that the energy of the electron will be lowered, as it can occupy the lower in energy of the two molecular orbitals formed, which now becomes the SOMO (Figure 4.36). A singly occupied orbital will also be stabilized by interaction with a filled orbital, however, as two of the electrons enter the lower-energy molecular orbital, and only one enters the higher (now the SOMO) (Figure 4.37).

Because of this, the most important factor contributing to the stability of radicals is conjugation. The allyl radical, described above, is more stable than simple alkyl radicals because of the conjugation with the adjacent double bond. Benzyl radicals are still more stable, and increasing conjugation increases the stability further, so the

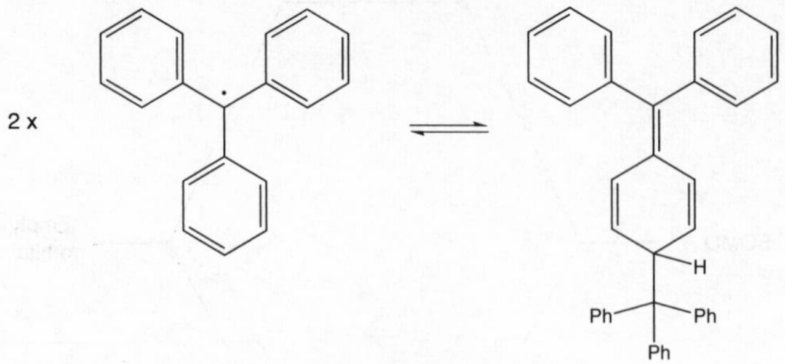

Figure 4.38

triphenylmethyl, or trityl, radical is stable enough to exist in solution in equilibrium with its dimer. (The dimer has the structure shown in Figure 4.38, rather than the symmetrical structure, for steric reasons.) The order of stability of simple alkyl radicals is the same as for the cations, namely tertiary>secondary>primary, because of the greater hyperconjugation possible in a tertiary radical, although this effect is less marked than for carbocations.

Because radicals are stabilized by interaction with a filled orbital, they are also stabilized by an adjacent lone pair. For example, a radical next to an ether oxygen is more stable than a simple alkyl radical. This interaction with an adjacent lone pair has a stabilizing effect on cations as well. However, radicals are also stabilized by factors that stabilize anions; for example, radicals adjacent to ketone groups are stabilized. The resulting delocalized electron system is somewhat similar to the allyl radical, except that the oxygen removes the symmetry. Radicals are thus stabilized by many of the factors that stabilize cations and also by those that stabilize anions.

We have seen that the electronic configuration of cations is normally sp_2 and that of anions is normally sp_3. What about radicals? In keeping with their intermediate nature, they may be either sp_2 or sp_3, depending on the circumstances. The sp_2 configuration is normally favoured, as is confirmed by the well-known observation that the stereochemistry at an asymmetric carbon is normally lost if that carbon forms a radical. This configuration will be even more favoured if conjugation is possible. However, it is also possible to form radicals with sp_3 geometry. For example, while the impossibility of forming the sp_2 configuration precludes the formation of a cation from the bridged alkyl bromide shown in Figure 4.39, it is perfectly possible to form a radical. This radical must have an sp_3 configuration because of the shape of the molecule.

4.3.2 Formation of radicals

As radicals are mainly reactive intermediates, they are not normally present at the start of a reaction. There are two general ways in which they may be formed:

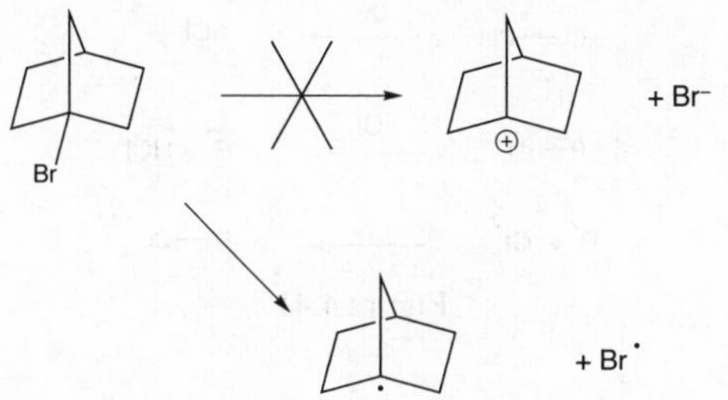

Figure 4.39

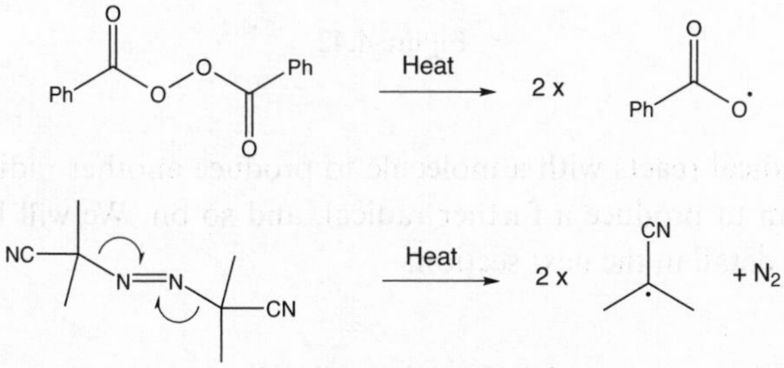

Figure 4.40

(1) by homolytic cleavage of a paired-electron molecule;
(2) by reaction of a radical with a paired-electron molecule.

It may appear that the second of these is a bit of a cheat, because it presupposes that a radical be present, but this is the way in which carbon radicals are normally formed. When carrying out radical reactions, it is common practice to use what is known as a radical initiator. A radical initiator is a substance that readily undergoes homolytic cleavage to supply a source of radicals, which can then react with paired-electron molecules, as in (2) above, to produce the radicals of interest. Benzoyl peroxide and 2,2'-azobisisobutyronitrile (AIBN) are commonly used radical initiators, both of which cleave to give radicals on heating (Figure 4.40). Halogens may also be used as radical initiators, although these are more easily cleaved photochemically than thermally.

Let us consider the free radical chlorination of alkanes as an example. A mixture of an alkane and chlorine is treated with ultraviolet (UV) light, which splits the chlorine into radicals. A chlorine radical then reacts with the alkane, abstracting a hydrogen to produce a carbon radical. This carbon radical can react with a chlorine radical to give the alkyl halide (Figure 4.41).

The chlorine also plays the part of a reagent here, as well as that of a radical initiator, although radical initiators are often used in just catalytic amounts, and serve only to start a radical chain which can then propagate itself. By 'radical chain', we

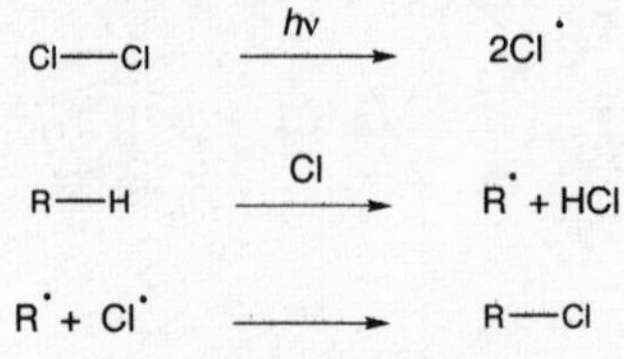

$$\text{Cl—Cl} \xrightarrow{h\nu} 2\text{Cl}^\bullet$$

$$\text{R—H} \xrightarrow{\text{Cl}} \text{R}^\bullet + \text{HCl}$$

$$\text{R}^\bullet + \text{Cl}^\bullet \longrightarrow \text{R—Cl}$$

Figure 4.41

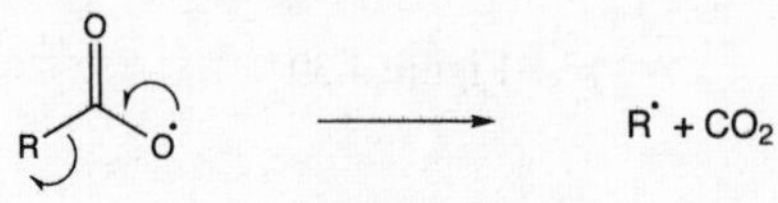

$$\longrightarrow \text{R}^\bullet + CO_2$$

Figure 4.42

mean that a radical reacts with a molecule to produce another radical, which can then react again to produce a further radical, and so on. We will look at radical chains in more detail in the next section.

4.3.3 Reactions of radicals

Many radical reactions are chain reactions, in which new radicals are constantly being formed. The steps in a radical chain reaction can be divided into three types.

(1) Initiation

The initial formation of a radical, normally by homolytic cleavage as described above:

$$X-Y \rightarrow X^\bullet + Y^\bullet$$

(2) Propagation

The step in which a radical reacts with a paired-electron molecule, giving rise to a further radical. This may be abstraction of a hydrogen or other group, as in the chlorination reaction discussed above, or addition to a multiple bond:

$$X^\bullet + H-R \rightarrow X-H + R^\bullet$$
$$X^\bullet + A = B \rightarrow X-A-B^\bullet$$

(3) Termination

The step in which two radicals combine to form a paired-electron molecule:

$$X^\bullet + R^\bullet \rightarrow X-R$$

A further possible fate of a radical is decomposition or rearrangement, which is another kind of propagation step. An example is the decomposition of carboxylate radicals (Figure 4.42).

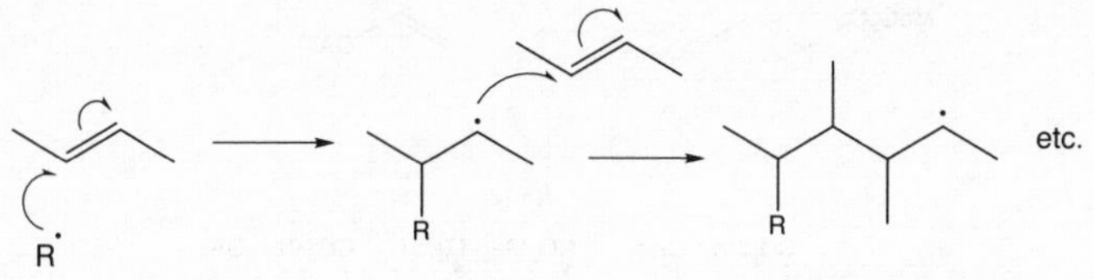

Figure 4.43

Because the concentration of radicals is usually low, the collision of two radicals will be less frequent than the collision of a radical with a paired-electron molecule, so there will normally be many propagation reactions for every termination reaction. This gives rise to what is known as a chain reaction. An example is the radical-initiated polymerization of olefins. Initial attack of a radical on an olefin gives a radical, which can then react with another molecule of olefin, and so on, leading to a long hydrocarbon chain (Figure 4.43). Eventually, this chain will react with a radical in a termination reaction, but it is quite possible for molecules with molecular weights in the range of many thousands to be produced before this happens.

In looking at the pattern of reactivity of radicals, it is helpful to consider the SOMOs. In Section 4.3.1 we saw that a radical could be stabilized if its unpaired electron interacts with either a filled or an empty orbital. However, when the interaction is with an empty orbital, the energy of the unpaired electron itself (the SOMO) is lowered, while interaction with a filled orbital leads to a SOMO of raised energy.

In a reaction, a SOMO can interact with the HOMO or the LUMO of a molecule, in the same way that an unpaired electron can be stabilized by interactions with both filled and empty orbitals. For any given reaction, one of these two interactions is likely to be dominant. If the SOMO has a low energy, then the dominant interaction will be with the HOMO of the molecules with which it is reacting, and if the SOMO has a high energy, the interaction with the LUMO will predominate. We can therefore regard radicals with low-energy SOMOs (normally those stabilized by interaction of the unpaired electron with a vacant orbital) as electrophilic radicals, and those with high-energy SOMOs (normally those stabilized by interaction of the unpaired electron with a filled orbital) as nucleophilic radicals. We should remember that these terms are relative, and a nucleophilic radical will not normally be as nucleophilic as a nucleophile in the conventional sense; the same applies to electrophilic radicals.

Nonetheless, consideration of the energy of the SOMO of a radical allows us to make useful predictions about the reactivity. As an example, let us consider the radical copolymerization of dimethyl fumarate and vinyl acetate. This polymerization gives a regular arrangement of alternating monomers with the vinyl acetate unit in the same orientation throughout the chain (Figure 4.44). Clearly, some sort of control is operating, as we do not get a random mixture.

This control can easily be explained by considering the SOMOs of the intermediate species. If the radical initiator, Ra·, reacts first with the dimethyl fumarate, this

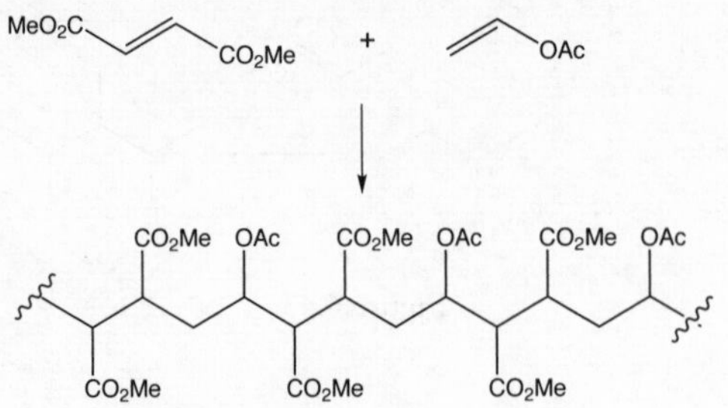

Figure 4.44

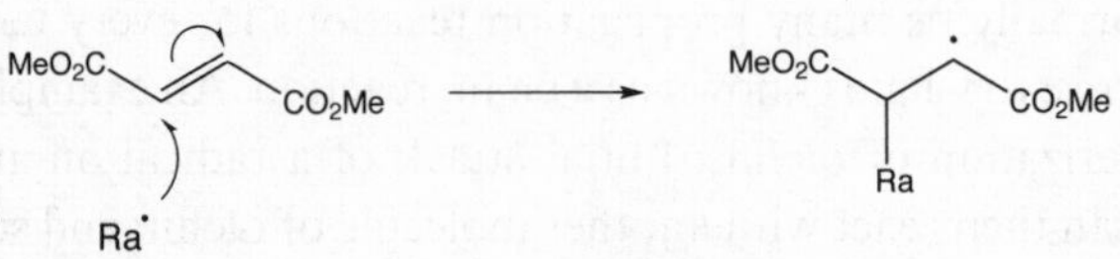

Figure 4.45

forms the radical shown in Figure 4.45, in which the unpaired electron is stabilized by interaction with the vacant carbonyl π^* (antibonding) orbital. The SOMO is therefore of low energy and the radical is electrophilic. This radical has a choice of two double bonds to attack: the electron-poor bond of fumarate (electron-withdrawing effect of the carbonyl groups), or the electron-rich bond of vinyl acetate (electron-donating effect of the oxygen lone pair). Since the radical is electrophilic, it is obvious that it will react with vinyl acetate. It is possibly more satisfactory to explain this in terms of the molecular orbitals: the radical derived from fumarate has a low-energy SOMO, and will therefore have a stronger interaction with the HOMO of the reacting molecule than with its LUMO. Vinyl acetate has a high-energy HOMO because of the interaction between the double bond and the oxygen lone pair. In contrast, the HOMO of dimethyl fumarate is of much lower energy, because of the interaction with the vacant carbonyl π^* orbitals. This tells us that the frontier orbital interaction will be stronger with vinyl acetate than with dimethyl fumarate; this is indeed how the radical reacts.

Frontier orbitals can also explain the orientation of this reaction. Remember that the vinyl acetate units are all aligned in the same direction in this polymerization, so this orientation is controlled. We have already seen that the important frontier orbital of the vinyl acetate is its HOMO; to explain the orientation of the reaction we need to consider what this HOMO looks like. This system is reminiscent of the allyl anion; the double bond is conjugated with a further two electrons, the only difference is that here the extra electrons come from an oxygen lone pair, and in the allyl anion they come from a negatively charged carbon.

The structure of the allyl anion was discussed in Section 4.1.1. The HOMO is the non-bonding orbital, in which the molecular orbital coefficient on the central

104

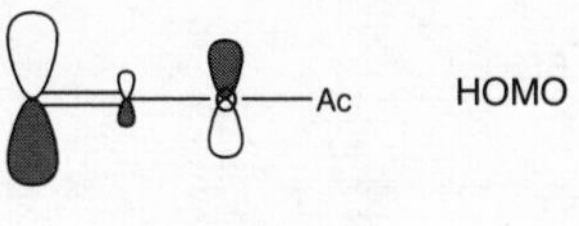

Figure 4.46

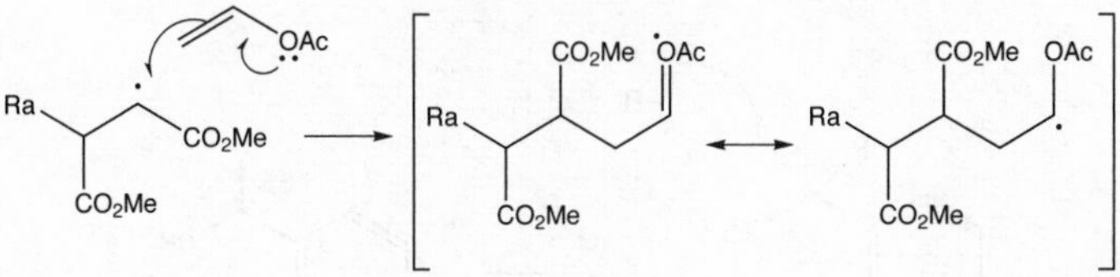

Figure 4.47

carbon atom is zero. In vinyl acetate, the central carbon no longer has a coefficient of zero because the system is not symmetrical. Nonetheless, it is still much smaller than the coefficient on the end (Figure 4.46). Because this is a reaction of soft species and is therefore controlled by the frontier orbital interactions, it is the coefficient of the molecular orbital that determines the site of the reaction, and the molecule therefore reacts at the terminal carbon.

We could also have explained this by drawing curly arrows (recall from Section 1.3.1 that we use single-headed arrows to show the movement of single electrons in radical reactions), and saying that the oxygen lone pair is the source of electrons leading to reaction with the electron-deficient radical (Figure 4.47). Although this has some validity, it is not as rigorous as the molecular orbital approach. Curly arrows, drawn with electron sources and electron sinks in mind, can be very useful in discussions of two-electron reactions, when it is normally clear which reactant is the nucleophile and which the electrophile. However, the molecular orbital approach can always be used instead (although it may be conceptually more complicated), and there are times when we may be led astray by a simplistic reliance on curly arrows. The danger of this is greater in radical reactions, when we no longer necessarily have a simple nucleophile–electrophile interaction, so it is safer to rely on frontier orbital considerations when discussing these reactions.

The reaction of the electron-deficient radical with vinyl acetate leads to a new radical. The SOMO of this new radical results from the interaction of the unpaired electron with the oxygen lone pair, and so this time is of high energy, or in other words the radical is nucleophilic. Using analogous arguments to those above, we can see that the radical reacts with the electron-deficient double bond of dimethyl fumarate in preference to the electron-rich double bond of vinyl acetate. Here we do not have to worry about orientation, because dimethyl fumarate is symmetrical. The regular alternating structure of the polymer can thus be explained by a knowledge of the frontier orbitals.

We must remember in radical reactions that some radicals are very reactive, and

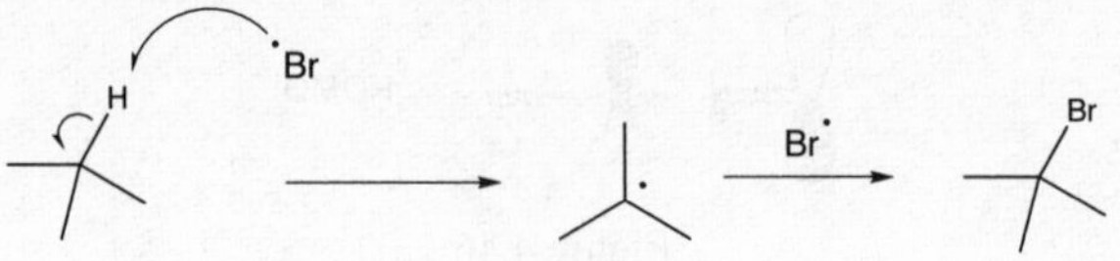

Figure 4.48

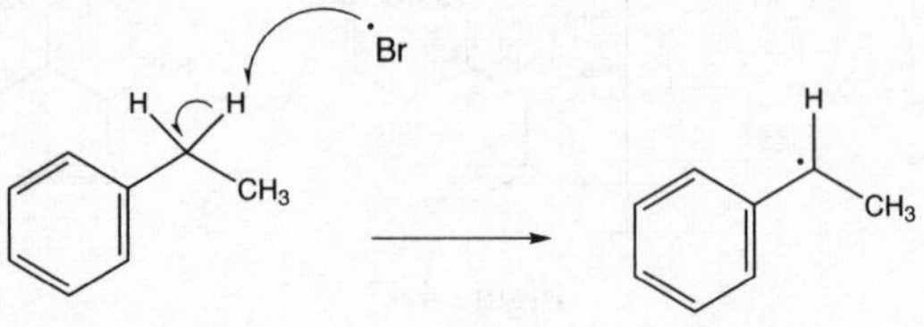

Figure 4.49

will react with more or less the first molecule they encounter, showing little selectivity. In general, the more reactive a species, the lower its selectivity. A reactive radical such as a methyl radical will rapidly abstract hydrogen atoms or add to double bonds, and is unlikely to give a controlled reaction.

In contrast, a bromine radical is comparatively stable, and shows a considerable degree of selectivity. For example, free radical bromination of isobutane gives a high yield of t-butyl bromide (Figure 4.48). The high selectivity of bromine means that it preferentially abstracts the tertiary hydrogen to give the more stable tertiary radical rather than a primary hydrogen to give a less stable primary radical. Similarly, bromine radicals react with ethylbenzene to give a radical exclusively at the more stable benzylic position (Figure 4.49).

We have already mentioned that termination reactions are normally rare compared with propagation reactions, because of the low concentration of radicals. Sometimes, termination reactions will become more frequent as the reaction proceeds, because the molecule that reacts with the radical (e.g. the olefin in a polymerization reaction) is used up, and recombination of radicals is better able to compete.

Termination reactions are also more likely if the radicals are of low reactivity, because their reactions with molecules will be less efficient, and again, radical recombination is able to compete. Radical reactions can be terminated deliberately by making use of this principle, through the use of radical inhibitors. These are molecules that react with radicals to produce stable radicals, which react poorly with molecules and so are more likely to react with other radicals, leading to termination reactions. An example is hydroquinone, which readily loses a hydrogen atom by radical abstraction to give the stable radical shown in Figure 4.50.

Radical inhibitors are frequently added to suppress unwanted radical reactions. One important reaction of this type is the oxidation of hydrocarbons by oxygen in the air (autoxidation), which can be particularly troublesome at high temperatures.

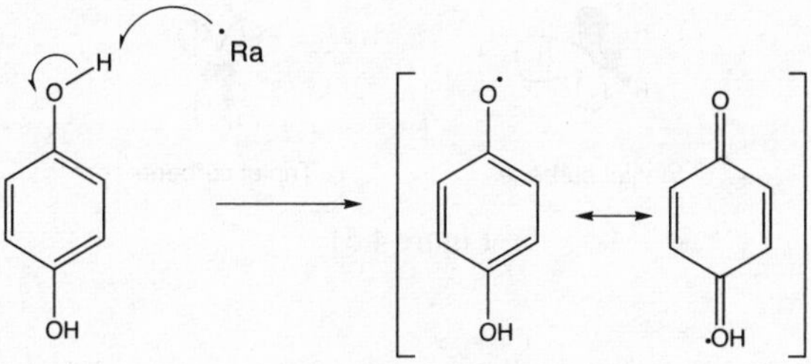

Figure 4.50

This reaction is the reason why cooking oils (which contain esters of fatty acids) become rancid. Radical inhibitors can also be used as mechanistic probes; if a reaction is slowed, or even stopped completely, by the addition of a radical inhibitor, then this provides good evidence that the reaction proceeds by a radical mechanism (see Section 7.5).

4.4 Carbenes

4.4.1 Structure and stability of carbenes

A carbene is a divalent carbon species. It has no electric charge and only six electrons in the valence shell. Carbenes are extremely reactive and short-lived, and never exist as stable species. They can be trapped in frozen inert gas matrices, but otherwise their existence is only fleeting.

The simplest and one of the most common carbenes is methylene, CH_2. Another common carbene is dichlorocarbene, CCl_2. Whilst many carbenes have been detected, those with simple structures such as the above two examples are far more likely to be encountered in practice than the more complicated ones.

Carbenes can exist in either of two electronic states: singlet or triplet. These terms refer to the magnetic properties of the species; in the triplet state, the nuclear magnetic resonance (NMR) signal of the carbon nucleus is split into three lines by interaction with the electrons, whereas in the singlet state this splitting is not present. In the singlet state, the two electrons are paired and occupy one orbital, and the fourth orbital is vacant, whereas in the triplet state there are two unpaired electrons in two separate orbitals. Thus a singlet carbene can be regarded as being both a carbanion and a carbocation combined (with the charges cancelling out) and a triplet carbene can be regarded as a diradical (Figure 4.51).

The triplet state is the ground state for simple alkyl carbenes such as methylene, although these species are often formed in the singlet state and react as such before they have a chance to decay to the triplet state. However, for dihalocarbenes, the singlet state is the one of lower energy.

Because carbenes are so reactive and short-lived, it is not always possible to tell

107

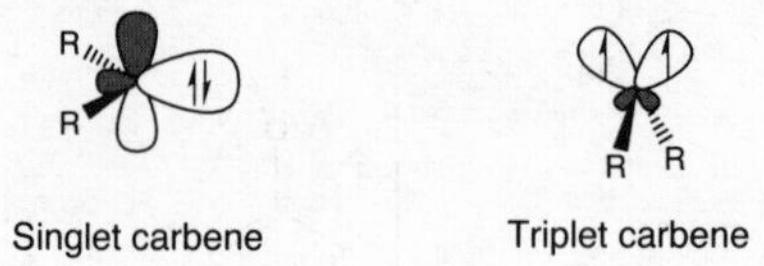

Singlet carbene   Triplet carbene

Figure 4.51

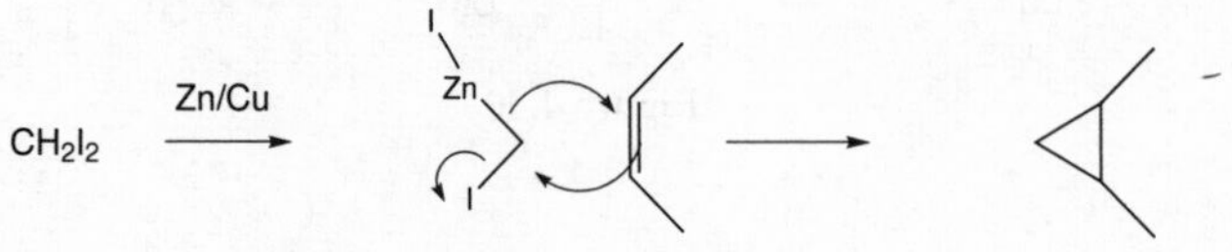

CH_2I_2 $\xrightarrow{\text{Zn/Cu}}$

Figure 4.52

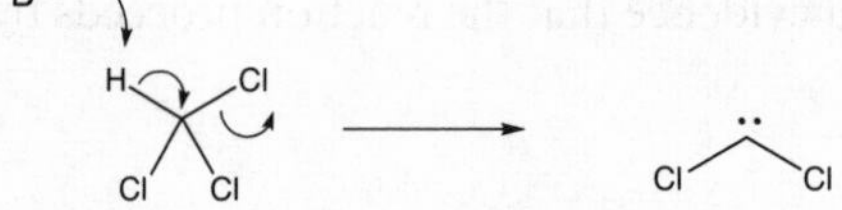

Figure 4.53

whether a free carbene is present in a reaction, or whether the reaction merely passes through some intermediate which has some properties in common with a carbene, but in which the carbon never actually becomes divalent. The term carbenoid is sometimes used to describe intermediates which have this reactivity, but are not necessarily true carbenes. For example, in the Simmons–Smith reaction (Figure 4.52), an alkene is treated with diiodomethane and a zinc–copper couple. The organozinc compound is an intermediate in this reaction, and behaves like a carbene, although the free carbene is not present at any stage. This intermediate can be called a carbenoid.

It is perhaps inadvisable to talk about features that stabilize carbenes, since they are invariably very unstable. However, dihalocarbenes tend to be less reactive than simple alkyl carbenes, and hence more stable.

4.4.2 Formation of carbenes

There are two common ways to form carbenes. The first of these is α-elimination, and the second is decomposition of compounds containing a double bond between carbon and another atom.

The most common α-elimination reaction used to form carbenes is the elimination of HCl from chloroform by treatment with base to form dichlorocarbene (Figure 4.53). Other carbenes may also be formed in this manner, for example dichloromethane may be treated with base to form chlorocarbene, CHCl.

The second way of forming carbenes is by the breakdown of certain compounds

Figure 4.54

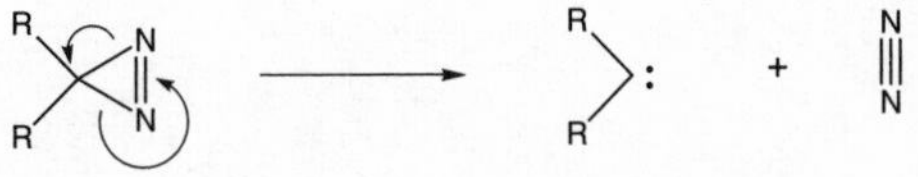

Figure 4.55

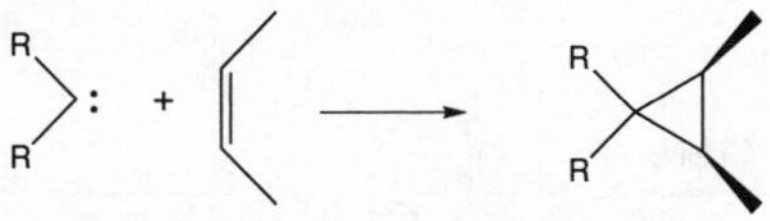

Figure 4.56

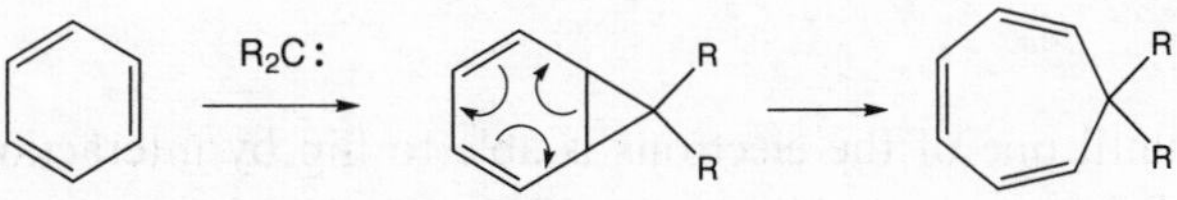

Figure 4.57

with double bonds, such as diazomethane, which may be induced thermally or photochemically (Figure 4.54). The decomposition of diazirines is similar, although here two single bonds are broken instead of one double bond (Figure 4.55).

4.4.3 Reactivity of carbenes

The most useful reaction of carbenes is their insertion into double bonds (Figure 4.56). This is in fact the most common way of preparing cyclopropanes. It is a very efficient reaction, and takes place even with electron-poor double bonds such as tetracyanoethylene, a reagent inert to many of the reactions of olefins, and with the double bonds of aromatic compounds. The products of reactions between carbenes and aromatic compounds normally rearrange by an electrocyclic mechanism (see Section 6.5) to give a ring-expanded product (Figure 4.57).

The stereospecificity of this reaction varies depending on whether the carbene is singlet or triplet. For a singlet carbene, the two bonds are made more or less simultaneously, and so the stereochemistry of the double bond is conserved (see Figure 4.56). However, when a triplet carbene reacts only one bond can form initially, because the two free electrons of the carbene have the same spin. (Electrons must have opposing spin to form bonds, so when the first carbene electron has formed a bond with one of the π electrons of the olefin, the remaining π electron must have the same spin as the second carbene electron.) The second bond therefore

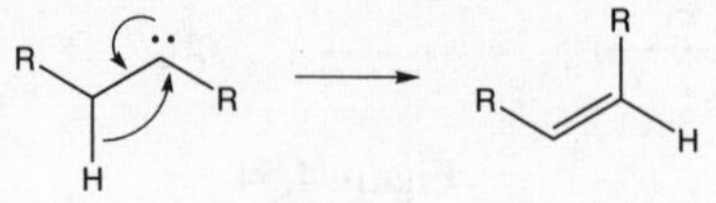

Figure 4.58

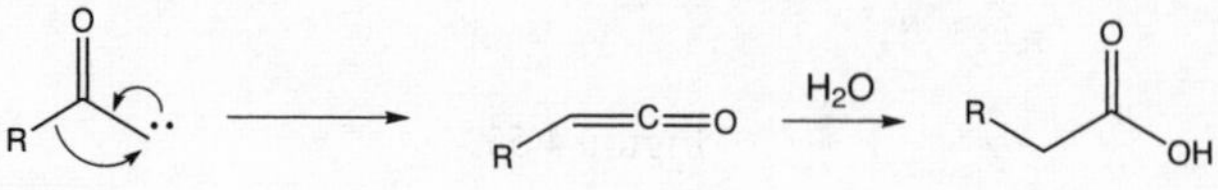

Figure 4.59

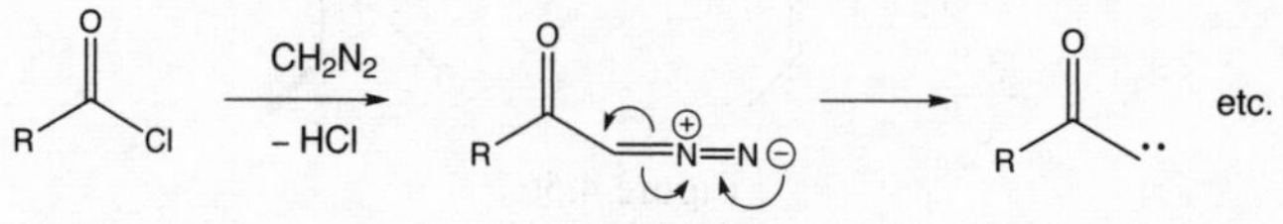

Figure 4.60

cannot form until one of the electrons is able to flip by interaction with its surroundings. This interaction will in general be a slower process than rotation about a single bond, so the stereochemistry of the olefin will be lost and a mixture of products will result. This is a useful, although not infallible, guide that can be used in mechanistic investigations to discover whether a suspected carbene is singlet or triplet.

Carbenes can also insert into C−H bonds. Thus propane treated with methylene gives a mixture of butane and isobutane. This reaction is seldom of any use preparatively, as it is not selective. It can be a troublesome side reaction with alkyl carbenes, although it is rarely a problem for dihalocarbenes, as they undergo this reaction much less readily.

Carbenes can rearrange to give stable products. Any carbene with an α-hydrogen rearranges rapidly to give an olefin (Figure 4.58). This rearrangement is so rapid that it is not generally possible to use these carbenes for any other purpose, which is one of the reasons why the carbenes most often encountered have such simple structures. A similar reaction is the rearrangement of α-carbonylcarbenes, which is known as the Wolff rearrangement (Figure 4.59). The initial product is a ketene, but this is readily hydrolysed to a carboxylic acid in the presence of water. It can also be converted to other acid derivatives if suitable alternatives to water are present; for example, esters are produced if alcohols are used as the solvent. This reaction is an important step in the Arndt–Eistert synthesis (see also Section 5.2.4), used for extending the carbon chain of a carboxylic acid by one unit (Figure 4.60).

Carbenes can also abstract radicals, leading to the formation of two radicals (Figure 4.61). This reaction is not unexpected for triplet carbenes, as these are a kind of free radical. However, under some circumstances singlet carbenes can also react

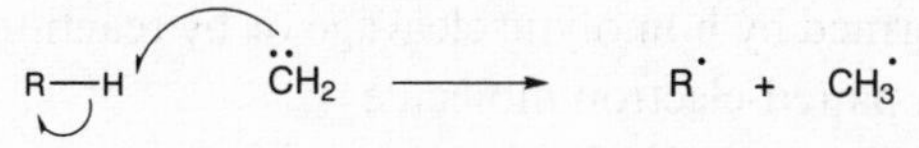

Figure 4.61

in this way. Radical abstraction and recombination has been postulated as a possible mechanism for the insertion of carbenes into single bonds, although it is known that this mechanism does not always operate, if at all.

Summary

- Molecules with a negatively charged carbon atom are called carbanions.
- Carbanions are stabilized by the same factors that stabilize the conjugate bases of any acids.
- One of the most important ways of stabilizing carbanions is by delocalization of the negative charge, particularly onto atoms such as oxygen.
- The most common way of forming carbanions is by removal of a proton from a carbon atom.
- Carbanions can also be formed from alkyl halides with metals or organometallic compounds.
- Grignard reagents (organomagnesium compounds) are a common type of carbanion-like molecule.
- Most carbanions are strong bases, and so must be handled with strict exclusion of proton sources such as water.
- Carbanions are nucleophilic, and undergo many of the typical reactions of nucleophiles, such as reacting with carbonyl groups.
- Grignard reagents react with esters to produce tertiary alcohols, but do not react with alkyl halides.
- Carbanions may give elimination reactions if a leaving group is present.
- Molecules with a positively charged carbon atom are called carbocations.
- Among simple alkyl carbocations, tertiary carbocations are the most stable and primary carbocations the least stable.
- Carbocations are stabilized by resonance interactions and by adjacent lone pairs.
- Carbocations are formed by the loss of a leaving group from a carbon atom or by addition of an electrophile to a carbon–carbon double bond.
- Carbocations may react by combination with a nucleophile, loss of a proton, or rearrangement.
- Free radicals are molecules with unpaired electrons.
- The frontier orbital of a free radical is the called the SOMO.
- Radicals are stabilized by resonance and by interaction with either filled or empty orbitals.

111

- Radicals are formed by homolytic cleavage or by reaction of a radical initiator with a paired-electron molecule.
- Reactions of radicals can be divided into initiation, propagation, and termination steps.
- Consideration of SOMOs can tell us much about the reactivity of radicals.
- More reactive radicals tend to be less selective in their reactions.
- Radical reactions can be slowed by radical inhibitors.
- Carbenes are divalent carbon species, and are two electrons short of their octet.
- Most carbenes are very reactive.
- Carbenes may exist in singlet or triplet states.
- Carbenes are formed by α-elimination or by breakdown of molecules such as diazo compounds.
- Carbenes can insert into double or single bonds, rearrange, or abstract radicals.

Problems

1. Arrange the following carbanions in order of stability.

2. Enolates are often made by treatment of the carbonyl compound with LDA, itself prepared from diisopropylamine and butyllithium. Why is butyllithium not used directly to form the enolate?

3. Which of the following molecules will react with phenyl magnesium bromide, and what will the products be?

112

4. Suggest a synthesis of the following alcohol.

5. Arrange the following carbocations in order of stability.

6. Draw resonance structures to show how the following carbocation is stabilized.

7. When esters are hydrolysed in acid, it is normally the acyl–oxygen bond that is cleaved. However, acidic hydrolysis of t-butyl esters results in cleavage of the alkyl–oxygen bond. Why?

8. Arrange the following radicals in order of stability.

9. Draw a molecular orbital diagram for the relevant interaction of the unpaired electron in the peroxide radical (HOO˙), showing which orbital is the SOMO. Would you expect this radical to be nucleophilic or electrophilic?

10. Treatment of benzene with sodium in liquid ammonia in the presence of an alcohol produces 1,4-cyclohexadiene; this is known as Birch reduction. Treatment of anisole gives 1-methoxy-1,4-cyclohexadiene, not 3-methoxy-1,4-cyclohexadiene. Why?

113

11. Bromination of cyclohexene with N-bromosuccinimide in the presence of benzoyl peroxide gives a monobrominated product. What is the product, and what is the mechanism of its formation?

12. A reaction between methylene and *cis*-2-butene gave *trans*-1,2-dimethylcyclopropane. What does this tell us about the electronic configuration of the carbene in this reaction?

5

The effect of heteroatoms

Heteroatoms in organic chemistry are defined as those atoms other than carbon or hydrogen. In theory, this leaves over a hundred possibilities, but in practice the number of atoms likely to be encountered in organic molecules is much more limited. We shall look at some of the most important heteroatoms in this chapter.

Heteroatoms have a profound effect on the reactivity of organic molecules. Hydrocarbons have only a limited range of reactions, and so although organic chemistry is often defined as the chemistry of carbon, it is the other elements that make it interesting. In this chapter we shall look at how heteroatoms affect the reactivity of organic molecules. While the great variety of organic molecules may at first seem bewildering, when molecules are considered in terms of their heteroatoms and the effects that these have, then the reactivity patterns observed are in fact quite logical, and consequently easy to study.

5.1 Oxygen

Oxygen has two main properties that shape its reactivity in organic chemistry: it is electronegative and has electron lone pairs. The electronegativity of oxygen means that organic molecules containing oxygen are polarized such that the oxygen atom usually bears a partial negative charge. This is particularly marked in carbonyl compounds, which have a double bond between oxygen and carbon.

This property also makes it possible for oxygen to form anions (recall from Section 2.2.2 that more electronegative atoms form anions more easily). Oxygen anions are intermediates in many of the reactions of oxygen-containing organic molecules.

The lone pairs of oxygen enable it to act as a nucleophile. This nucleophilicity is of course enhanced if the oxygen is in the form of an anion, but the electron lone pairs on neutral oxygen atoms can also be good nucleophiles. Like all nucleophiles, oxygen can also act as a base, being protonated on a lone pair. This is also a common

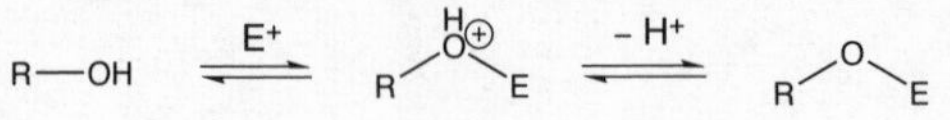

R—OH $\rightleftharpoons$ $\overset{E^+}{}$ $\underset{R}{} \overset{\overset{H}{\underset{\oplus}{O}}}{\diagup} \underset{E}{}$ $\rightleftharpoons$ $\overset{-H^+}{}$ $\underset{R}{} \overset{O}{\diagup} \underset{E}{}$

Figure 5.1

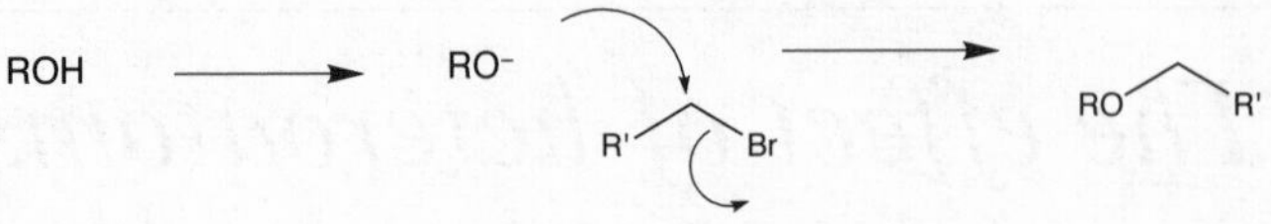

ROH $\longrightarrow$ RO$^-$ $\qquad$ $\underset{R'}{\diagup} \underset{Br}{}$ $\longrightarrow$ $\underset{RO}{} \overset{}{\diagup} \underset{R'}{}$

Figure 5.2

process in oxygen-containing molecules; many of their reactions under acidic conditions have protonated intermediates.

The lone pair also allows oxygen to stabilize an adjacent carbocation, as mentioned in Section 4.1.1.

5.1.1 Alcohols

The properties of alcohols are largely predictable from the properties described above. Whilst the electronegativity of oxygen means that the carbon–oxygen bond in alcohols is polarized in the direction of the oxygen, this effect is not nearly as pronounced as in carbonyl compounds, and is of only minor importance in the reactivity of alcohols.

Of far greater importance are the other two major properties of oxygen, its nucleophilicity and its ability to form anions. Alcohols act as nucleophiles in many of their reactions. A schematic reaction between an alcohol and an electrophile is shown in Figure 5.1. After reaction between the oxygen and the electrophile, the resulting oxonium ion loses its proton to give a stable molecule. Of course, these steps will in general be reversible and the position of the equilibrium depends on the nature of the electrophile.

Oxygen's ability to form anions is also important in the reactions of alcohols; the alcoholic proton may be removed by bases. This proton is slightly less acidic than that of water, so aqueous bases are not normally used to effect this deprotonation. With lower alcohols, such as methanol and ethanol, a common procedure is to dissolve sodium metal in the alcohol, which gives the sodium alkoxide directly. Higher alcohols are often deprotonated with sodium hydride in an inert solvent such as THF or 1,2-dimethoxyethane (DME).

The deprotonated alcohol, or alkoxide, is of course a far better nucleophile than the neutral alcohol, and can undergo many more reactions. A common example of this is the reaction of alkoxides with alkyl halides (Figure 5.2). The product of this reaction is an ether. Whilst it is sometimes possible to prepare ethers by reacting neutral alcohols with alkyl halides, the reaction is extremely sluggish and of little practical value for all but the most reactive alkyl halides, and so alkoxides are gener-

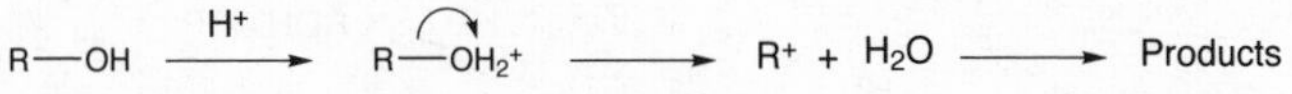

R—OH $\xrightarrow{\text{H}^+}$ R—OH$_2^+$ $\longrightarrow$ R$^+$ + H$_2$O $\longrightarrow$ Products

Figure 5.3

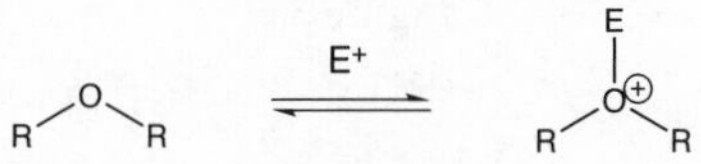

Figure 5.4

ally used. This is the most useful way of preparing ethers and is known as the Williamson ether synthesis.

Like water, alcohols can be either deprotonated or protonated. Under acidic conditions, alcohols are protonated readily. This process is of course reversible, and if the proton simply comes off again, then the reaction is of no use. However, while the alcohol is protonated, the oxygen becomes part of a much better leaving group. Neutral water can leave from the protonated molecule, whereas the leaving group from the unprotonated molecule would be hydroxide ion. Under acidic conditions, therefore, the carbon–oxygen bond in alcohols can be cleaved (Figure 5.3), leading to elimination and substitution reactions (of which more in Chapter 6).

5.1.2 Ethers

Ethers undergo far fewer reactions than alcohols, but when they do react, their reactivity is analogous to that of alcohols.

Despite their lone pairs, ethers do not act as effective nucleophiles, because there is now no proton to be lost after initial attack of the oxygen on an electrophile. Recall from the discussion of alcohols above that it is this deprotonation that makes the nucleophilic attack of alcohols on electrophiles irreversible. In general, if an ether does act as a nucleophile, the reaction will rapidly revert to starting materials (Figure 5.4).

Because there is no proton attached to the oxygen, the deprotonation that is avail-able to alcohols is no longer a possibility. Whilst this considerably narrows the range of reactions that ethers may undergo compared with alcohols, this is not all bad news for ethers. Frequently in organic chemistry, reactions are carried out using strong bases such as sodium hydride, butyllithium, or LDA. Many of these reactions need a polar solvent to dissolve the resulting ion pairs of organic bases and metal ions, and the relative inertness of ethers is an advantage here. Alcohols cannot be used as solvents for such reactions, as they would react with the bases. Many other common polar solvents, such as acetone and acetonitrile, also have acidic hydrogens and so cannot be used. However, the electronegativity of oxygen means that ethers are also polar molecules (although not generally as polar as alcohols), and so they make ideal solvents for reactions involving strong bases. Diethyl ether, THF, and DME are all commonly used.

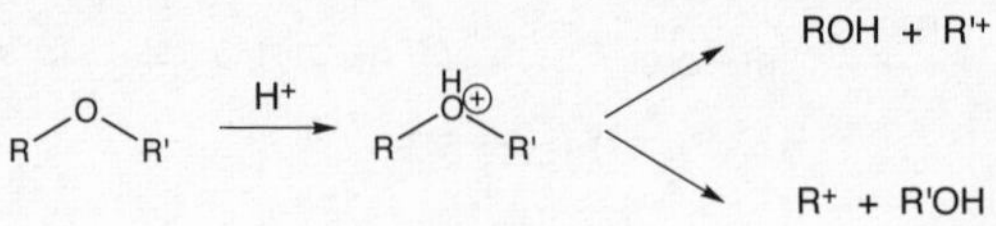

ROH + R'$^+$

R$^+$ + R'OH

Figure 5.5

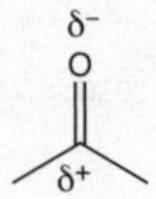

Figure 5.6

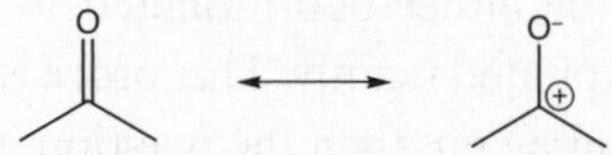

Figure 5.7

Ethers may be protonated under acidic conditions, in much the same way as alcohols. They can thus be cleaved, if the protonated oxygen acts as a leaving group, although this is not normally a useful reaction for unsymmetrical ethers, because either bond may be cleaved, and so a mixture of products results (Figure 5.5). Sometimes one bond may be cleaved more readily than the other to give a good yield, although a far more reliable way of cleaving ethers is to use trimethylsilyl iodide (see Section 8.2.4).

5.1.3 Carbonyl groups

The carbonyl group is of fundamental importance in organic chemistry. Carbonyl groups take part in a wide variety of reactions, although these can all be explained by straightforward consideration of the properties of the group.

The carbonyl group has a significant dipole moment, the oxygen bearing the partial negative charge. Partial charges are often written as δ^+ and δ^-, as shown in Figure 5.6. The reason why the carbon–oxygen bond is considerably more polarized in carbonyl groups than in alcohols or ethers is that the π electrons, being less tightly bound to the nuclei, are more polarizable, and so tend to congregate towards the electronegative oxygen more than σ electrons do.

Carbonyl groups can be written as a resonance hybrid as shown in Figure 5.7, where one resonance form contains formally separated charges. However, the uncharged double-bonded representation is the way carbonyl groups are normally written.

This polarization is the dominant feature in the reactivity of the carbonyl group. Whilst carbonyl chemistry is a huge subject, and entire books have been devoted to

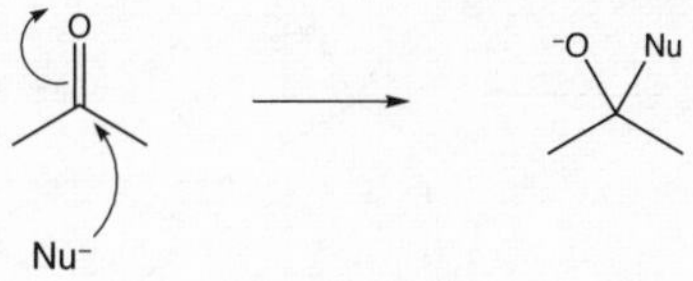

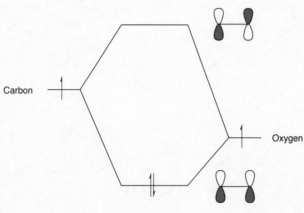

Figure 5.8

Figure 5.9

it, the reactivity of the carbonyl group can be summed up by the following simple statement: the carbonyl group contains an electron-deficient carbon atom.

This knowledge easily allows us to predict the most important property of carbonyl groups, namely that they are electrophilic and so react with nucleophiles. A generalized carbonyl reaction is shown in Figure 5.8. Note that the oxygen acts as the electron sink, as shown by the curly arrows.

It is also easy to consider the HOMOs and LUMOs that take part in this reaction. As carbonyl groups are electrophiles, the frontier orbital of interest is the LUMO. A molecular orbital diagram of the π electrons of the carbonyl group (Figure 5.9) shows that the LUMO is the antibonding π orbital, or π^* orbital. When the carbonyl group reacts with a nucleophile, the electrons from the HOMO of the nucleophile enter this π^* orbital. Because this is an antibonding orbital, the double bond in the starting material becomes a single bond in the product, which is of course consistent with the curly arrow way of looking at the reaction.

This discussion of carbonyl chemistry has been very abstract so far, and it is time to illustrate it with an example, for which we will use the hydration of acetaldehyde (Figure 5.10). For the moment, we shall look at the reaction at neutral pH, although we shall see below that it is actually much faster in acid or base. The nucleophile is water, and the initial attack on the carbonyl group is followed by simple proton transfer to give the hydrate. The reaction is readily reversible, and acetaldehyde in aqueous solution exists partly as its hydrate. This example serves to illustrate the

119

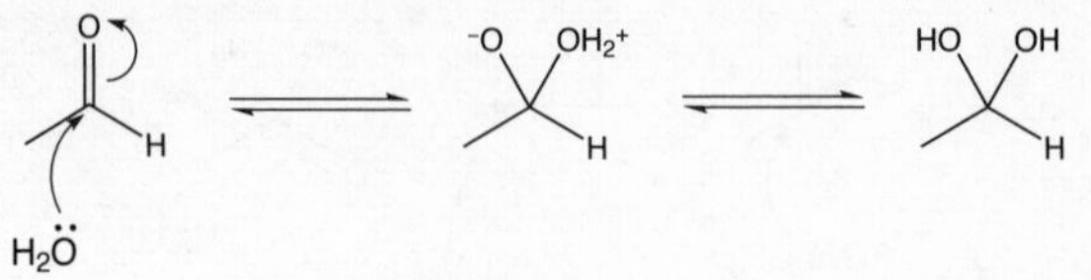

Figure 5.10

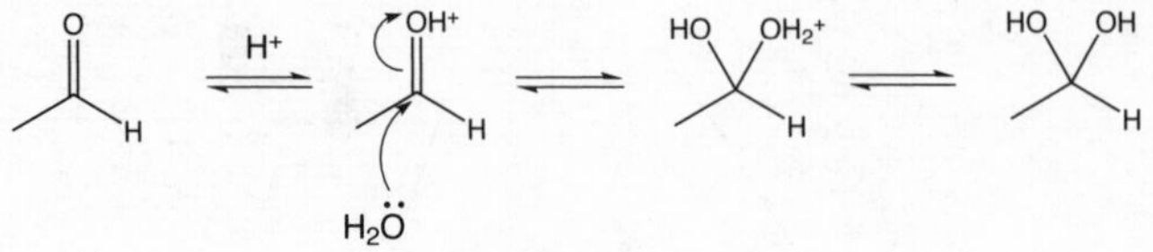

Figure 5.11

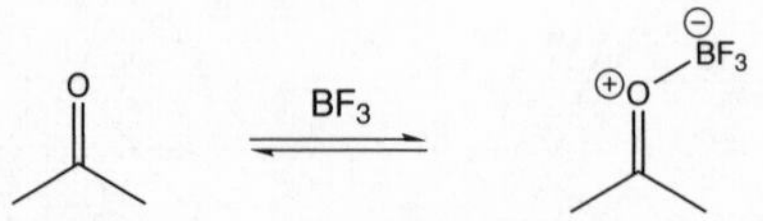

Figure 5.12

point that the initial attack on the carbonyl group does not normally give a stable product, and further steps must occur before the reaction is complete. Here, there is just a simple protonation step, although the subsequent steps are frequently more complex.

We may also use this example to introduce another important feature of carbonyl group reactions, namely that they are sensitive to the effects of pH. We assumed that the hydration reaction shown in Figure 5.10 is proceeding at somewhere near neutral pH. Let us now consider how this reaction will behave in acid (Figure 5.11). We know that the lone pair of oxygen is easily protonated under acidic conditions, and carbonyls are no exception to this behaviour. This protonation has the effect of speeding up the reaction because the protonated carbonyl group is a better electro-phile than its neutral counterpart (see Section 2.2.4).

Protonation of carbonyl groups is often a prerequisite for their reactions; many carbonyl groups are not sufficiently reactive when neutral to react with the desired nucleophiles. Lewis acids may also be used to speed up carbonyl reactions in much the same way. Figure 5.12 shows this for boron trifluoride.

Going back to the hydration of acetaldehyde, let us now consider what happens under basic conditions. Here too the reaction is faster than at neutral pH, this time because hydroxide can act as the nucleophile instead of water (Figure 5.13).

Many carbonyl reactions are dependent on catalysis by acids or bases, but they are not always interchangeable as in this simple example. For example, if water is replaced in this reaction by an alcohol such as ethanol, a product analogous to the hydrate, an acetal, is formed, but only under acidic conditions. The mechanism of

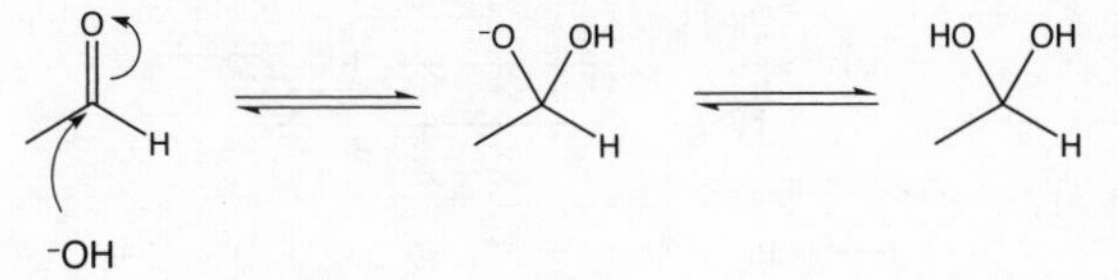

Figure 5.13

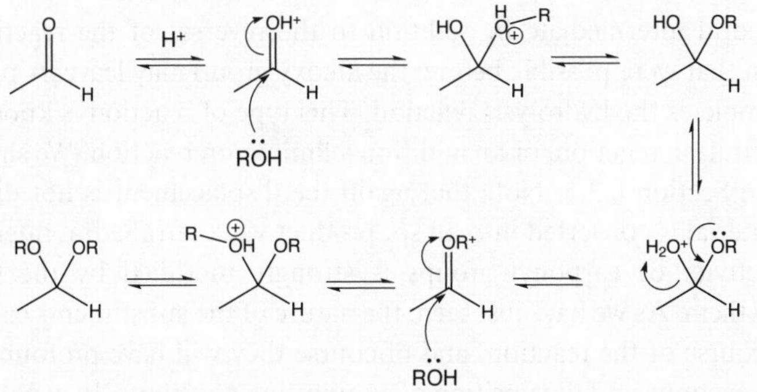

Figure 5.14

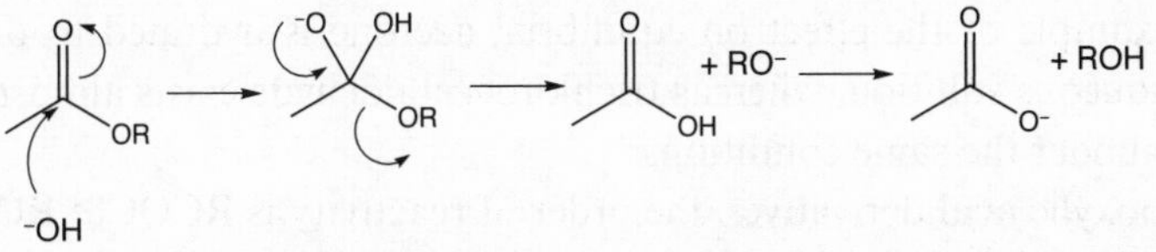

Figure 5.15

this reaction is shown in Figure 5.14. Note that all the steps are reversible, and the reaction is an equilibrium. The position of the equilibrium will depend on the relative amounts of water and alcohol present.

The step in this reaction that requires the acid catalysis is the loss of water from the protonated intermediate. The water may leave readily because of the adjacent oxygen lone pair, which stabilizes the positive charge on the carbon atom. However, the oxygen that leaves must be protonated before it can leave; hydroxide ion is far too poor a leaving group for this reaction to be possible without acid catalysis. It is important to remember that the water is replaced by the alcohol through an sp_2 hybridized intermediate. Direct S_N2 displacements (see Section 6.3.2) *never* occur in the formation or hydrolysis of acetals.

The reactivity pattern of carbonyl groups is slightly different if a leaving group is attached to the carbonyl group, as in esters or acid chlorides. If we consider the base-catalysed hydrolysis of an ester (Figure 5.15), the initial attack of hydroxide gives a tetrahedral intermediate in much the same way as the first step of the base-catalysed hydration of aldehydes. Here, however, a third possibility is available to

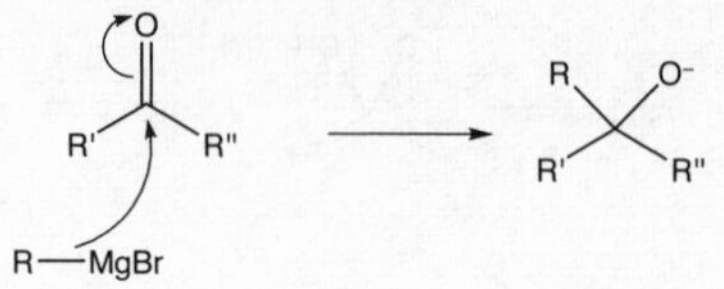

Figure 5.16

this tetrahedral intermediate in addition to the reversal of the reaction and the protonation that were possible before: the alkoxy group may leave. A proton transfer step completes the hydrolysis reaction. This type of reaction is known as a carbonyl substitution reaction, or an addition–elimination reaction. We shall return to this topic in Section 6.3.5. Note that again the displacement is not direct; an sp_2 starting material is converted into an sp_2 product via a tetrahedral intermediate.

The reactivity of carbonyl groups is strongly modified by the substituents attached to them. As we have just seen, the nature of the substituents can determine the entire course of the reaction, and of course they will have profound effects on the rate of reactions and the position of equilibrium constants. In general, electron-withdrawing substituents make carbonyl groups more reactive, in other words they speed up rates of reactions and shift equilibria in favour of the product.

As an example of the effect on equilibria, acetone is hydrated to only a small extent in aqueous solution, whereas trichloroacetaldehyde exists almost entirely as its hydrate under the same conditions.

For carboxylic acid derivatives, the order of reactivity is $RCOCl > RCO_2COR' > RCOSR' > RCO_2R' \approx RCO_2H > RCONH_2$. Note that esters and carboxylic acids are of similar reactivity only if the acid is protonated. If the acid is present as its carboxylate salt its reactivity is of course completely different, because the molecule is then negatively charged. There is a large difference between the extremes of this reactivity series: acetyl chloride is hydrolysed readily, even violently, with neutral water, whereas the hydrolysis of acetamide requires prolonged heating under strongly acidic conditions.

Carbonyl groups can be attacked by many other nucleophiles besides water and alcohols. Most carbonyl compounds react readily with Grignard reagents and other organometallic compounds (Figure 5.16); this is a useful reaction for forming carbon–carbon bonds. The products of such reactions will of course be different depending on whether a leaving group is attached to the carbonyl group. We saw in Section 4.1.3 how the reaction of Grignard reagents with esters produces ketones as intermediates, which then react with more Grignard reagent to give tertiary alcohols.

Another important feature of the reactivity of carbonyl groups is that they can stabilize an adjacent negative charge. The resulting anion is known as an enolate ion, which we have already encountered in Chapters 2 and 4. Enolate anions undergo many reactions, as we would expect of a carbanion. One reaction is worthy of special mention, namely the reaction of enolates with the carbonyl compounds from which

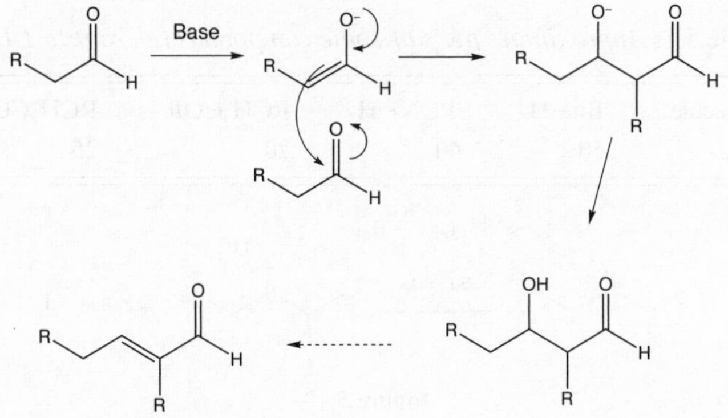

Figure 5.17

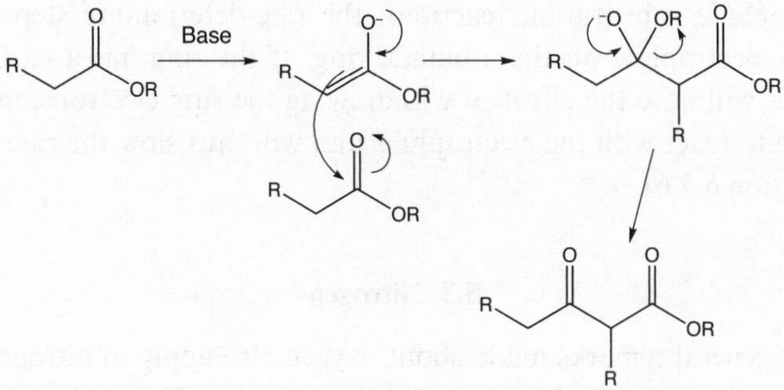

Figure 5.18

they are derived. When the carbonyl compound is an aldehyde, the reaction is known as the aldol reaction, and generally proceeds very readily when an aldehyde is treated with base. It is so favourable, in fact, that it can be difficult to carry out any other reactions with aldehydes under basic conditions, because the aldol reaction is normally dominant.

The aldol reaction may lead to a product in which the aldehyde is β to a hydroxyl group (this is known as an aldol), or this may dehydrate to give an α,β-unsaturated aldehyde, depending on the conditions of the reaction (Figure 5.17). Ketones give a similar reaction, although this does not normally happen as readily as with aldehydes. If the carbonyl compound is an ester, then the reaction gives a β-keto ester; this reaction is known as the Claisen condensation (Figure 5.18).

Although these examples are all of a carbonyl compound reacting with itself, it is of course entirely possible for the enolate of one carbonyl compound to react with a different compound. In general, crossed carbonyl condensations of this sort will result in a mixture of products, although there are ways to control the outcome of such reactions.

Finally, the electron-withdrawing effect of carbonyl groups can be important in

Table 5.1. *Approximate pKₐs of some compounds relevant to LDA*

Molecule	Bu—H	^{i}Pr${}_2$N—H	RCH${}_2$COR	RCH${}_2$CO${}_2$R
pK_a	50	40	20	25

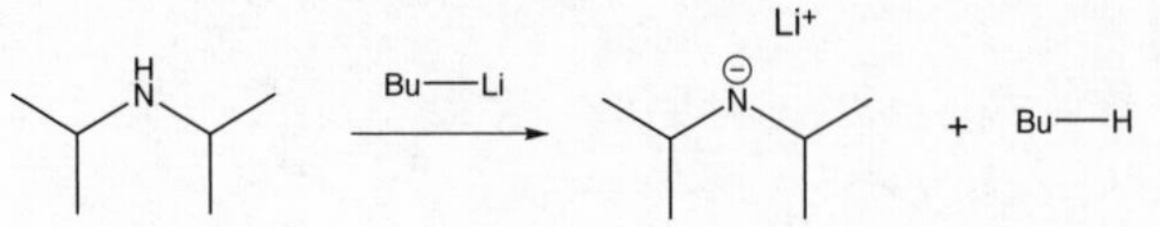

Figure 5.19

many other situations besides helping to stabilize an anion. For example, in aromatic electrophilic substitution reactions, the rate-determining step is normally attack of an electrophile on the aromatic ring. If the ring has a carbonyl group attached, this will have the effect of withdrawing the ring electrons, making them less available to react with the electrophile, and will thus slow the rate of the reaction (see Section 6.3.6).

5.2 Nitrogen

Many of the general remarks made about oxygen also apply to nitrogen, although we must remember that nitrogen is less electronegative. This means that although carbon–nitrogen bonds are polarized towards nitrogen, the effect is considerably less marked than for oxygen. Nitrogen is also much more reluctant to form negatively charged species than oxygen, so that lithium diisopropylamide (LDA), a base commonly used in synthetic organic chemistry with a negatively charged nitrogen atom, is strong enough to remove protons irreversibly from carbonyl compounds to form enolates. LDA is normally prepared from diisopropylamine and the exceedingly strong base butyllithium (Figure 5.19). Some illustrative pK_as are shown in Table 5.1.

The lone pair is much more important in the chemistry of nitrogen than in that of oxygen. This is again because of the difference in electronegativity; the lone pair of nitrogen is held less tightly and is therefore more available for reactions or protonation. An illustration of this is that ammonia is normally thought of as a base, as its lone pair is so easily protonated, whereas the oxygen analogue, water, is not generally regarded as a base (although it is of course a weak base, and can accept a proton to form H${}_3$O${}^+$).

5.2.1 Amines

Firstly, let us be clear about the nomenclature of amines, which differs from that of alcohols in a rather confusing way. We talk of primary, secondary, and tertiary alco-

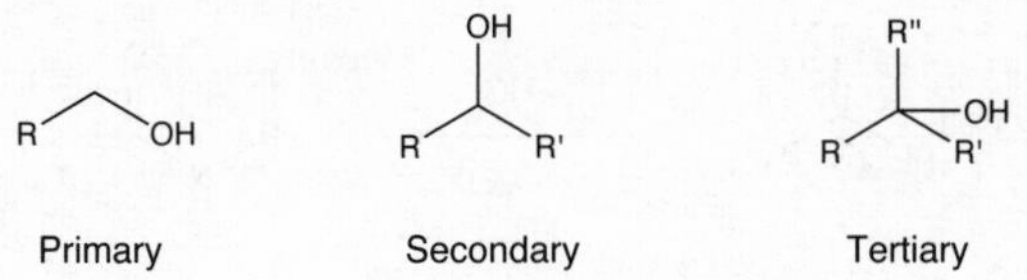

Figure 5.20

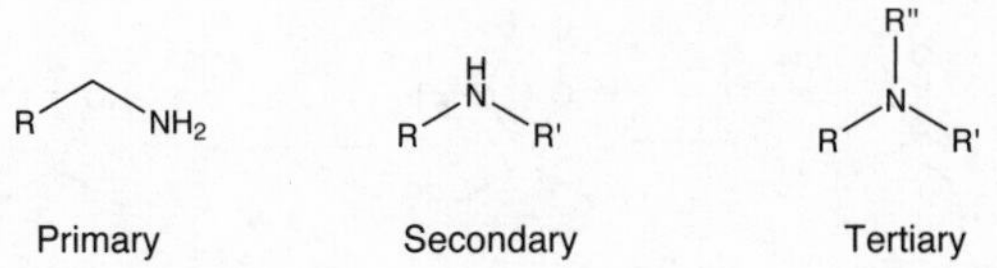

Figure 5.21

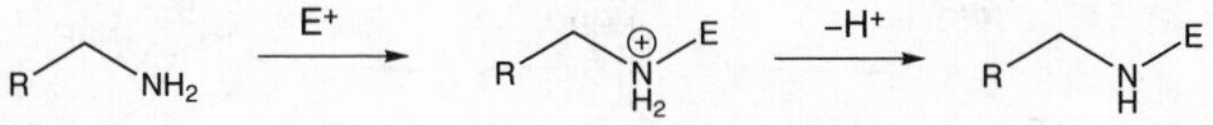

Figure 5.22

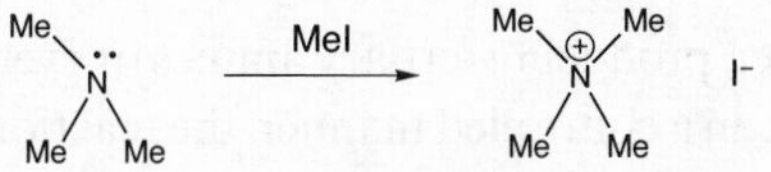

Figure 5.23

hols; a series of these is shown in Figure 5.20. However, the terms primary, secondary, and tertiary amines do not apply in the same way, as shown in Figure 5.21.

Amines have two important properties, as expected from the above description of nitrogen lone pairs: they are basic and nucleophilic. The properties of amines as bases were discussed at some length in Section 2.2.2.5; to recap briefly, the basicity of amines tends to increase with increasing alkyl substitution, although tertiary amines are not always more basic than secondary amines.

Amines are excellent nucleophiles. A typical reaction of a primary amine is shown in Figure 5.22. This could apply just as well to a secondary amine, although not to a tertiary amine, because these have no proton to lose. In general, tertiary amines do not undergo as many reactions as primary and secondary amines, because this final deprotonation step is not possible. However, tertiary amines may still react with electrophiles to form stable quaternary ammonium salts. For example, trimethyl-amine is readily alkylated by methyl iodide to form the stable salt tetramethyl-ammonium iodide (Figure 5.23).

Alkylation of amines with alkyl halides presents a special problem: the products of these reactions are often more reactive than the starting materials, and so the reactions are poorly controlled. An attempt to prepare primary, secondary, or even tertiary amines by this method will frequently be unsuccessful, because the desired

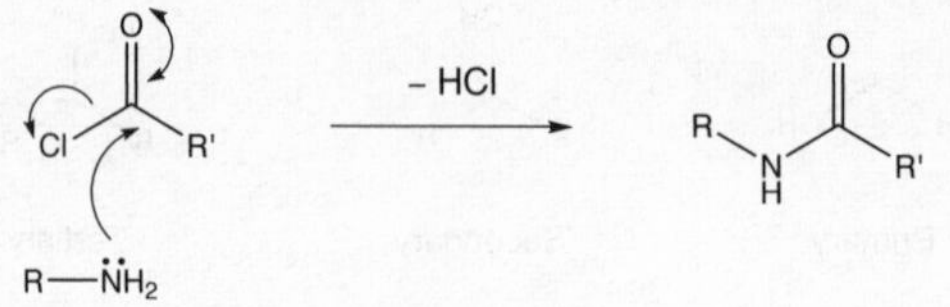

Figure 5.24

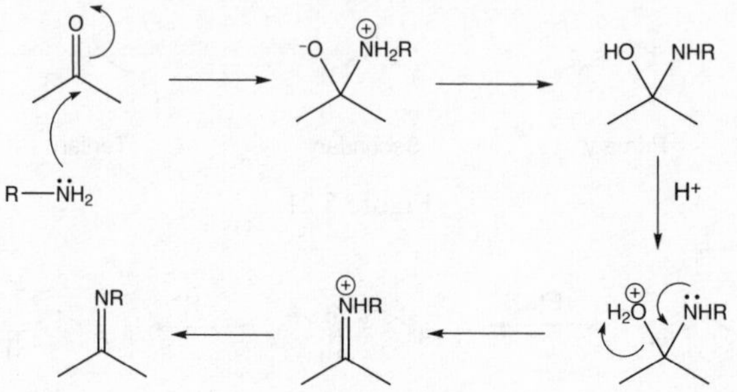

Figure 5.25

product will react further, producing tertiary amines or even quaternary ammonium salts. To alkylate amines in a controlled manner, the reaction must be done indirectly with amides or imines as intermediates (see below).

Amines react readily with carbonyl compounds. If the reaction is an addition–elimination (see Section 6.3.5), as with acyl halides or anhydrides, then the product is an amide (Figure 5.24). Otherwise, an imine is formed, as shown in Figure 5.25. Amides and imines are discussed separately below.

Of course, a prerequisite for all these reactions is that the amine is not protonated. A protonated amine is not nucleophilic, as it has no lone pair. These reactions therefore generally fail under acidic conditions, although weakly acidic conditions can be tolerated, as the protonated amine will be in equilibrium with the free base. Weak acid may even be beneficial in carbonyl reactions, as the amine can react more efficiently with a protonated carbonyl group. Even though acid will increase the proportion of the protonated amine, there will still be some unprotonated amine in equilibrium with it, which will react with the carbonyl group. Under the right conditions, the extra reactivity of the protonated carbonyl group will more than compensate for the lower concentration of unprotonated (and therefore reactive) amine.

This requirement for neutral or basic conditions is important, because acid is generated in many of these reactions. For example, the reaction between diethylamine and acetyl chloride produces one equivalent of HCl (Figure 5.26). This will protonate the remaining amine, rendering it unreactive. When simple and cheap amines such as this are used, it is common practice to use a twofold excess of the amine, so that the extra amine can act as a base. If more valuable amines are used, an extra base must be added. Sometimes a tertiary amine is used for this purpose,

126

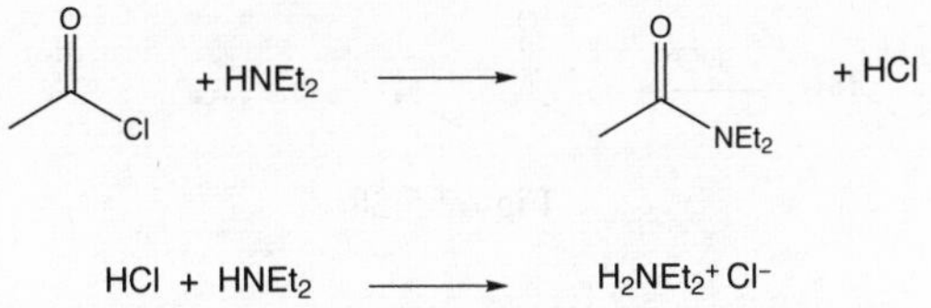

Figure 5.26

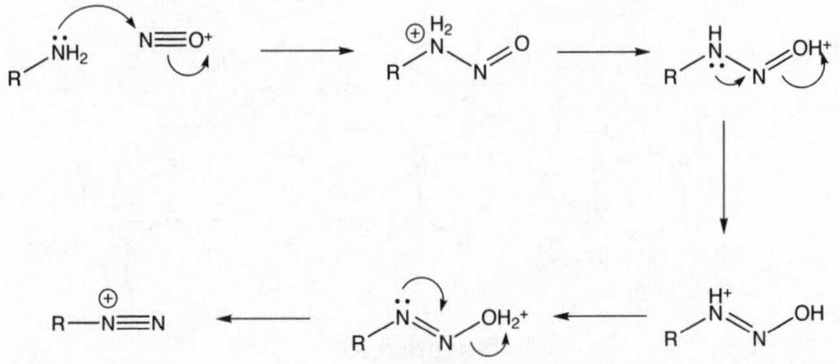

Figure 5.27

such as triethylamine or diisopropylethylamine. The latter is sometimes known as Hünig's base, and has the advantage in this situation of being unreactive as a nucleophile because of its steric crowding. It is much loved by peptide chemists for this reason.

Inorganic bases such as sodium carbonate or sodium hydroxide can also be used. You might think that sodium hydroxide would not be appropriate here, because it is itself a strong nucleophile and would react with the acyl halide. In practice this is not usually a problem, because this reaction is usually done with hydrophobic acid chlorides, such as benzoyl chloride, which do not dissolve in water, but instead form oily droplets. They therefore cannot mix efficiently with the hydroxide ions, as these are dissolved in the water. However, the amine will normally be miscible with the droplets and will therefore react efficiently. The reaction of acid chlorides with amines in this way is known as the Schotten–Baumann procedure.

Another interesting reaction of amines is with nitrous acid. The electrophile here is probably the N_2O_3 formed by the dehydration of nitrous acid (HNO_2). This serves as a donor of NO^+. The rest of the reaction is straightforward (Figure 5.27). This reaction must take place under acidic conditions, and it is the small amount of unprotonated amine that reacts. The reaction is faster for aromatic amines, as they are less basic, and so at a given pH more of the free base is present. The products of this reaction are normally diazonium salts, although if the reaction takes place next to a carbonyl group diazoalkanes are formed (Figure 5.28).

Aliphatic diazonium salts are extremely reactive, since N_2 is such a powerful leaving group. They are so reactive that this reaction is not normally useful synthetically, as the cations produced react in an uncontrolled manner by a mixture of

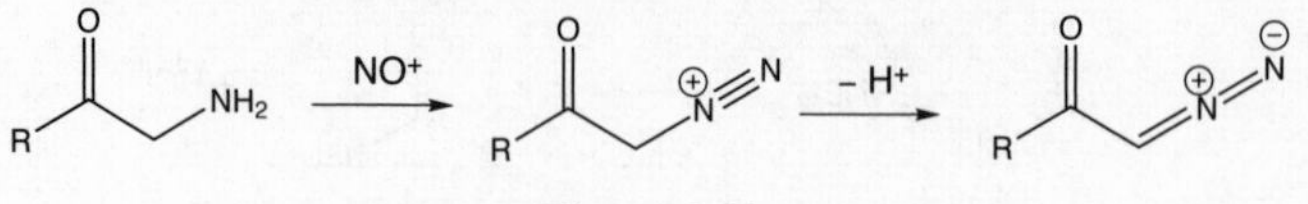

Figure 5.28

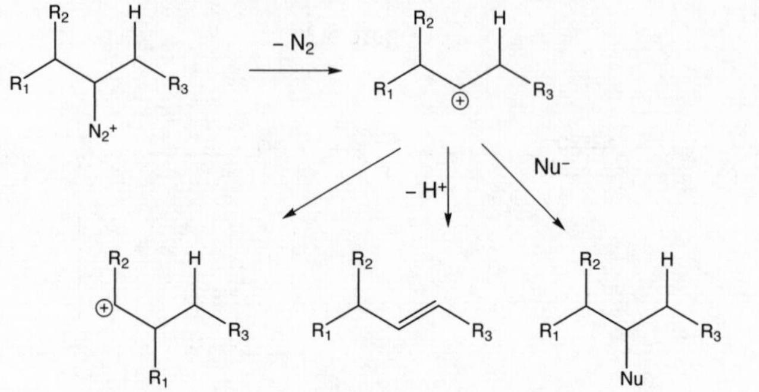

Figure 5.29

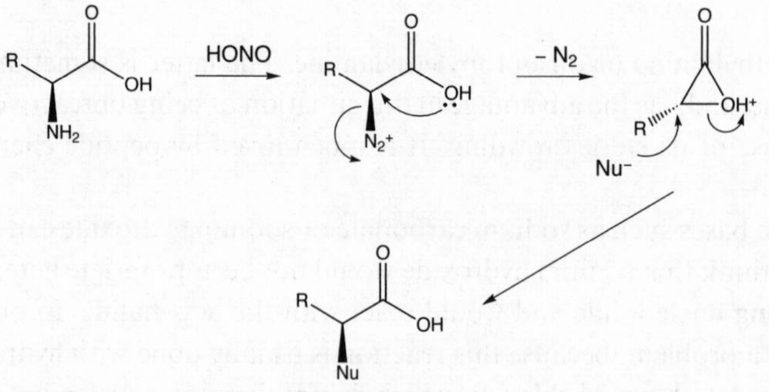

Figure 5.30

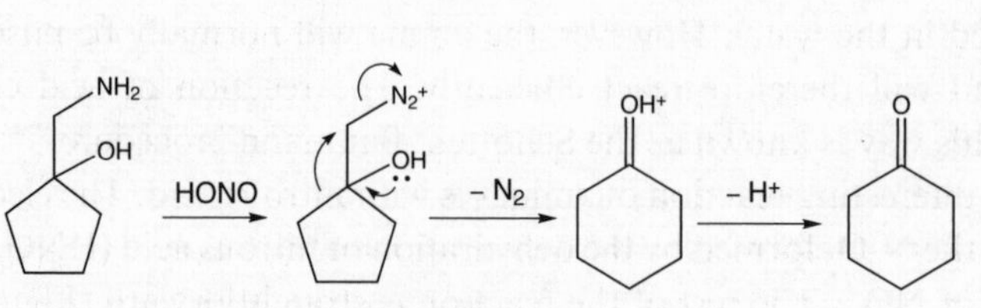

Figure 5.31

rearrangements, eliminations, and substitutions (Figure 5.29). However, there are occasions in which this reaction can be used synthetically, the two most common of which are shown in Figures 5.30 and 5.31.

When the amine is an α-amino acid, the carboxylate group is placed such that it can react rapidly with the α-carbon in an S_N2 displacement, before the nitrogen has time to leave spontaneously. Not only is the carboxylate group close, and therefore

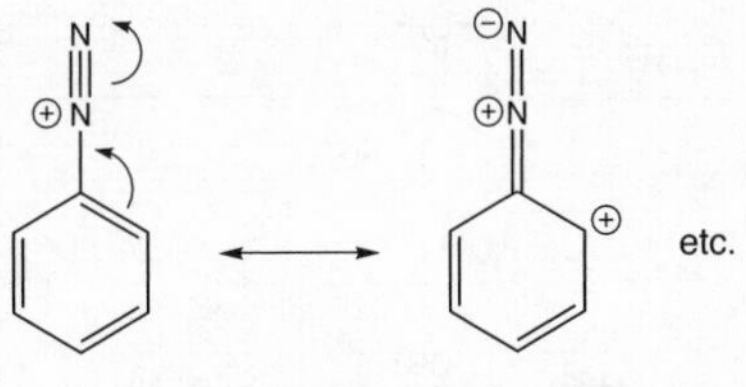

Figure 5.32

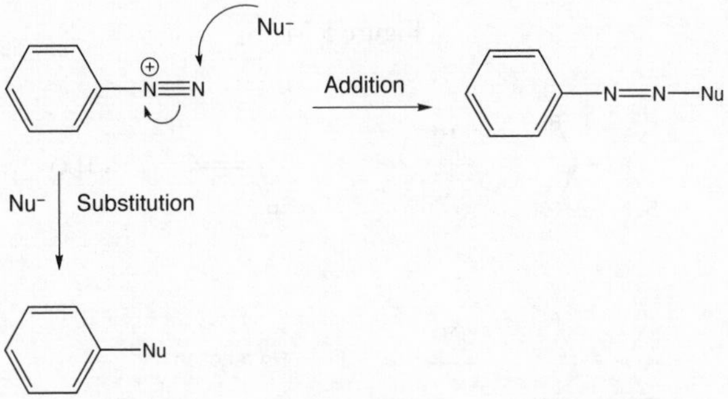

Figure 5.33

in a good position to act as a nucleophile, but the electron-withdrawing effect of the carbonyl group means that the normal S_N1 reaction is inhibited (formation of a carbocation at the α-position is an unfavourable process). Moreover, the α-position is activated towards S_N2 displacement, as we saw in Section 3.3.1. The resulting α-lactone is itself unstable, but lasts long enough to react with any nucleophiles present in another S_N2 reaction usually displaying some selectivity for better nucleophiles. Halides are commonly used. The overall effect of this reaction is displacement of the amino group with net retention of configuration.

The second common use for this reaction is in the ring expansion of α-hydroxy amines. This is known as the Tiffeneau–Demjanov ring expansion. The rearrangement shown is favoured because of the formation of the cation stabilized by the oxygen. This is shown here with a concerted mechanism; in other words the nitrogen leaves at the same time as the alkyl group migrates. However, it is also possible that the nitrogen departs first, leaving a primary carbocation which is quickly stabilized by rearrangement.

Aryl diazonium salts are considerably more stable, although must generally be kept at ice-bath temperatures. The extra stability is due to resonance stabilization (Figure 5.32). They undergo a range of reactions, the most important of which are shown in Figure 5.33. The most straightforward is S_N1 substitution by nucleophiles. Substitution with halides or cyanide may also be carried out with their copper (I) salts. This is known as the Sandmeyer reaction. It is generally a more efficient procedure, and is thought to take place by a free radical mechanism. Diazonium salts

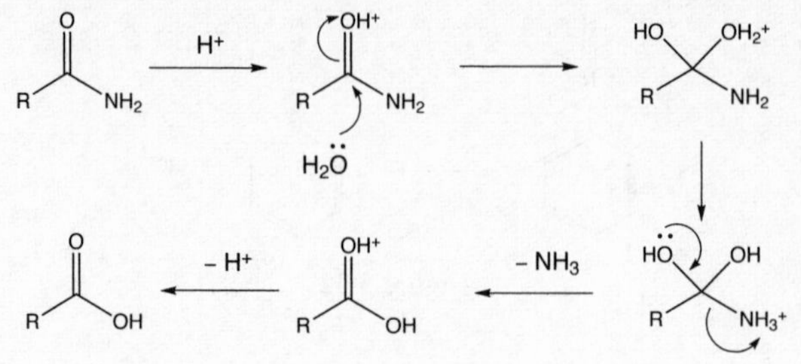

Figure 5.34

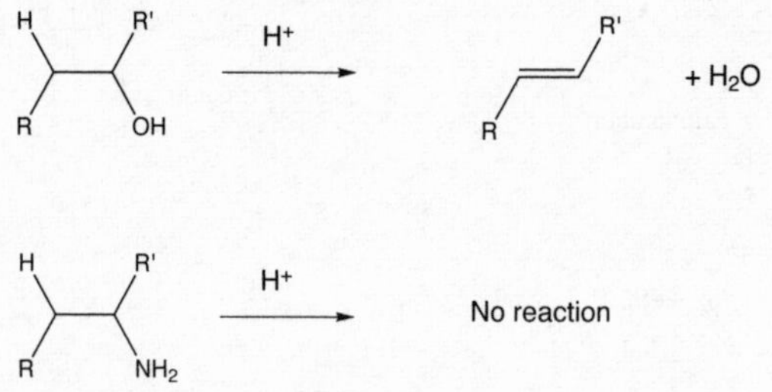

Figure 5.35

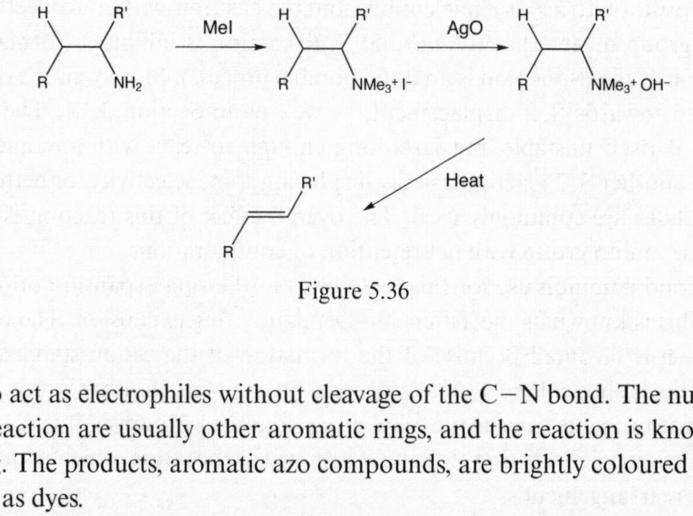

Figure 5.36

may also act as electrophiles without cleavage of the C–N bond. The nucleophiles in this reaction are usually other aromatic rings, and the reaction is known as azo coupling. The products, aromatic azo compounds, are brightly coloured and many are used as dyes.

Unlike oxygen, nitrogen is very reluctant to act as a leaving group. The most common reaction in which it does is the hydrolysis of amides, and this must normally take place under harsh acidic conditions (Figure 5.34). Amino groups cannot be eliminated by acid in the same way that hydroxyl groups can (Figure 5.35). However, they can be eliminated in a process known as the Hofmann degradation (Figure 5.36). The amine is first converted into a quaternary ammonium compound with methyl iodide, and then into its hydroxide salt by treatment with silver oxide.

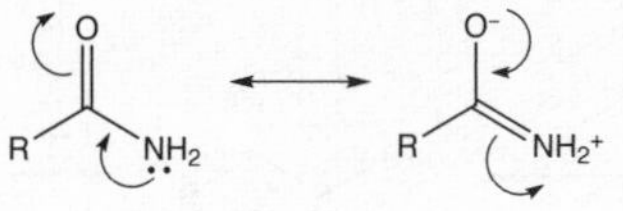

Figure 5.37

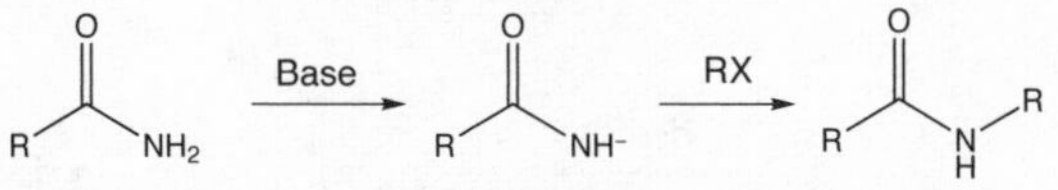

Figure 5.38

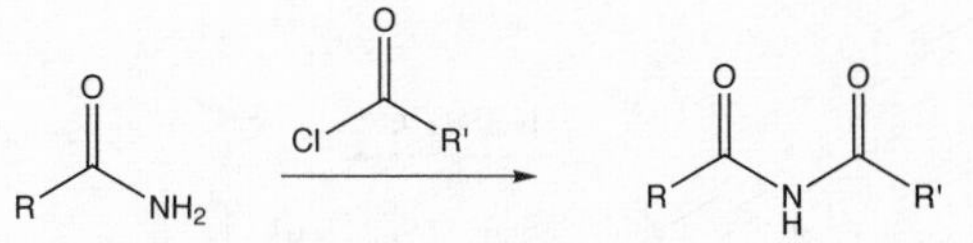

Figure 5.39

Heating this hydroxide salt leads to an E2 elimination (see Section 6.2.2). This reaction is largely of historical interest; it is little used synthetically, but enjoyed great popularity in the structural elucidation of natural products, particularly alkaloids, in the days before modern spectroscopic techniques were available.

5.2.2 Amides

Not surprisingly, the carbonyl group has a considerable influence on the reactivity of the nitrogen in amides. The lone pair of the nitrogen interacts with the π^* orbital of the carbonyl group, as we can see from the resonance structures shown in Figure 5.37. This interaction makes the nitrogen much less nucleophilic and less basic, so that it no longer undergoes many of the reactions of amines.

However, this electron-withdrawing effect of the carbonyl group makes the nitrogen substantially more acidic, and a proton can be removed from the amide nitrogen under relatively mild conditions. The pK_a of a normal primary amide is about 17, which means that an appreciable quantity of the anion will be formed in equilibrium with a base such as sodium hydroxide. Of course, this pK_a is higher than that of water, so most of the amide will not be deprotonated at equilibrium, but the deprotonated amide that is present will readily take part in reactions.

Amides can be alkylated on the nitrogen with alkyl halides after conversion to the base in this way (Figure 5.38), although they are not normally reactive enough to be alkylated in the neutral form. However, it is sometimes possible to acylate amides with acid chlorides (Figure 5.39).

The problem of cleanly alkylating amines was described above. The lower reactivity of amides provides a solution, because treating an amine with one equivalent of

131

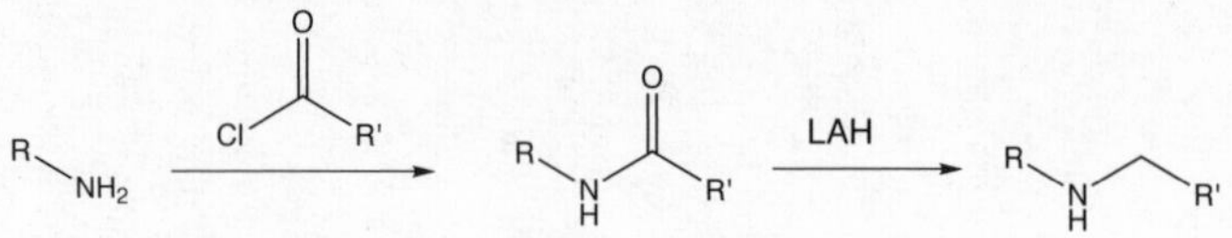

Figure 5.40

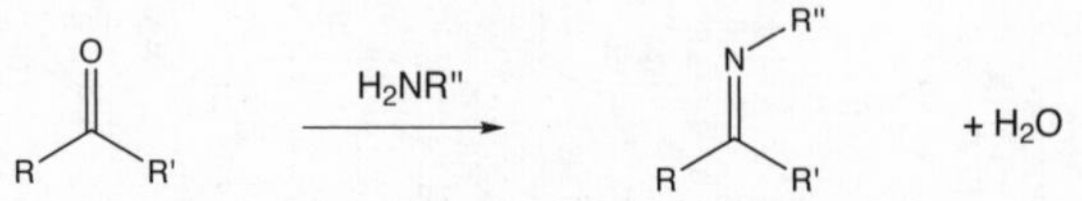

Figure 5.41

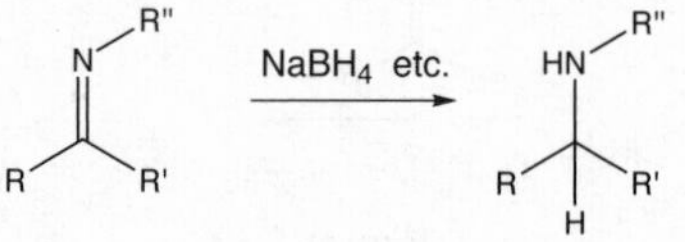

Figure 5.42

an acid chloride will normally produce the corresponding amide in good yield. Although it may be possible for the amide thus produced to react with a second molecule of acid chloride, the much greater reactivity of amines means that this seldom occurs in practice. The desired amine can then be prepared by reduction of the amide with lithium aluminium hydride (LAH) (Figure 5.40). Of course, not only does the carbonyl group modify the reactivity of the nitrogen, but the nitrogen also affects the reactivity of the carbonyl group, making it less reactive (see Section 5.1.3). This is why a powerful reagent such as LAH is needed to carry out this reduction.

5.2.3 Imines, oximes, and hydrazones

The product of the reaction between ammonia or a primary amine and a ketone is known as an imine (Figure 5.41). Unsubstituted imines (i.e. those derived from ammonia), are unstable, being rapidly hydrolysed even under neutral conditions. Alkyl-substituted imines, sometimes known as Schiff bases, are more stable, and can normally be isolated without too much difficulty. Aryl substitution improves the stability still further because of conjugation between the aryl group and the carbon–nitrogen double bond. However, all imines are readily hydrolysed in acid.

Imines can be easily reduced to amines with a variety of reducing agents (Figure 5.42). Sodium borohydride is a common choice, while the milder reagent sodium cyanoborohydride provides a way of reducing imines selectively in the presence of ketones. Imines themselves are not normally useful compounds. They are most often encountered as intermediates in the synthesis of secondary amines.

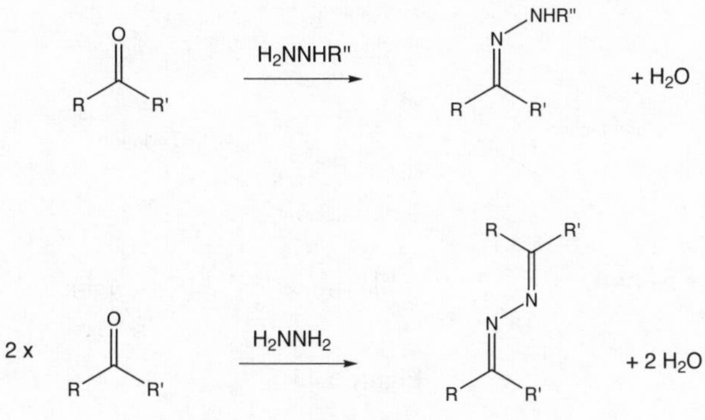

Figure 5.43

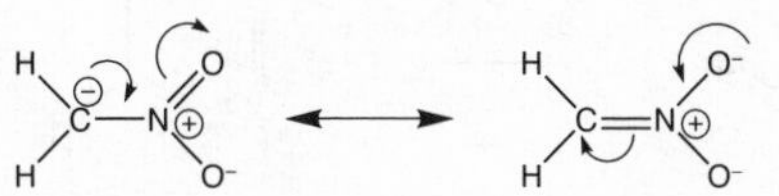

Figure 5.44

Oxygen substitution on the imine nitrogen produces a compound known as an oxime. These are prepared by reaction of ketones with hydroxylamine, and are generally stable compounds, being less easily hydrolysed than imines. Oximes undergo an interesting reaction known as the Beckmann rearrangement (see Section 3.2).

Hydrazones can also be regarded as derivatives of imines. These are prepared from ketones and a hydrazine, although if hydrazine itself is used, the product is often an azine (Figure 5.43). Hydrazones are usually stable crystalline compounds, and are mainly of historical importance, in that their main use was for the character-ization of ketones (or occasionally hydrazines) by the preparation of crystalline hydrazone derivatives. Modern spectroscopic methods have rendered this procedure almost, but not completely, obsolete.

5.2.4 Other nitrogen functions

There are many other functional groups containing nitrogen, although some of them are rather obscure. Perhaps the most important is the *nitro* group. Its main feature is that it is a powerful electron-withdrawing group, and so it profoundly modifies the properties of any group to which it is attached. The most common example of this is its well-known property as a directing group in aromatic rings (see Section 6.3.6). However, nitro groups are not restricted to aromatic compounds, and have the effect of substantially increasing the acidity of adjacent protons in aliphatic compounds. For example, nitromethane has a pK_a of 10, and so can easily form a carbanion. Figure 5.44 shows how this is stabilized by resonance. Carbanions from

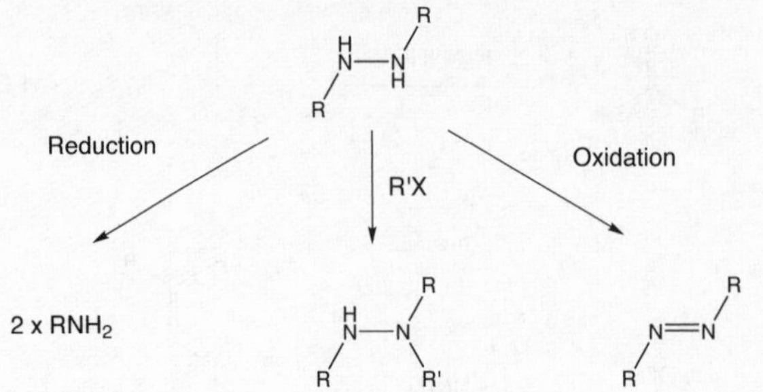

Figure 5.45

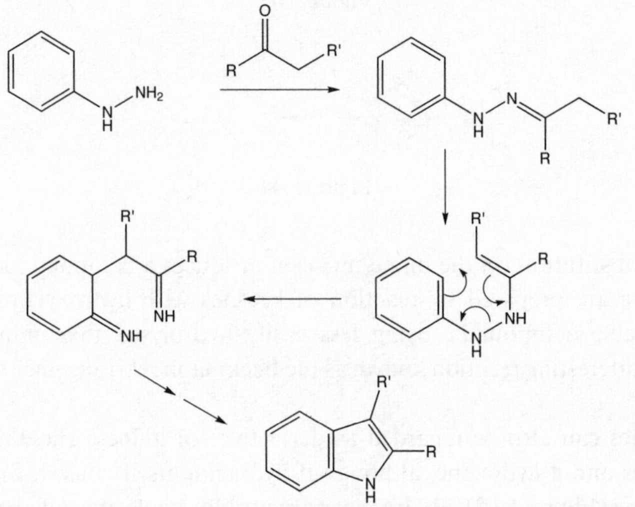

Figure 5.46

nitroalkanes can of course take part in the usual carbanion reactions such as addition to carbonyl groups. The nitro group itself is rather inert, and remains unchanged in most reactions. However, it can be reduced to an amino group by a number of reducing agents, such as LAH.

Hydrazines are another important class of nitrogen compounds. They have three principal modes of reactivity (Figure 5.45). Firstly, because of the alpha effect they are very nucleophilic, more so than amines, and are therefore easily alkylated and acylated, although they are less basic than amines. Secondly, they may be cleaved, because of the weak N−N bond. This may take place with reducing agents such as zinc metal to give two amine molecules, or may alternatively occur in a rearrangement. An example of such a rearrangement is found in the Fischer indole synthesis (Figure 5.46). In the key step of this synthesis, the N−N bond of hydrazine is cleaved in a sigmatropic rearrangement (see Section 6.5). This is followed by ring closure and elimination of ammonia to give the indole.

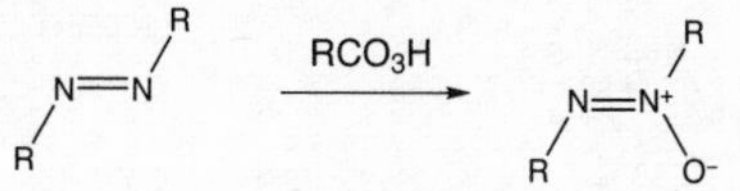

Figure 5.47

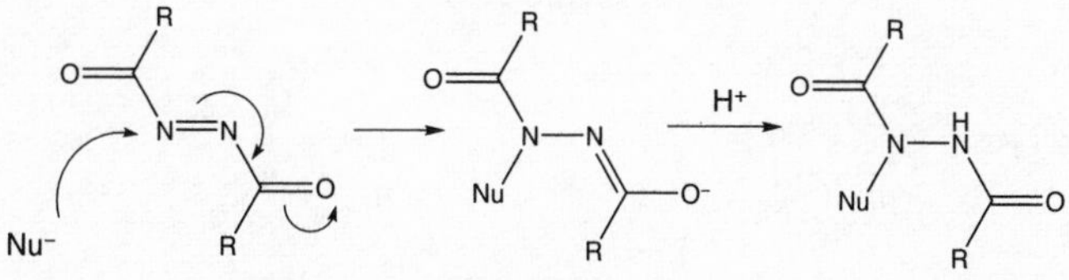

Figure 5.48

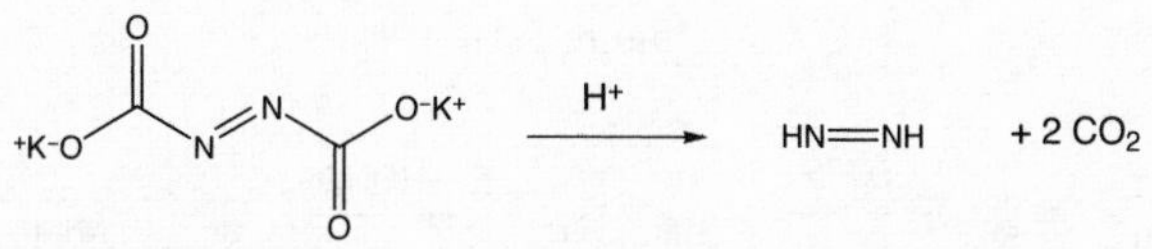

Figure 5.49

The third reactivity pattern of hydrazines is that they can be oxidized to *azo compounds*. This is particularly easy if the resulting azo compound is conjugated; for example, hydrazobenzene is oxidized to azobenzene on standing in air. Azo compounds are extremely weak bases, partly because they are sp_2 hybridized (see Section 2.2.2.3), and partly because of the electron-withdrawing effect of the adjacent nitrogen. They are weak nucleophiles, and the only reaction of any importance in which they behave as nucleophiles is their oxidation to azoxy compounds by peracids (Figure 5.47).

Acyl azo compounds can act as electrophiles, being attacked by nucleophiles in a reaction analogous to Michael addition to α,β-unsaturated carbonyl compounds (Figure 5.48). Another use of azo compounds is that they can be decomposed to nitrogen and two radicals. We saw in Section 4.3 that this is a commonly used method for generating radicals. Azo compounds can also be reduced to hydrazines or even directly to amines.

The parent azo compound, $HN=NH$, which is called diimide (or diazine), is a fascinating molecule. It is unstable and must be prepared *in situ*. The most common way of doing this is the acid-catalysed decomposition of potassium azodicarboxylate (Figure 5.49). Left to its own devices diimide quickly decomposes mainly to nitrogen, hydrogen, and hydrazine. However, it can be used to reduce $C=C$ or $N=N$ double bonds, proceeding by the cyclic mechanism shown in Figure 5.50. This reaction is driven by the thermodynamically very favourable formation of nitrogen. The reaction works best for symmetrical double bonds, and generally fails if the bonds are significantly polarized. Monosubstituted azo compounds are also unstable.

135

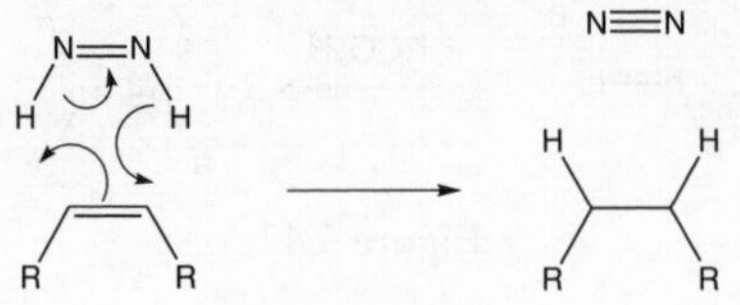

Figure 5.50

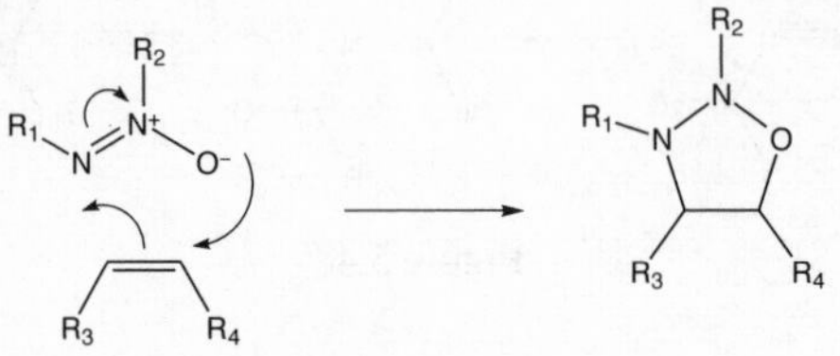

Figure 5.51

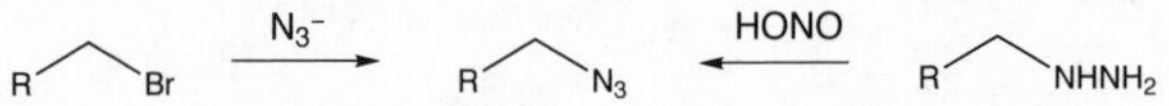

Figure 5.52

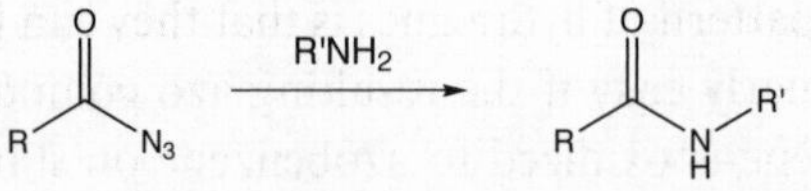

Figure 5.53

Azoxy compounds are relatively unimportant, although they are normally reasonably stable. They may be reduced to azo compounds, hydrazines, or amines, but their most useful reaction is cycloaddition to olefins (Figure 5.51).

Azides are compounds containing the N_3 group. They are normally prepared by reaction of the azide anion with appropriate electrophiles (Figure 5.52). This anion is only weakly basic, but is a good nucleophile, so these reactions are usually rather efficient. Azides may also be prepared by treatment of hydrazines or hydrazides with nitrous acid, in a reaction analogous to diazotization.

Organic azides are not normally stable indefinitely, although they can usually be isolated. They have a tendency to decompose by losing nitrogen, and low molecular weight azides may even do this explosively when heated. The azide group is moderately electron-withdrawing and can act as a leaving group. The best-known example of this is the use of acyl azides in peptide synthesis (Figure 5.53). Azides also take part in cycloaddition reactions, and can be reduced to amines.

The most commonly encountered *diazo compound* is diazomethane, which is often used for the preparation of methyl esters (Figure 5.54). Diazomethane is unstable, has been known to explode, and is prepared *in situ* by the reaction shown in Figure

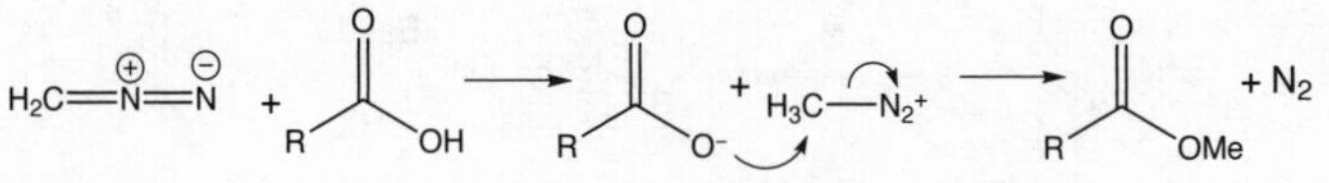

Figure 5.54

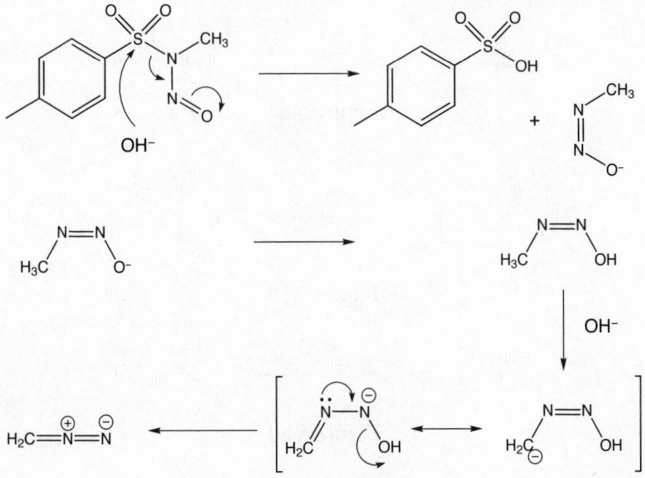

Figure 5.55

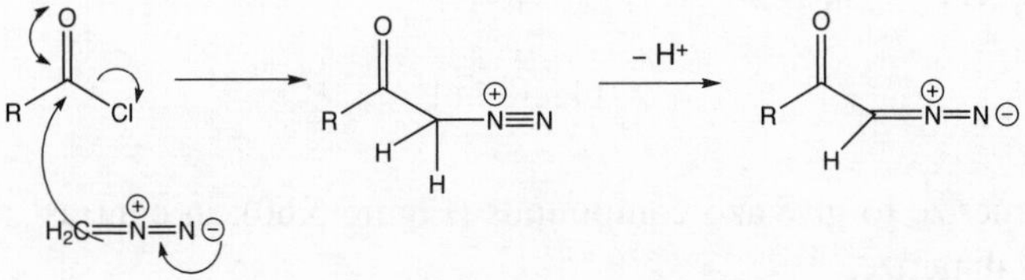

Figure 5.56

5.55. Diazomethane is nucleophilic, and can react with acyl chlorides. This is the first step in the Arndt–Eistert synthesis, shown in Figure 5.56. The subsequent step is a rearrangement involving a carbene (see Section 4.4). Diazo compounds can also take part in cycloaddition reactions.

5.2.5 Nitrenes

Nitrenes are the nitrogen equivalents of carbenes. Like carbenes, they are unstable intermediates with only a fleeting existence. There are two common ways in which they are formed: α-elimination of appropriately substituted amines and decomposition of azides (Figure 5.57).

Nitrenes undergo much the same reactions as carbenes. They may insert into single bonds (Figure 5.58), add to double bonds (Figure 5.59), or rearrange. They

137

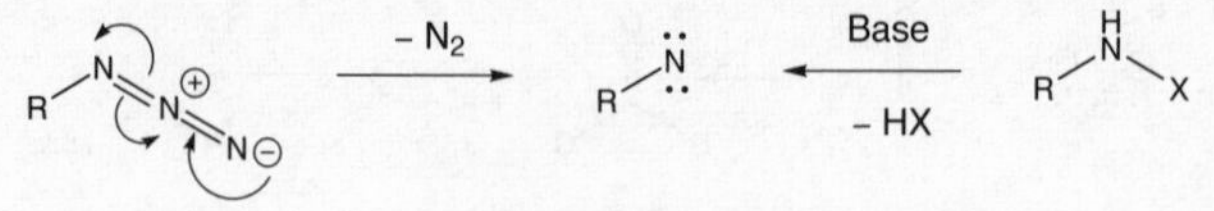

Figure 5.57

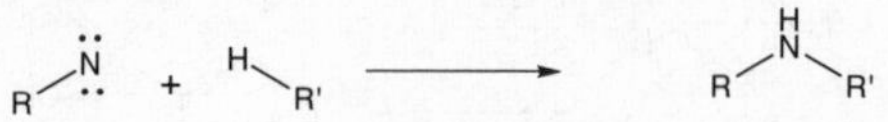

Figure 5.58

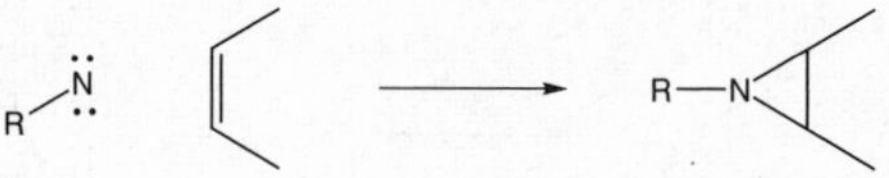

Figure 5.59

Figure 5.60

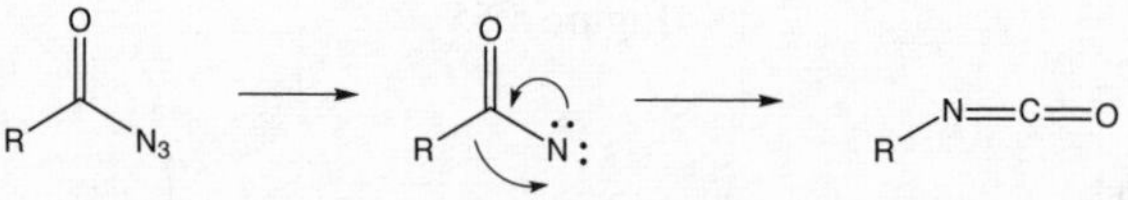

Figure 5.61

may also dimerize to give azo compounds (Figure 5.60), in contrast to carbenes, which rarely dimerize.

One nitrene rearrangement in particular is worth mentioning. This actually goes by four different names, depending on how the nitrene is formed, but the rearrangement is always the same, an acyl nitrene rearranging to an isocyanate. In the Curtius rearrangement, the starting material is an acyl azide (Figure 5.61). This decomposes thermally to give the nitrene, which then rearranges to the isocyanate. Isocyanates can be isolated, although they normally react further, either by hydrolysis to the amine (via a carbamic acid, which decarboxylates) or by alcoholysis to a carbamate (Figure 5.62).

The Schmidt reaction is similar, but here the starting material is a carboxylic acid. This reacts under rather harsh conditions with hydrazoic acid (HN_3) to form the azide, which then reacts further as in the Curtius reaction. Here, however, the conditions used are such that the final product is almost always the amine.

In the Hofmann rearrangement (Figure 5.63), a primary amide is treated with bromine and sodium hydroxide (or sodium hypobromite). This initially forms an N-bromo amide which then undergoes α-elimination to give the nitrene, which reacts

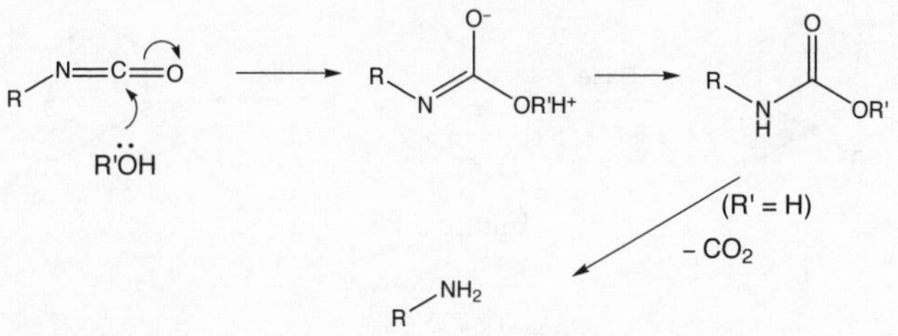

Figure 5.62

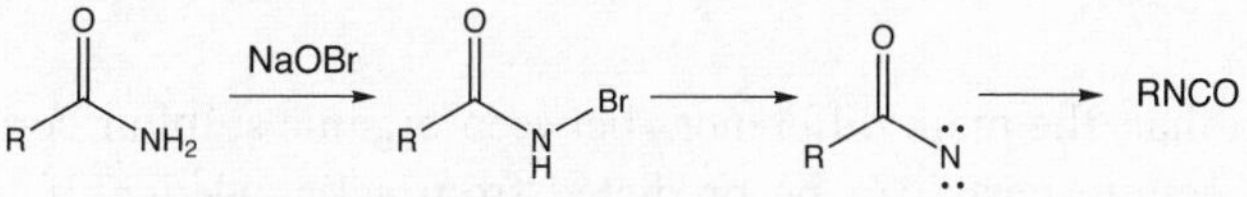

Figure 5.63

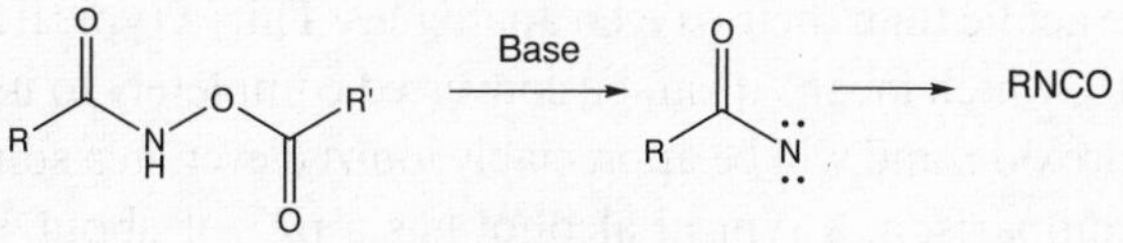

Figure 5.64

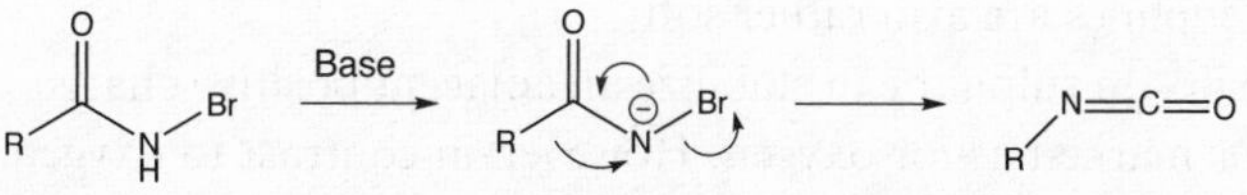

Figure 5.65

as before. The Lossen rearrangement is similar; here the starting material is an O-acyl hydroxamic acid (Figure 5.64). This is treated with base to give the nitrene.

These reactions have all been described as proceeding via free nitrenes, although this has not been proved experimentally. It is also entirely possible that some or all of these reactions may be concerted, the rearrangement taking place at the same time as the other groups leave the nitrogen (for example Figure 5.65).

5.3 Sulphur

Sulphur is directly below oxygen in the periodic table, so we would expect there to be some similarities in their chemistry. Indeed there are, although there are also many differences. One of the most important of these is that whereas the carbonyl group is of pivotal importance in the chemistry of organic oxygen compounds, thio-carbonyl groups play almost no part in the chemistry of sulphur, owing to their instability. Most thioketones decompose rapidly, and some of the simpler thioalde-hydes are so unstable that they have never been isolated.

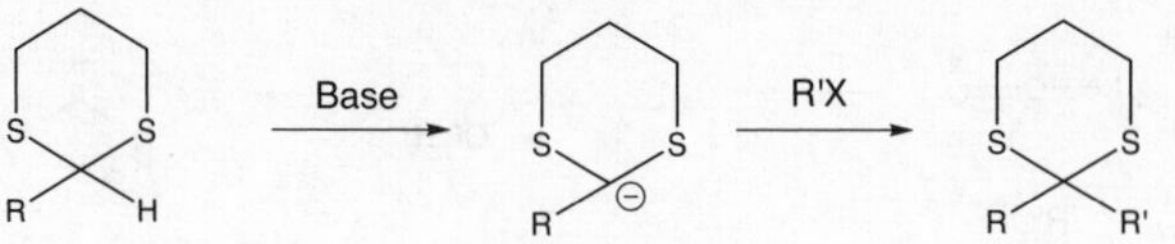

Figure 5.66

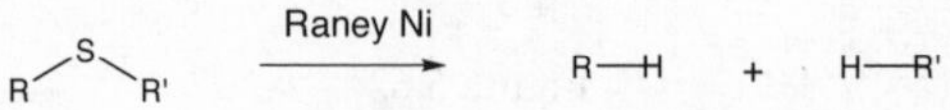

Figure 5.67

Other than that, the main differences between organic sulphur compounds and their oxygen counterparts can be predicted from a knowledge of trends within groups of the periodic table. As discussed in Section 2.2.2, elements further down the periodic table tend to be more acidic than those above them, so sulphur compounds are more acidic than their oxygen analogues. Thus a typical alkane thiol has a pK_a of about 10, which means it can be converted completely to its anion in water with sodium hydroxide, and will be appreciably ionized even in a solution of sodium carbonate. By comparison, a typical alcohol has a pK_a of about 17, so thiols are considerably more acidic. Sulphur compounds are also more nucleophilic than their oxygen analogues. Thiolate anions are some of the strongest nucleophiles known. Sulphur nucleophiles are also rather soft.

The lone pairs on sulphur can stabilize an adjacent positive charge, although this effect is not as marked as for oxygen. However, in contrast to oxygen, sulphur can also stabilize an adjacent negative charge. For example, thioacetals can be deprotonated and alkylated by treatment with a suitable base (normally butyllithium), as shown in Figure 5.66.

Unlike oxygen, sulphur can form stable compounds with valences higher than 2, as in compounds such as sulphoxides and sulphones. This has been attributed to the availability of d orbitals for extra bonding, although the evidence for this is not entirely clear.

The carbon–sulphur bond is weaker than the carbon–oxygen bond. The main consequence of this is that sulphur can be reductively removed from most organosulphur compounds by treatment with Raney nickel (Figure 5.67).

Sulphur-based radicals are formed more easily than their oxygen analogues.

5.3.1 Thiols

Thiols are the sulphur analogues of alcohols, and as mentioned above they are both more acidic and more nucleophilic. As nucleophiles, they undergo many of the same reactions as alcohols, and often faster. The reactions are not necessarily always more efficient, as the products may be less stable. This applies mainly to attack on an acyl carbon, where the product is a thioester (see below).

140

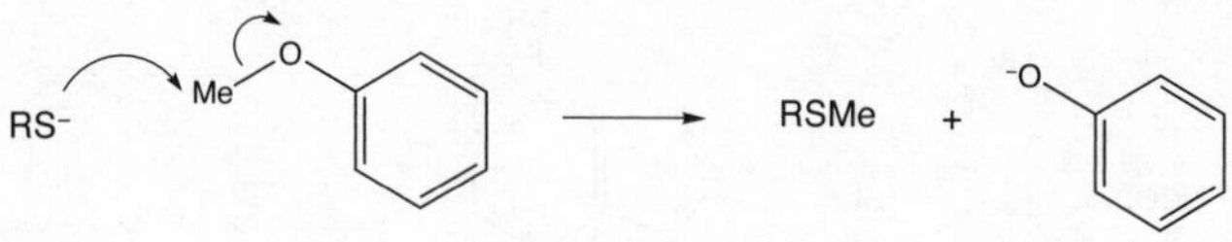

Figure 5.68

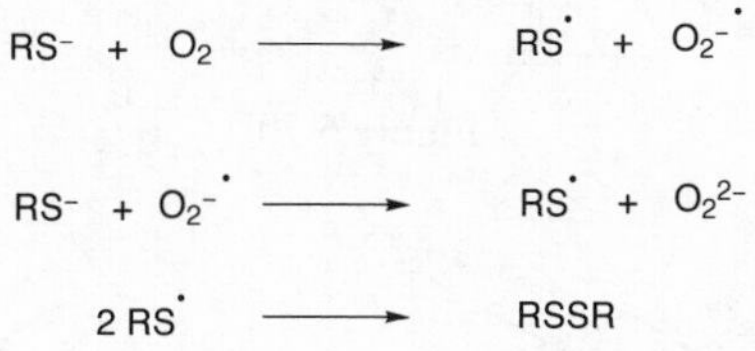

$$RS^- + O_2 \longrightarrow RS^{\cdot} + O_2^{-\cdot}$$

$$RS^- + O_2^{-\cdot} \longrightarrow RS^{\cdot} + O_2^{2-}$$

$$2\,RS^{\cdot} \longrightarrow RSSR$$

Figure 5.69

Thiolate anions are such powerful nucleophiles that they undergo some reactions that are unknown for alcohols or alkoxides. One such reaction is the cleavage of aryl methyl esters by an S_N2 mechanism (Figure 5.68).

Thiols are easily oxidized to disulphides. This reaction frequently takes place on standing in air, and can be a nuisance when working with thiols. It is a lot faster if a base is present; the mechanism is shown in Figure 5.69. The reaction can be carried out very efficiently if desired with a number of reagents such as hydrogen peroxide or iodine.

Thiols may also be oxidized to other products, most notably sulphonic acids and sulphonyl halides.

5.3.2 Sulphides

Sulphides, also known as thioethers, are more reactive than their oxygen counterparts. The sulphur is still nucleophilic, despite not being able to lose a proton, and undergoes a number of reactions as a nucleophile. One of the most important of these is oxidation. A common reagent for this is hydrogen peroxide, and the product is either a sulphoxide or a sulphone, depending on the conditions.

Sulphides can be alkylated to give trialkylsulphonium salts. These are stable compounds and can be isolated, although they are usually reasonably easy to cleave by treatment with nucleophiles. A trialkylsulphonium salt of considerable biological importance is S-adenosylmethionine (SAM), which acts as a methyl group donor in many biological systems (Figure 5.70). SAM is often described as nature's equivalent of methyl iodide.

Sulphides can be cleaved with iodine if one of the groups attached to the sulphur can form a stable cation (Figure 5.71). This is frequently used to regenerate a thiol that has been protected as a sulphide during a synthesis (the product of the reaction is actually a disulphide, although this can easily be reduced to the thiol if required).

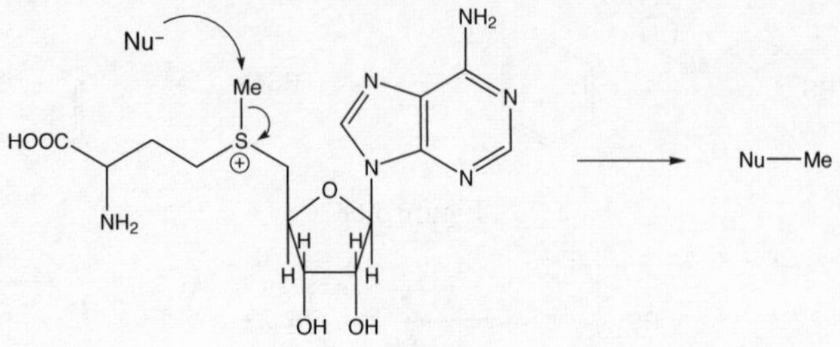

Figure 5.70

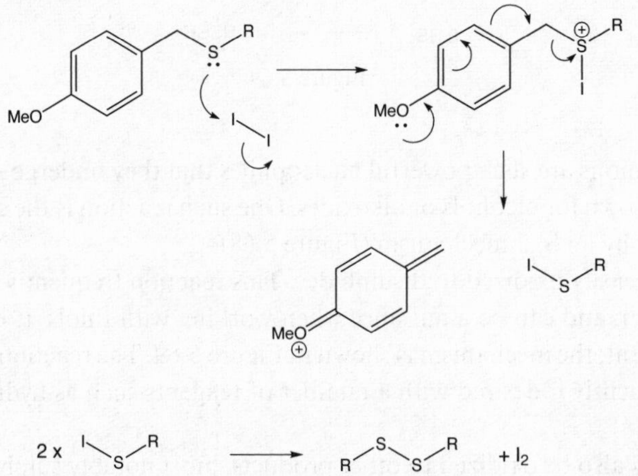

Figure 5.71

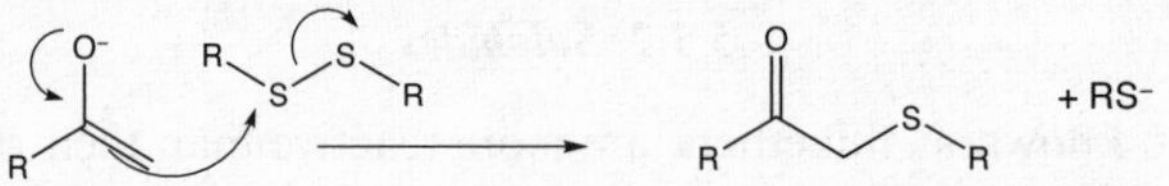

Figure 5.72

Sulphides can sometimes be deprotonated at their α-position. This is easier if there is another group present to stabilize the anion, although it can sometimes be done even if there is not.

5.3.3 Disulphides

Disulphides can be readily reduced to thiols by a number of reagents, such as lithium in liquid ammonia or zinc in dilute acid.

Disulphides are also susceptible to cleavage by nucleophiles, for example enolates can be thiolated in this way (Figure 5.72). However, this reaction is generally better

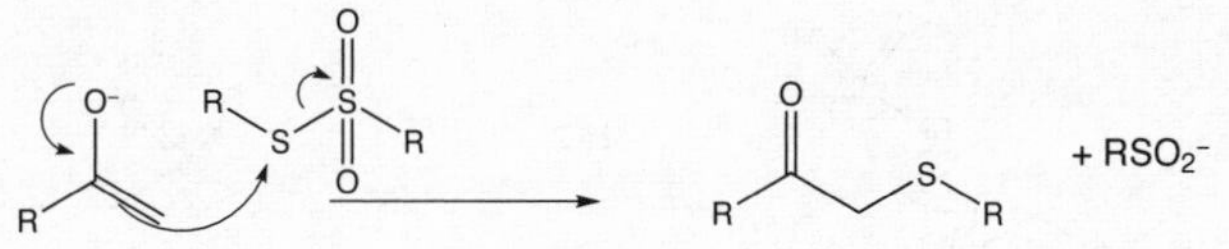

Figure 5.73

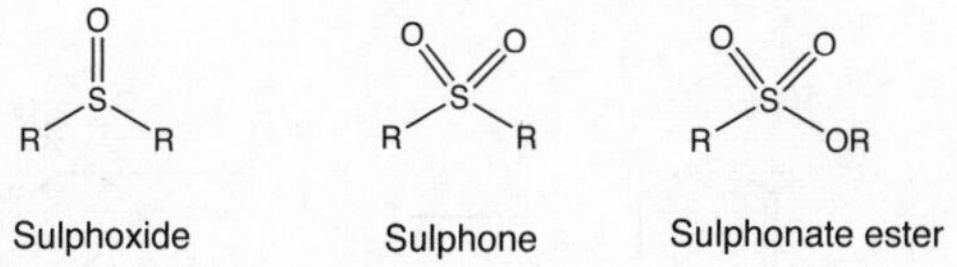

Sulphoxide   Sulphone   Sulphonate ester

Figure 5.74

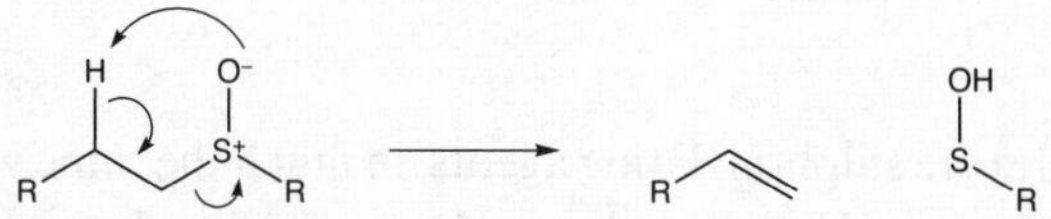

Figure 5.75

accomplished with a thiosulphonic ester, as this has a better leaving group (Figure 5.73).

5.3.4 Thioesters

Thioesters are moderately reactive carboxylic acid derivatives. They are more reactive than esters, but not as reactive as anhydrides.

They are of rather limited use in synthetic chemistry, although they are important biologically. In the same way that SAM is nature's methyl iodide, acetyl coenzyme A (acetyl CoA) is a ubiquitous thioester that acts as nature's acetyl chloride.

5.3.5 Oxidized sulphur functions

There are a number of functional groups in which sulphur is combined with oxygen, of which the most important are sulphoxides, sulphones, and sulphonic acids and their derivatives (Figure 5.74).

Sulphoxides can stabilize an adjacent negative charge better than sulphides, and sulphones can do so better still. The α-protons of sulphones are of comparable acidity to those of ketones. Oxidized forms of sulphur are reasonably good leaving groups, and both sulphoxides and sulphones may undergo β-elimination to give olefins. Sulphoxides are eliminated by a concerted cyclic mechanism (Figure 5.75).

Sulphonic acids are strong acids, comparable in strength to mineral acids. Trifluoromethanesulphonic acid (triflic acid) is one of the strongest acids known.

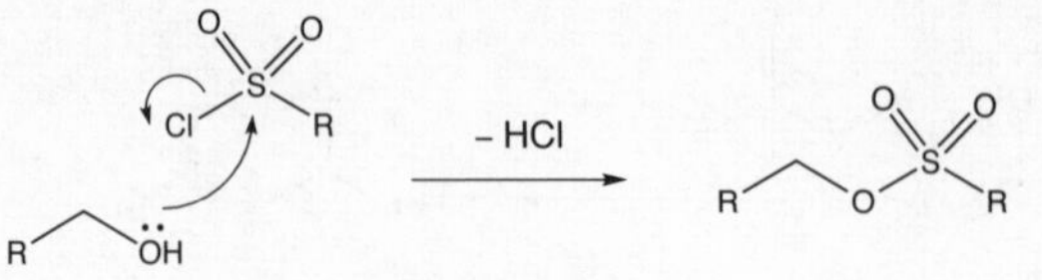

Figure 5.76

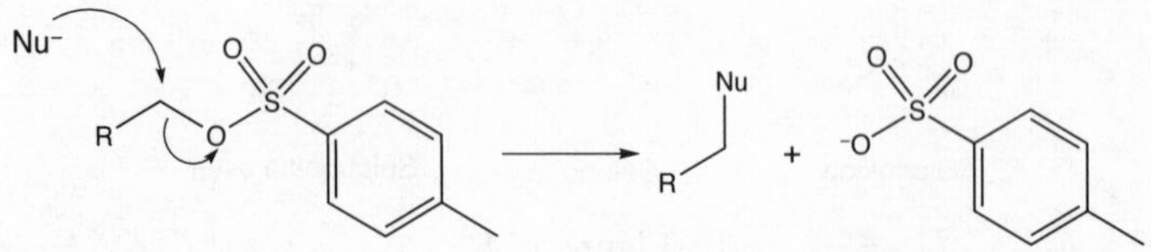

Figure 5.77

Sulphonyl chlorides are sulphonylating agents, in much the same way that acyl chlorides are acylating agents. However, the mechanism of displacement of the chloride is different here, being a direct displacement (Figure 5.76). Toluenesulphonyl chloride (tosyl chloride) is commonly used to prepare sulphonate esters from alcohols. The sulphonate group of this can act as a leaving group, so the method can be used as an indirect way of displacing alcohols with other nucleophiles (Figure 5.77).

5.4 Phosphorus

The differences between oxygen and sulphur are similar to those between nitrogen and phosphorus. Thus phosphorus compounds are less basic than their nitrogen analogues, but are still excellent nucleophiles if they have free lone pairs. Phosphorus can form compounds with valences higher than 3; these higher-valency compounds play a large part in its chemistry.

One recurring feature in the chemistry of phosphorus is that it forms particularly strong bonds with oxygen, and the conversion of triphenylphosphine to triphenylphosphine oxide, for example, is a common thermodynamic driving force for reactions.

Phosphines are the phosphorus analogues of amines. However, they are not very frequently encountered, with the exception of tertiary phosphines and triphenylphosphine in particular. Triphenylphosphine has only weakly basic properties (the pK_a of its conjugate acid is 2.7), but is a good nucleophile. It is often encountered in complexes with various metals, but its most important use is in the Wittig reaction (see below).

Phosphoric acid (H_3PO_4) may be esterified with one, two, or three alkyl groups to give phosphate esters. Phosphotriesters are reasonably easily hydrolysed in base, whereas phosphodiesters and phosphomonoesters are much more stable. They are all hydrolysed readily in acid. Phosphodiesters are extremely important in nature –

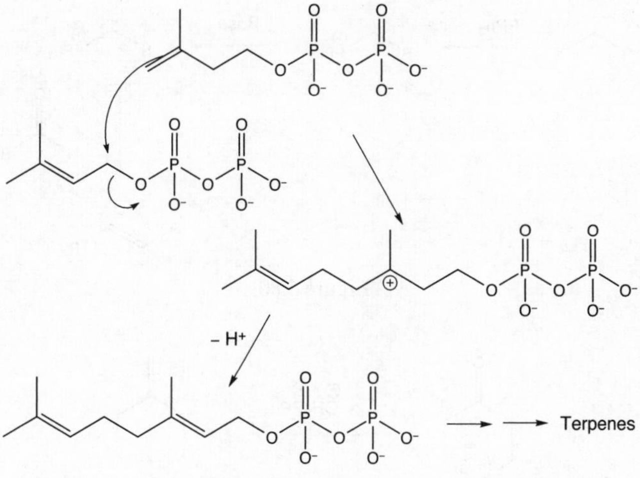

Figure 5.78

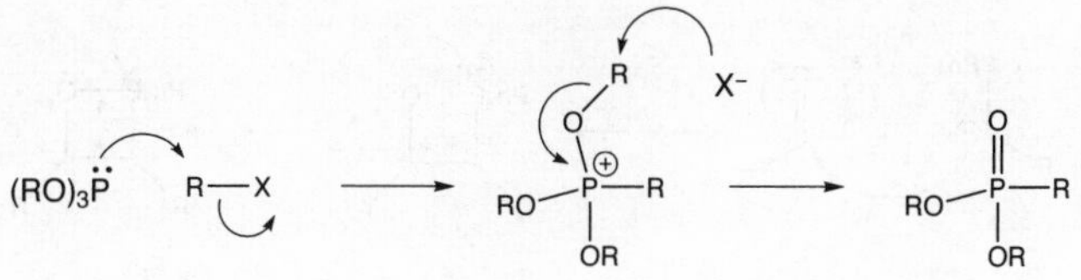

Figure 5.79

DNA, the molecule that contains genetic information, is a polymer in which the monomers are held together by phosphodiester linkages.

Phosphate groups are good leaving groups, and under acidic conditions phosphate esters are cleaved at the $C-O$ bond. The pyrophosphate group, the anhydride of phosphoric acid, is an even better leaving group. This is a common leaving group in biosynthetic reactions; terpenes (a group of compounds which among other things are responsible for the characteristic odours of herbs and many other plants) are formed from dimethylallyl pyrophosphate and isopentenyl pyrophosphate (Figure 5.78).

Trialkyl phosphites are another class of phosphorus compounds. They are most often encountered in the Arbuzov reaction, in which they are converted to dialkyl phosphonates. The mechanism is shown in Figure 5.79. This demonstrates two of the properties of phosphorus compounds: the trialkyl phosphite is able to act as a nucleophile by virtue of its lone pair, and the reaction is driven by the formation of a phosphorus–oxygen double bond.

The most important reaction of organophosphorus compounds is the Wittig reaction, which is an extremely useful synthetic method for the preparation of double bonds. This may use triphenylphosphine, or in a modification known as the

145

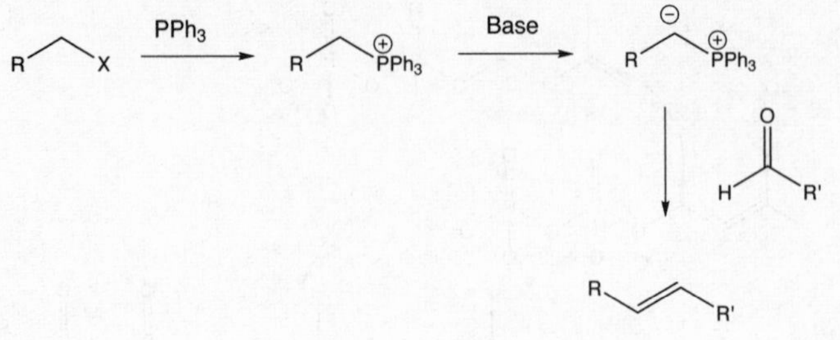

Figure 5.80

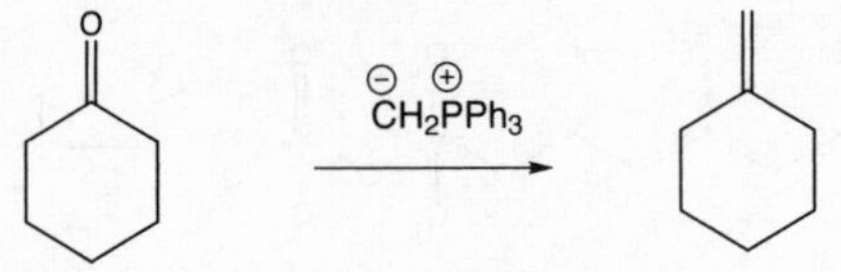

Figure 5.81

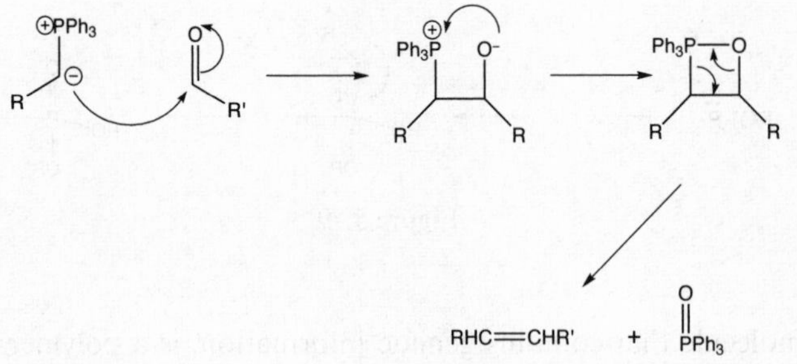

Figure 5.82

Horner–Emmons or Wadsworth–Emmons reaction dialkyl phosphonates are used. The reaction joins an alkyl halide with an aldehyde (or sometimes a ketone) to produce an olefin (Figure 5.80). The great advantage of the reaction is that the double bond is produced in a controlled manner, always between the two carbon atoms that are joined. It is often the only way of preparing molecules in which the double bond is in a thermodynamically unfavourable position, such as the exocyclic compound in Figure 5.81.

The mechanism of the key step is shown in Figure 5.82. The carbanion is stabilized by the adjacent positive charge of the phosphonium salt. This type of compound, in which a negative charge is stabilized by an adjacent positive charge, is known as an ylide. Vacant d orbitals on phosphorus may play a part in this stabilization, although quantum mechanical calculations suggest that the bonding may be more complicated than this. Whatever the reason, phosphorus ylides can be formed far more easily than their nitrogen analogues. If the carbanion has no further means

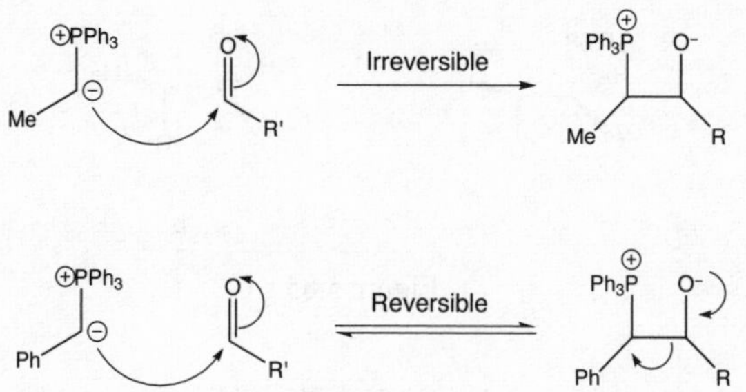

Figure 5.83

Figure 5.84

of stabilization, then a strong base such as butyllithium is normally needed to form the ylide. However, if another stabilizing group is present, such as an adjacent aromatic ring or carbonyl group, then a much milder base can be used, for example sodium ethoxide. The ylide reacts with the aldehyde, and the resultant zwitterion cyclizes to form a four-membered oxaphosphetane. This breaks down to give triphenylphosphine oxide, which provides a thermodynamic driving force for the reaction, and the olefin.

This reaction is usually stereoselective, in that unstabilized ylides normally lead to *cis* double bonds (i.e. with both substituents on the same side of the double bond), whereas stabilized ylides (i.e. those with extra carbanion stabilization in addition to the phosphorus) normally lead to *trans* double bonds. This can be remembered by the mnemonic that unstabilized ylides lead to the less stable product (*cis* double bond) and that stabilized ylides lead to the more stable product (*trans* double bond). The reason for this difference is connected with the initial attack of the ylide on the aldehyde. For unstabilized ylides this reaction is irreversible, whereas the reaction is an equilibrium process for stabilized ylides (Figure 5.83). This is because stabilized ylides can more easily leave the other carbon, as they are better leaving groups.

Let us first consider what happens with stabilized ylides. The zwitterion can exist as one of two diastereoisomers, as shown in Figure 5.84. Both of these are shown here in the conformation they must adopt for the oxygen and the phosphorus to join together. In isomer a, which leads to the *cis* product, the two R groups are on the same side of the bond, and in isomer b, which leads to the *trans* product, the two R groups are on opposite sides. Thus for steric reasons, it is easier for isomer b to adopt the conformation in which the oxygen and phosphorus can meet than it is for isomer a. Isomer b is

147

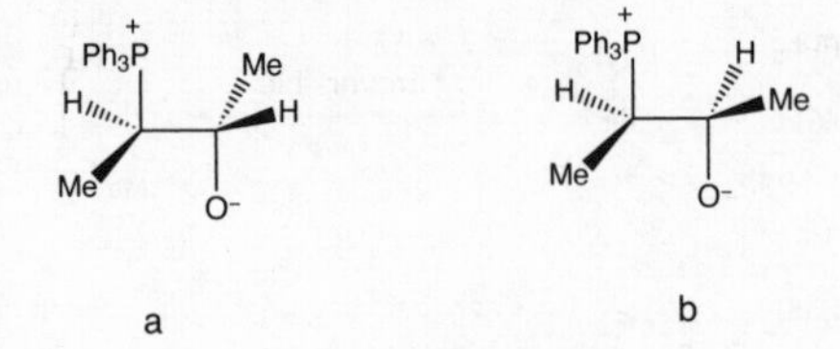

a   b

Figure 5.85

thus more likely to react to give the product than isomer a, and so isomer a is more likely to undergo the reverse reaction and have another chance to form isomer b.

For unstabilized ylides, this equilibration is not possible, and the isomer that is initially formed determines the geometry of the product. Although the conformations of the two isomers shown here are those that ultimately react, they are not the conformations that are formed initially. When the two reactants come together, the phosphorus will tend to point away from the oxygen since these are the two largest groups (the oxygen is actually very bulky, because it is charged and therefore highly solvated; the solvent molecules attached to it give it its steric bulk). In this conformation, which is similar to the transition state for the attack of the ylide on the carbonyl group, the two isomers a and b are as shown in Figure 5.85. Here, isomer a is the more stable, owing to less steric crowding between the R groups. Isomer a is therefore more likely to be formed, and so the *cis* product will predominate.

Note that the reaction of stabilized ylides is under thermodynamic control, whereas the reaction of unstabilized ylides is under kinetic control (see Section 3.1.6).

Needless to say, this description of the stereoselectivity is not an invariable rule, and is clearly less applicable if ketones are used instead of aldehydes or if the other starting material is a secondary rather than primary halide.

5.5 Halogens

The chemistry of organohalogen compounds is relatively straightforward, and can be summed up by saying that halogens attached to carbons are good leaving groups and have electron-withdrawing properties. The orders of these two properties down the group are in opposite directions, iodine being the best leaving group and the least electron-withdrawing.

Halides act as leaving groups in their well-known substitution and elimination reactions (Figure 5.86). Fluorine is such a poor leaving group that it rarely takes part in substitution reactions, although it can be eliminated.

Although the leaving group ability of the halides improves on going down the group, they also become more expensive. This is seldom a consideration in laboratory reactions, when reagents are used in only small quantities, and so alkyl chlorides are rarely used in displacement reactions, bromides and iodides being more popular, owing to their greater reactivity. However, many industrial processes use

Table 5.2. *pK$_a$s of various halogenated acetic acids*

Molecule	CH_3COOH	CH_2ICOOH	$CH_2BrCOOH$	$CH_2ClCOOH$
pK$_a$	4.8	3.2	2.9	2.9
Molecule	$CHCl_2COOH$	CCl_3COOH	CH_2FCOOH	CF_3COOH
pK$_a$	1.3	0.7	2.6	0.2

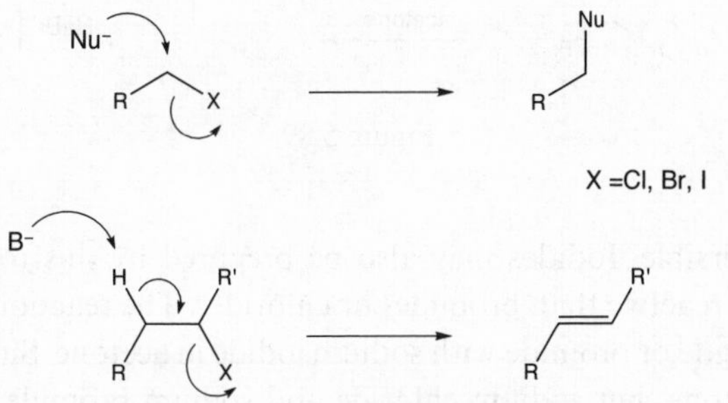

Figure 5.86

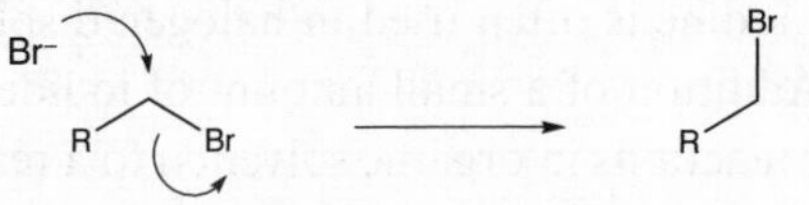

Figure 5.87

alkyl chlorides despite their more sluggish reactions, because the cost of the reagents is more important when large quantities are used.

The electron-withdrawing effect of the halogens is demonstrated by the pK$_a$s of various halogenated acetic acids as shown in Table 5.2. Note that trifluoroacetic acid is as strong as some mineral acids.

Halogens are also important as electron-withdrawing substituents in aromatic compounds. Aromatic rings substituted with halogens undergo electrophilic substitution reactions more slowly than their parent compounds. Halogenated aromatic rings can also undergo nucleophilic substitution under certain circumstances (see Section 6.3.6).

Since halides are both good leaving groups and good nucleophiles, alkyl halides can undergo halide exchange reactions. Sometimes this is undetectable, as in the reaction of a primary alkyl bromide with bromide ion (Figure 5.87). However, this reaction can sometimes lead to loss of stereochemistry at a chiral centre as a result of successive substitutions with inversion, particularly for iodides (Figure 5.88).

Halide exchange can also be used synthetically. Alkyl fluorides are often prepared by this route, because they do not easily undergo displacement reactions and so the

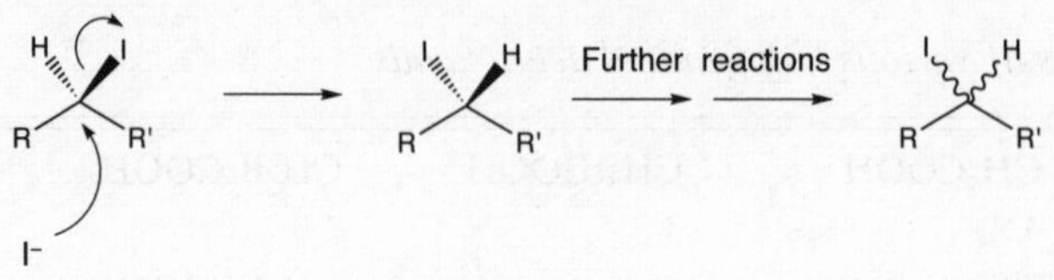

Figure 5.88

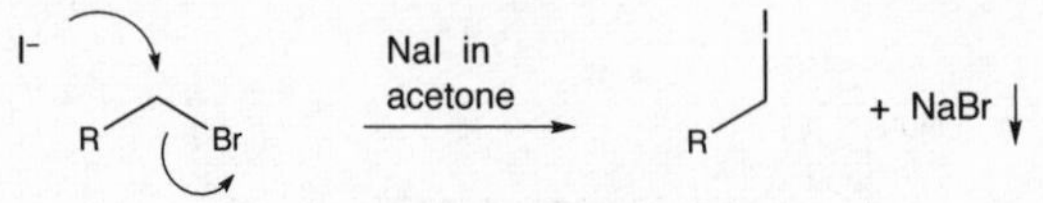

Figure 5.89

reaction is irreversible. Iodides may also be prepared by this route, even though iodides are more reactive than bromides or chlorides. The reaction is done by treating an alkyl chloride or bromide with sodium iodide in acetone. Since sodium iodide is soluble in acetone, but sodium chloride and sodium bromide are not, the precipitation of sodium chloride or bromide drives the reaction towards the formation of the alkyl iodide (Figure 5.89).

In a related process, iodine is often used in halogen displacement reactions as a nucleophilic catalyst. Addition of a small amount of iodide (tetrabutylammonium iodide is often used for reactions in organic solvents) to a reaction between an alkyl chloride or bromide and a nucleophile can often cause a considerable increase in the rate. The way this works is that the iodide rapidly displaces the other halide, because it is such a good nucleophile. Although in the absence of the nucleophile this would rapidly revert to the chloride (or bromide), because iodide is also a good leaving group, the small amount of alkyl iodide that is present at any one time can also react with the added nucleophile. Since alkyl iodides are much more reactive than chlorides or bromides, this has the effect of speeding up the reaction. This is similar to the use of pyridine to catalyse acylation reactions, which is discussed in Section 6.3.5.

Carbon–halogen bonds can also be cleaved homolytically. An example is their reduction with tributyltin hydride in the presence of a radical initiator (Figure 5.90).

5.6 Group I and II metals

A great many organometallic compounds are known, and a comprehensive discussion of their properties would fill several books. We shall restrict ourselves here to the most important properties of the metals most commonly encountered in organic chemistry, namely lithium, sodium, potassium, and magnesium.

These metals mainly act as counter-ions to carbanions, which were discussed in Section 4.1. However, the properties of the carbanions can differ markedly depending on the metal with which they are associated. The carbon–metal bonds in

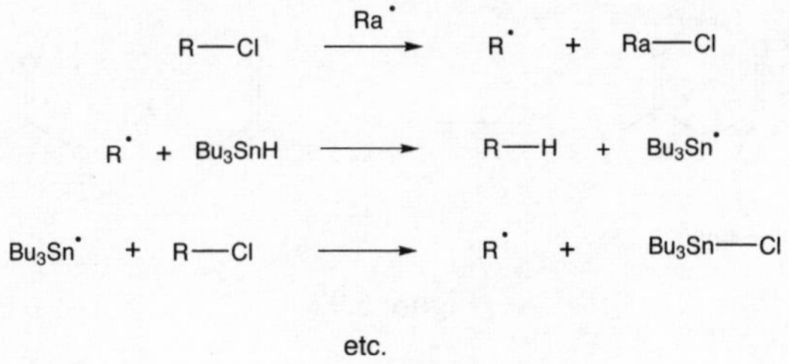

$$R\text{—}Cl \xrightarrow{\text{Ra}^{\bullet}} R^{\bullet} + Ra\text{—}Cl$$

$$R^{\bullet} + Bu_3SnH \longrightarrow R\text{—}H + Bu_3Sn^{\bullet}$$

$$Bu_3Sn^{\bullet} + R\text{—}Cl \longrightarrow R^{\bullet} + Bu_3Sn\text{—}Cl$$

etc.

Figure 5.90

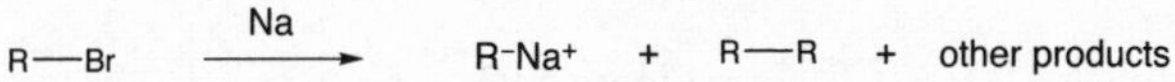

$$R\text{—}Br \xrightarrow{\text{Na}} R^-Na^+ + R\text{—}R + \text{other products}$$

Figure 5.91

organolithium and organomagnesium (Grignard) compounds have significant covalent character, whereas organosodium and organopotassium compounds are essentially ionic.

Sodium and potassium compounds are usually more reactive than organolithiums and Grignards. Thus, in general, Grignard reagents do not react in displacement reactions with alkyl halides (although they may do if the alkyl halide is sufficiently reactive, e.g. if it is benzylic), whereas organosodium and organopotassium compounds usually do give this reaction.

This greater reactivity of the sodium and potassium compounds can be a disadvantage, as it can make them difficult to handle. The standard way to prepare the lithium and magnesium compounds is by treatment of the corresponding alkyl halide with lithium or magnesium metal. Treatment with sodium or potassium, however, is usually unsuccessful because the Wurtz reaction (see Section 4.1.3) competes with the formation of the organometallic compound (Figure 5.91). For this reason, Grignards and organolithiums are used more frequently than the sodium and potassium compounds.

There are a number of other ways to prepare organosodium and organopotassium compounds, of which the most useful is metal exchange. A more readily available organometallic compound (often an organomercury compound) is treated with sodium or potassium metal, and an exchange takes place.

Another important feature of metal ions in organic chemistry is that they may act as Lewis acids. This is particularly relevant to lithium and magnesium. For example, in the reaction between a ketone and a Grignard reagent, the reaction is assisted by coordination of the ketone group to a magnesium atom (this makes the carbonyl group more electrophilic). We saw earlier (Section 5.1.3) that it is difficult to prepare ketones by the reaction of carboxylic acid derivatives with Grignard reagents. An exception to this is the reaction between acid anhydrides and Grignards. Here, the anhydride is perfectly set up to coordinate to a magnesium

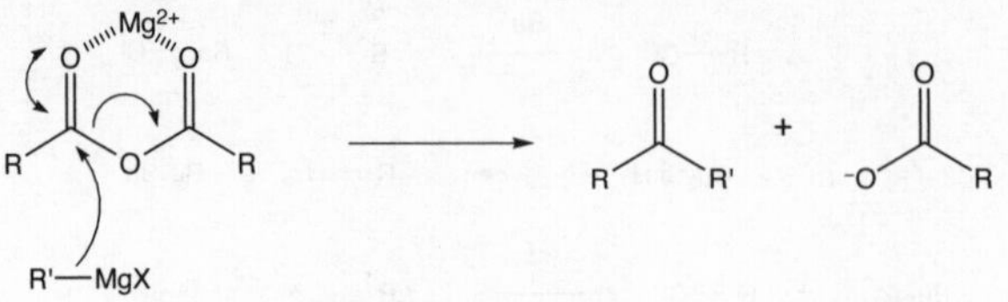

Figure 5.92

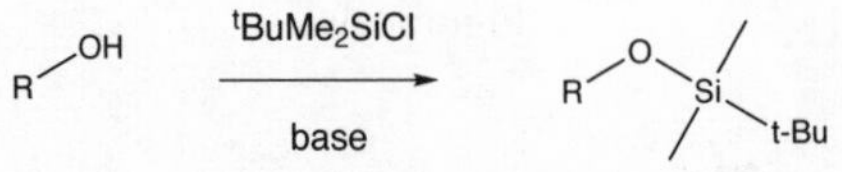

Figure 5.93

atom, which consequently is bound tightly (Figure 5.92). The anhydride is therefore much more reactive than the ketone (in addition to the normal difference in reactivity), so it is possible to stop the reaction at the ketone stage.

5.7 Silicon

Silicon is unlike the other heteroatoms discussed in this chapter (with the exception of the metals discussed in Section 5.6) in that it is more electropositive than carbon. The carbon–silicon bond is therefore polarized towards the carbon, although this effect is rather small and so silicon cannot meaningfully be described as a metal.

However, the organic silicon compounds most often encountered have silicon bonded to oxygen, rather than just carbon. Silicon has two common uses in organic chemistry, in protecting groups for alcohols, and in the related function of silyl enol ethers. There are of course many other types of organic compound containing silicon, but these will not be discussed here.

Silicon has an affinity with oxygen, and this feature is exploited by the use of tri-alkylsilyl groups as protecting groups for alcohol. An alcohol is treated with a tri-alkylsilyl chloride (t-butyldimethylsilyl chloride, or TBDMS-Cl, is often used) and a base to form the protected alcohol (Figure 5.93). This is stable to base, to many oxidizing and reducing agents, and to mild acid. It can be cleaved under mild conditions with fluoride ion (usually as tetrabutylammonium fluoride, or TBAF), which takes advantage of the particularly strong bond formed between silicon and fluorine (Figure 5.94).

Silyl enol ethers are prepared by treating a carbonyl compound with base and a trialkylsilyl chloride (usually trimethylsilyl chloride, or TMSCl). This may either be done by preforming the enolate with a strong base and then adding the TMSCl, or by using a weaker base (usually triethylamine) and adding the TMSCl at the same time, so that the TMSCl reacts with the small equilibrium concentration of the enolate. The choice of one method or the other can offer a certain amount of

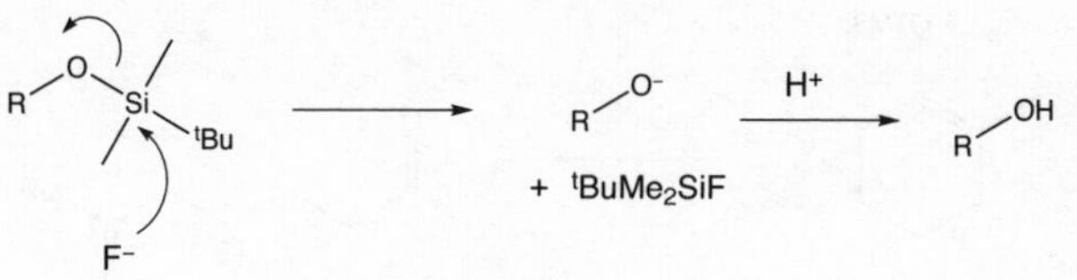

Figure 5.94

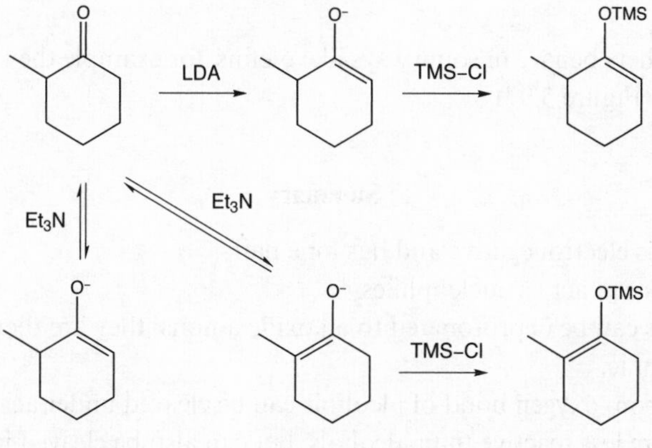

Figure 5.95

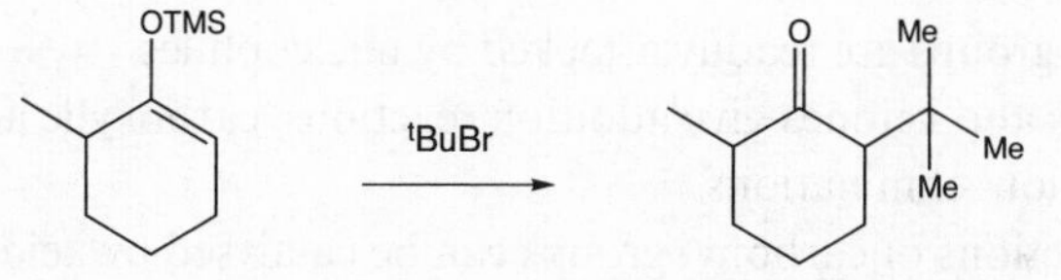

Figure 5.96

control, as shown in Figure 5.95. When the ketone is treated with LDA, the more acidic proton is lost, and the kinetic enolate is formed. However, if triethylamine is used, an equilibrium can establish, which leads to preferential formation of the more substituted and hence more stable double bond.

Silyl enol ethers are used as specific enolate equivalents. They undergo many of the reactions of enolates, such as alkylation with alkyl halides, acylation with acyl chlorides, and aldol reactions with aldehydes and ketones. They are of course not as reactive as enolates, and a catalyst is usually necessary, often a Lewis acid such as $TiCl_4$.

These compounds have the advantage over enolates that they offer greater control, for example by allowing a defined regiochemistry. Although silyl enol ethers are usually prepared and used straight away, they can sometimes be purified to separate a mixture of isomers. Their lower basicity means that they can be alkylated with tertiary halides (Figure 5.96). This reaction usually fails for enolates, since elimination predominates.

153

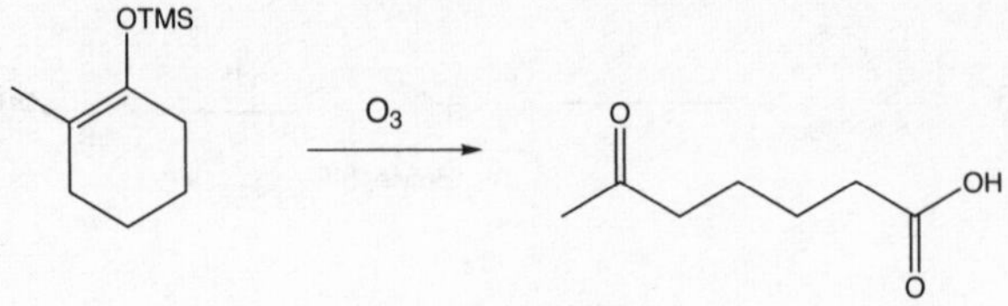

Figure 5.97

Silyl enol ethers behave in some ways like olefins, for example they can be cleaved by ozonolysis (Figure 5.97).

Summary

- Oxygen is electronegative and has lone pairs.
- Alcohols can act as nucleophiles.
- Alcohols can be deprotonated to alkoxide anions; they are then much more nucleophilic.
- The carbon–oxygen bond of alcohols can be cleaved under acidic conditions.
- Ethers are less reactive than alcohols, but can also be cleaved in acid.
- Carbonyl groups are polarized such that the carbon atom is electron-deficient.
- Carbonyl groups are readily attacked by nucleophiles.
- Aldehydes and ketones give addition reactions, carboxylic acid derivatives give addition–eliminations.
- Many reactions of carbonyl groups can be catalysed by acid or base.
- Carbonyl groups can stabilize an adjacent negative charge.
- Aldehydes readily give aldol reactions when treated with base; other carbonyl compounds can also react with themselves in a similar manner.
- Nitrogen is less electronegative than carbon, and so is more basic and more nucleophilic.
- Amines are excellent nucleophiles, and react with carbonyl groups and alkyl halides.
- Reaction of amines with alkyl halides can lead to multiple alkylation.
- Amines react best under basic conditions; they are readily protonated in acid, and so are no longer nucleophilic.
- Amines can be diazotized by treatment with nitrous acid.
- It is much more difficult to cleave the carbon–nitrogen bond in amines than it is to cleave the carbon–oxygen bond in alcohols.
- Amides are less nucleophilic and less basic than amines, but are more easily deprotonated.
- The carbonyl group of amides is less reactive than in many other carbonyl compounds.

- Unsubstituted imines are unstable. Their stability can be improved by substitution with alkyl or aryl groups (Schiff bases), with oxygen (oximes), or with nitrogen (hydrazones).
- Imines are readily reduced to amines.
- Oximes undergo the Beckmann rearrangement when treated with acid.
- Nitro groups are strongly electron withdrawing.
- Hydrazines are very nucleophilic, and contain a weak $N-N$ bond that can be cleaved. They can also be oxidized to azo compounds.
- Diimide is an unstable molecule that can reduce $C=C$ and $N=N$ bonds.
- Azoxy compounds take part in cycloaddition reactions.
- Acyl azides are used in peptide synthesis.
- Diazomethane is used to prepare methyl esters from carboxylic acids.
- Nitrenes are monovalent nitrogen species; they are analogous to carbenes.
- Nitrenes take part in the Curtius, Schmidt, Hofmann, and Lossen rearrangements.
- Sulphur compounds are more easily deprotonated and more nucleophilic than their oxygen analogues.
- Sulphur nucleophiles are soft.
- Carbon–sulphur bonds can be cleaved with Raney nickel.
- Sulphur readily forms radicals.
- Many sulphur compounds are easily oxidized.
- Sulphides and thioesters are important in nature.
- Many sulphur functions can stabilize an adjacent negative charge, particularly sulphoxides and sulphones.
- Phosphorus compounds are less basic and more nucleophilic than their nitrogen analogues.
- Phosphorus forms strong bonds with oxygen.
- Phosphoric acid can form esters; some of these are important in nature.
- Trialkyl phosphites take part in the Arbuzov reaction.
- Phosphorus plays an important part in the Wittig reaction.
- Halides are electron-withdrawing and are good leaving groups.
- Of the halides, iodide is the best leaving group, fluoride is the most strongly electron-withdrawing.
- Magnesium and lithium form bonds to carbanions with significant covalent character.
- Organosodium and organopotassium compounds are normally more reactive than their lithium and magnesium counterparts.
- Metal ions can act as Lewis acids.
- Silicon is slightly more electropositive than carbon.
- Silicon forms strong bonds with oxygen and fluorine.
- Trialkyl silyl groups are often used as protecting groups for alcohols.
- Silyl enol ethers can be used as specific enolate equivalents.

Problems

1. Ethyl bromide and ethanol can be interconverted by hydrolysis of ethyl bromide and reaction of ethanol with bromide ion. For each direction, should the reaction be run in acid or base, and why?

2. Arrange the following carbonyl compounds in order of their reactivity towards nucleophiles.

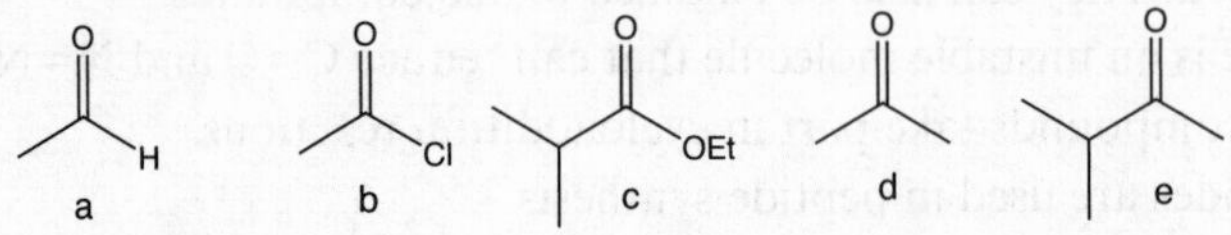

3. Write a mechanism for the reaction that ensues when the following aldehyde is treated with base.

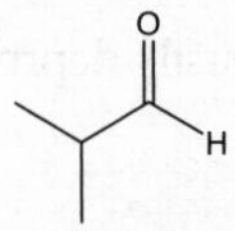

4. The first mixture of two carbonyl compounds shown below gives a mixture of four compounds when treated with base, but the second mixture gives a single product under the same conditions. Explain.

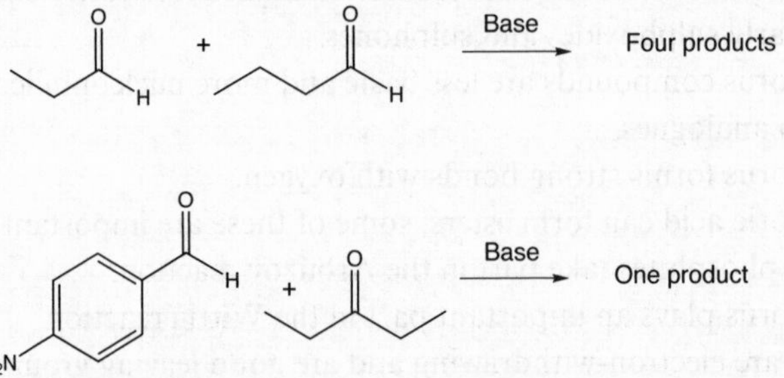

5. Reaction of the ketone shown below with hydroxylamine gives an oxime, A, that rearranges on treatment with acid to give a further product, B, hydrolysis of which gives the carboxylic acid and the amine shown. What are the intermediates, A and B, and what is the geometry of the oxime?

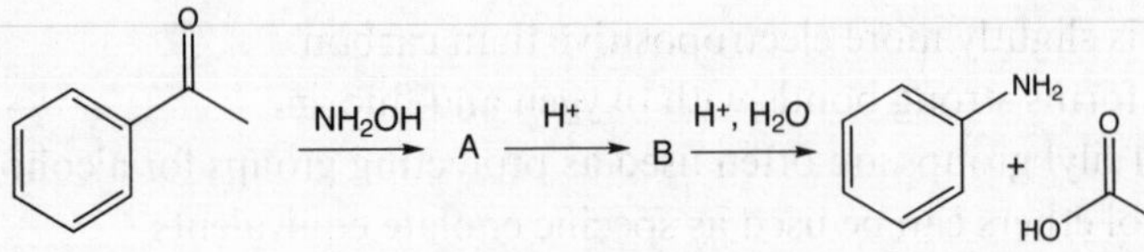

6. What are the most likely products of the reactions between ethylamine and the following compounds?

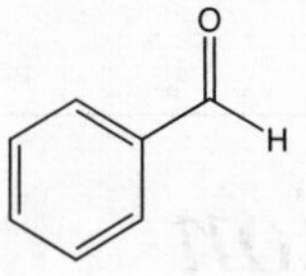

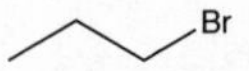

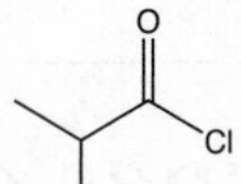

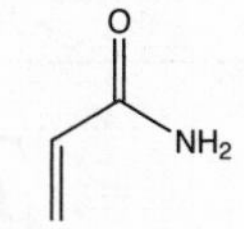

7. When the olefin shown below is treated with hydrazine and a small amount of copper sulphate, and air is bubbled through the solution, the double bond is reduced. Explain.

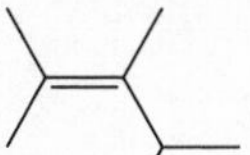

8. What is the product of the reaction between the olefin shown below and phenyl azide?

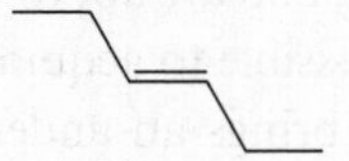

9. Sulphides are readily oxidized to sulphoxides on treatment with hydrogen peroxide. Write a mechanism for this transformation under acidic conditions. Sulphoxides can, in turn, be oxidized to sulphones. The mechanism of this reaction is the same as the oxidation of sulphides in neutral or acidic conditions, but in base a different mechanism operates. Suggest a likely mechanism.

10. A modification of the Wittig reaction (Section 5.4) is known as the Horner–Emmons reaction, in which a trialkyl phosphite is used instead of triphenylphosphine. Write a mechanism for this reaction, using benzyl bromide, triethyl phosphite, and acetaldehyde.

11. Why do phosphotriesters undergo hydrolysis in base more easily than phosphodiesters or phosphomonoesters?

12. What is the function of $TiCl_4$ when it acts as a catalyst in alkylations of silyl enol ethers with alkyl halides?

6

Types of reaction

So far, we have looked at the basic principles of chemical bonding and reactivity, the driving forces behind reactions, and the properties of molecules that affect the way they react. It is now time to consider the reactions themselves.

Newcomers to organic chemistry are often frightened by the seemingly huge array of reactions that take place, and are apt to think that a working knowledge of all, or even most of them is impossible to acquire. However, looking systematically at the various types of reaction brings an understanding of large numbers of the reactions of organic chemistry within easy reach. In fact, there are only four reactions in organic chemistry: additions, eliminations, substitutions, and rearrangements. For convenience, we discuss pericyclic reactions in this chapter as a fifth category, but any pericyclic reaction can also be grouped into one of the above four categories.

In the same way that many complex processes are made up of a number of simple ones, chemical reactions that may appear complicated are made up of combinations of these four basic reaction types strung together. Complicated reactions that are not easily understood are rare in organic chemistry; apparent complexity is almost always due to a large number of individual steps taking place in the same reaction.

We shall return to the theme of complex reactions in Chapter 8, but for now we shall look at the basic steps of which all reactions are made up.

6.1 Additions

6.1.1 Basic principles

One of the fundamental reactions in organic chemistry is addition to a multiple bond. This is normally a double bond, but triple bonds can undergo addition in much the same way. We can distinguish three main mechanisms for this reaction: electrophilic addition, nucleophilic addition, and free radical addition (Figure 6.1).

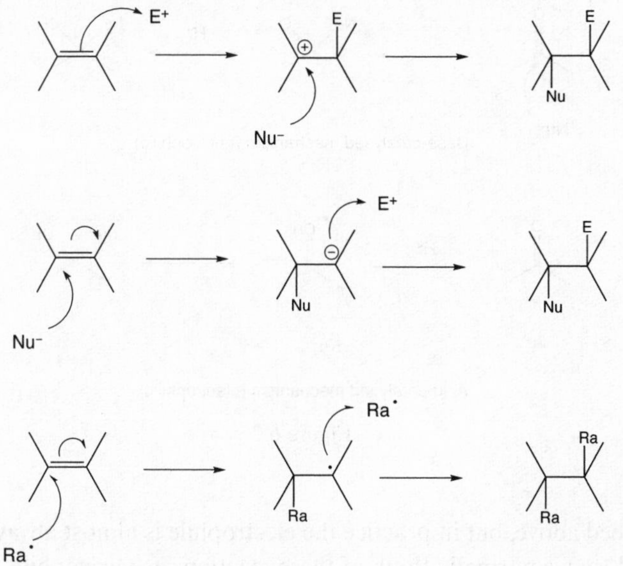

Figure 6.1

Addition may occasionally be concerted; this usually occurs only in pericyclic reactions (see Section 6.5).

The mechanism that occurs in any given reaction will of course depend on the nature of the double bond and the attacking species, and possibly also on the reaction conditions. Electrophilic and nucleophilic mechanisms are encountered considerably more often than radical ones. These two more common mechanisms can be thought of as variations of the same thing; the only difference between them is the sequence of the steps. In both mechanisms, a nucleophile and an electrophile add to a double bond. In the nucleophilic mechanism the nucleophile adds first, and in the electrophilic mechanism the electrophile adds first.

As yet, we have not specified what sort of a double bond we have. All multiple bonds can undergo addition, but the most commonly encountered examples in organic chemistry are carbon–carbon double bonds, and double bonds between carbon and a heteroatom, usually oxygen. Since there are important differences between these two types of double bond, we shall now consider each of them in turn.

6.1.2 Addition to double bonds between carbon and a heteroatom

We have already encountered additions to carbonyl groups in Section 5.1.3. Reactions in which carbonyl groups act as an electrophile, either in additions or substitutions, are very common. For a proper understanding of organic chemistry, it is essential to be familiar with the chemistry of the carbonyl group.

Additions to carbonyl groups may be either nucleophilic or electrophilic in the

159

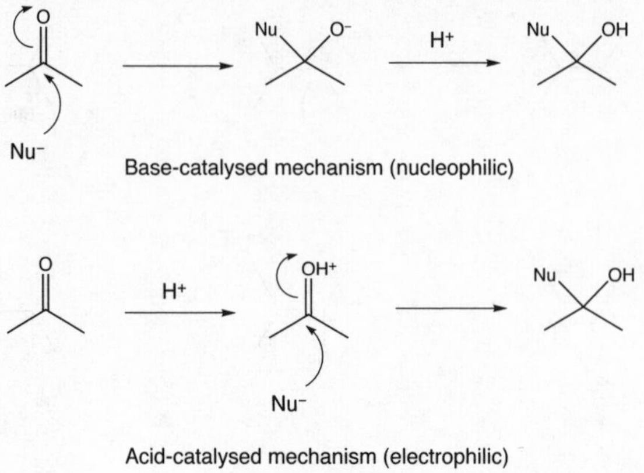

Base-catalysed mechanism (nucleophilic)

Acid-catalysed mechanism (electrophilic)

Figure 6.2

sense described above, but in practice the electrophile is almost always a proton or a Lewis acid, so we normally think of these reactions as nucleophilic. However, we should bear in mind that these reactions can proceed by either mechanism, depending on the reagents and frequently on the pH of the reaction medium. We saw in Section 5.1.3 how the mechanism of hydration of carbonyl groups differs depending on whether it is catalysed by acid or base.

Nonetheless, whether the reaction takes place under acidic or basic conditions, the overall result is the same: a nucleophile adds to the carbon, and an electrophile (almost always a proton) adds to the oxygen (Figure 6.2). Carbonyl groups can of course also undergo substitution reactions, which we shall discuss further in Section 6.3.5.

The reaction may or may not be reversible, depending on the attacking nucleophile. In general, the more reactive the nucleophile, the less likely the reaction is to be reversible. For example, if the nucleophile is water, then the reaction is readily reversible, whereas the reaction between a carbonyl group and a Grignard reagent is irreversible. The reason for this is simply that the group that has become attached to the carbonyl group has to act as a leaving group if the reaction is to be reversible, and more reactive nucleophiles are of course poorer leaving groups.

We should be careful here, however, as the nature of the nucleophile does not necessarily predict the nature of the leaving group that would be needed for the reverse reaction. We can see this by looking at the reduction of aldehydes and ketones with sodium borohydride (Figure 6.3). This is a mild reagent, and can even be used with water as a solvent (although it decomposes rapidly in acidic solutions). It acts by transferring hydrogen directly from boron to carbon; a free hydride ion is not an intermediate. However, if the reverse reaction were to occur, then hydride would be the leaving group. As hydride is a very poor leaving group, this is not generally observed.

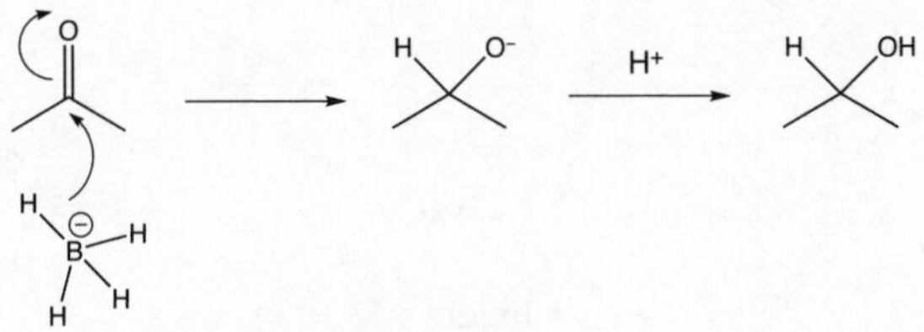

Figure 6.3

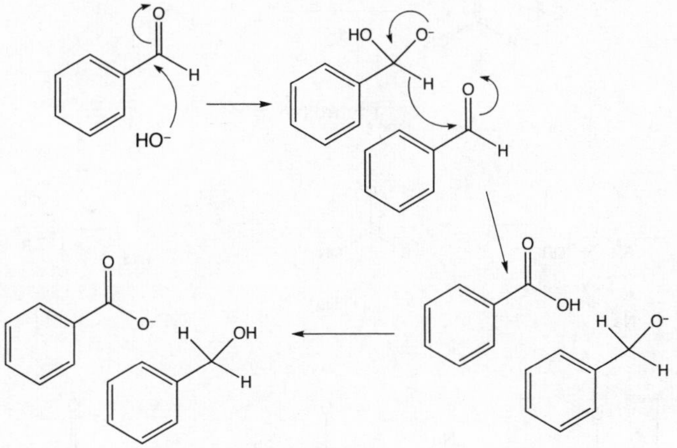

Figure 6.4

There is a reaction in which a hydride ion appears to leave a carbonyl group, although again it does not become free, but is transferred directly to another carbonyl group. This is known as the Cannizzaro reaction and is shown in Figure 6.4. Initial attack by hydroxide on the aldehyde group leads to the negatively charged tetrahedral intermediate, which can break down either by loss of hydroxide, to go back to the starting material, or by hydride transfer to a second aldehyde molecule. The driving force for the reaction is the formation of a molecule of the acid and one of the alcohol, which are more stable than the starting aldehyde. The reaction is limited to aldehydes with no α-hydrogen. In the presence of an α-hydrogen, aldol reactions would predominate.

The aldol reaction is in fact one of the most important additions to a carbonyl group (see Section 5.1.3). In this reaction, the nucleophile that attacks the carbonyl group is the enolate derived from either the same carbonyl compound or a different carbonyl compound. The reaction of enolates with carbonyl groups is very general, and may lead to substitution reactions (such as the Claisen condensation, Section 5.1.3) as well as simple additions.

The range of nucleophiles that add to carbonyl groups even includes electrons. If aldehydes or ketones are treated with sodium metal, an electron may add to the car-

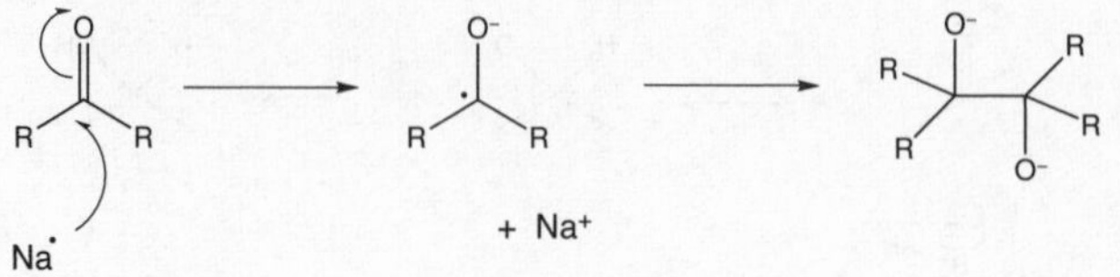

Figure 6.5

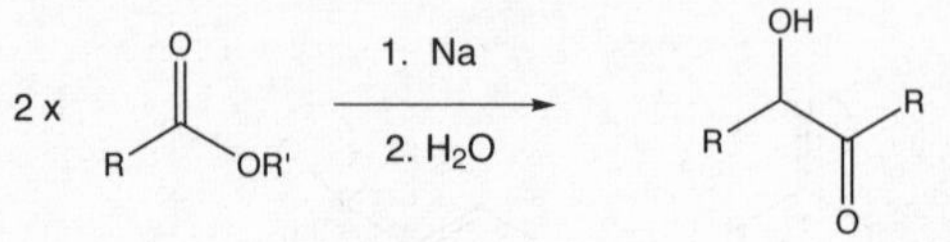

Figure 6.6

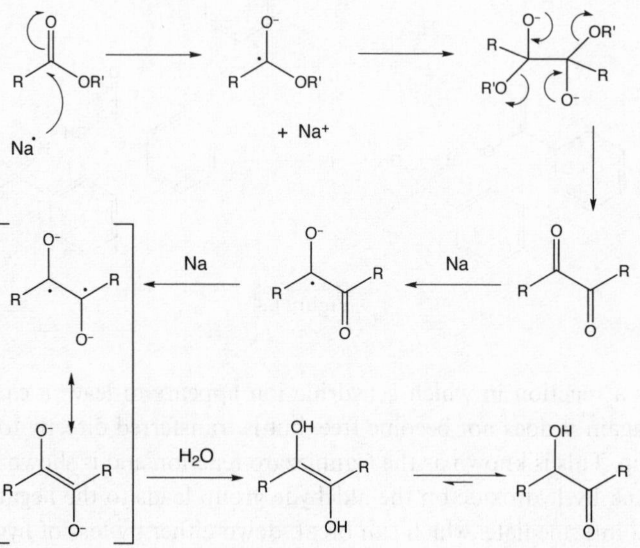

Figure 6.7

bonyl group to give a radical anion, as shown in Figure 6.5, which then dimerizes. The dianion thus formed gives a 1,2-diol on work-up.

A variation of this, which is more reliable synthetically, is the acyloin condensation. Here, two ester groups condense when treated with sodium metal to form an α-hydroxy ketone (Figure 6.6). The first stage of the mechanism is similar to the above reaction, with addition of an electron to the carbonyl group followed by dimerization (Figure 6.7). The resulting dianion can lose its alkoxy groups to give a diketone, which is then reduced further by sodium. This dianion gives the product on work-up.

In practice, by far the most common carbon–heteroatom double bonds to

162

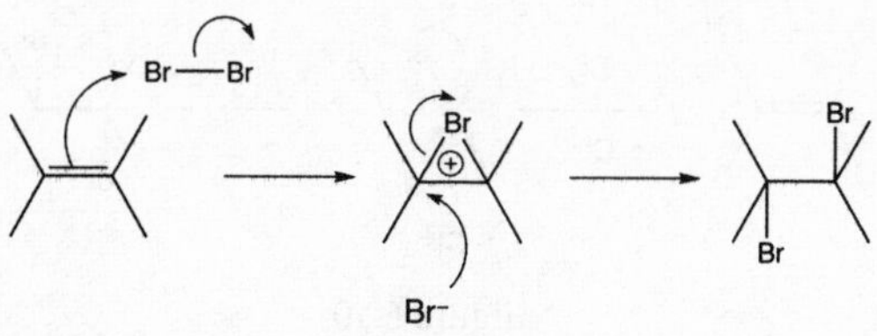

Figure 6.8

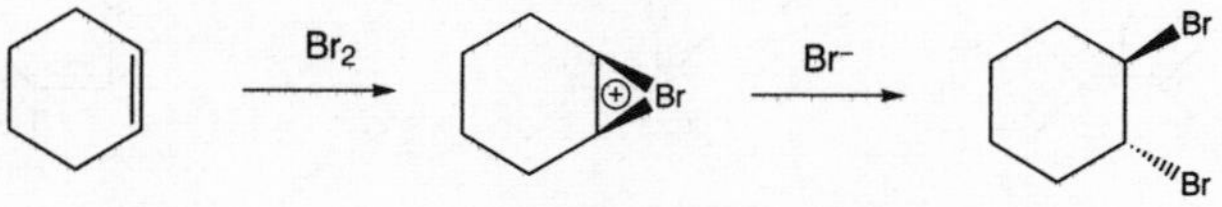

Figure 6.9

undergo addition reactions are carbonyl compounds, but compounds with C=N bonds and, more rarely, C=S bonds, can react similarly.

6.1.3 Addition to carbon–carbon double bonds

6.1.3.1 Electrophilic addition

A simple carbon–carbon double bond has areas of high electron density above and below the plane of the bond, and is therefore subject to attack by electrophiles. To put it another way, the double bond is a nucleophile.

One of the most important reactions of olefins, at least from a historical perspective, is the addition of bromine. This reaction is so efficient that it was commonly used as a test for double bonds in the days before spectroscopic techniques had found the widespread use that they enjoy today. An olefin rapidly decolorizes bromine water, thus giving a quick visual confirmation of its presence. The mechanism is shown in Figure 6.8.

The intermediate cyclic cation is known as a bromonium ion. This intermediate has a number of important consequences for the reaction. Firstly, it directs the stereochemistry. The bromide ion attacks the bromonium ion from the opposite side to the bromine, because this step is a sort of S_N2 displacement (see Section 6.3.2) and so takes place with inversion. The addition is therefore *trans* overall. Whilst this may be irrelevant for some simple olefins, there are many molecules for which this distinction is important, such as cyclohexene (Figure 6.9).

Another consequence of the bromonium ion is that the product of the reaction need not be a dibromide, as other nucleophiles can attack the bromonium ion. For example, if the reaction is run in the presence of a chloride salt, some of the mixed halide will be formed (Figure 6.10). The relative amounts of the two products depend, among other things, on the concentration of chloride present. However,

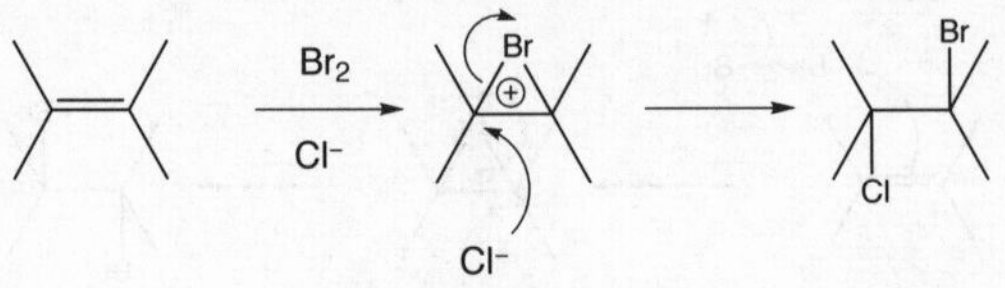

Figure 6.10

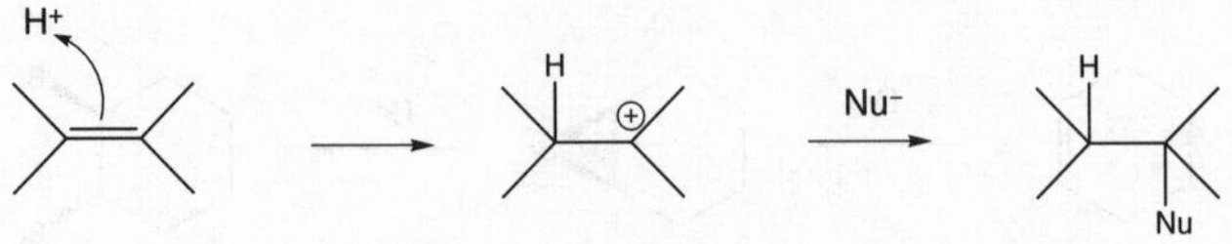

Figure 6.11

other things being equal, the dibromo compound will predominate, as bromide is a better nucleophile than chloride.

Iodine will add to double bonds in the same manner, although this reaction is of limited use preparatively as it is often readily reversible. However, the reaction can be used to isomerize double bonds, and can be useful synthetically as a method for the conversion of *cis* double bonds to their more stable *trans* isomers.

One of the most common electrophiles in organic chemistry is the proton. Since double bonds react with electrophiles, we would expect that they can be protonated, and indeed they are. This provides a carbocation, which can then react with a nucleophile (Figure 6.11). Thus we may carry out a number of additions to olefins by treating them with a nucleophile under acidic conditions. Of course this limits the range of nucleophiles we can use – we cannot, for example, use Grignard reagents in the presence of acid – but this is nonetheless a useful reaction.

Hydrogen halides are among the most commonly used reagents for this reaction. There is an extra complication here that was not present for bromination, namely that we now have to consider the orientation of addition in unsymmetrical olefins, in other words which end of the olefin receives the proton and which end receives the halogen. This can easily be predicted from our knowledge of carbocations (Section 4.2). The initial step in this reaction is the formation of a carbocation. We know that cations are stabilized by alkyl substituents, and the cation will therefore form in the position where it is more substituted. For example, isobutene will add a proton at the terminal position giving a tertiary carbocation, in preference to adding the proton at the central carbon, which would form the much less stable primary carbocation. The overall orientation of the addition will therefore be as shown in Figure 6.12. This is known as Markovnikov addition; the carbon that starts out with more hydrogens gains the extra hydrogen.

There is of course nothing special about the number of hydrogens attached to a carbon. Markovnikov addition arises because of the relative stabilities of the possible intermediate carbocations. With simple alkyl-substituted olefins, Markovnikov

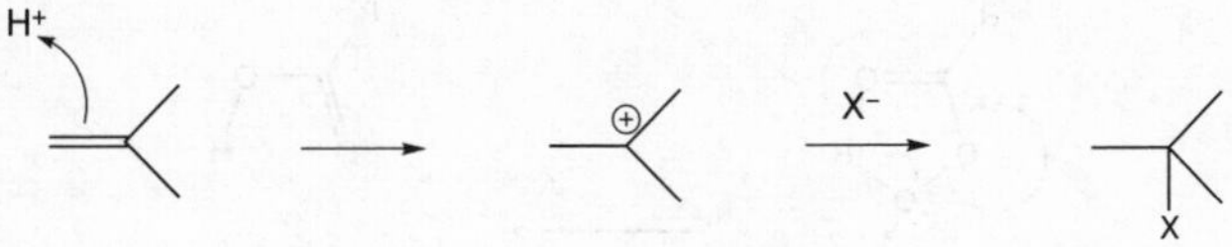

Figure 6.12

Figure 6.13

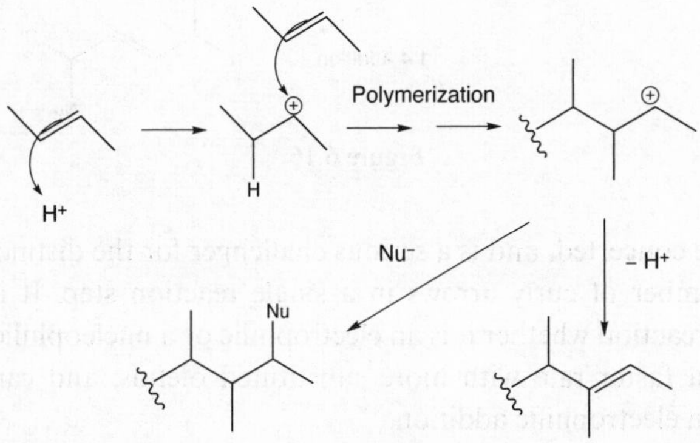

Figure 6.14

addition occurs because the carbon with more hydrogens is less substituted, and therefore forms the less stable carbocation. This will therefore be the site of attack by the proton, since the carbocation will then form at the other end of the double bond.

The presence of other substituents that can stabilize a double bond may also dictate the orientation of the reaction. For example, an enol ether will always protonate at the other end to the oxygen, as the oxygen is so good at stabilizing a carbocation (Figure 6.13).

The carbocation formed by addition of a proton will, in general, react with a nucleophile. Olefins are themselves nucleophiles, so the carbocation can react with the starting material (see Section 4.2.2). This is the basis of cationic polymerization: initial reaction of an olefin with an electrophile can start a chain reaction in which each addition of an olefin to a carbocation forms another carbocation. The reaction will eventually terminate by addition of a different nucleophile or loss of a proton (Figure 6.14).

One addition to an olefin is epoxidation with peracids (Figure 6.15). The reaction

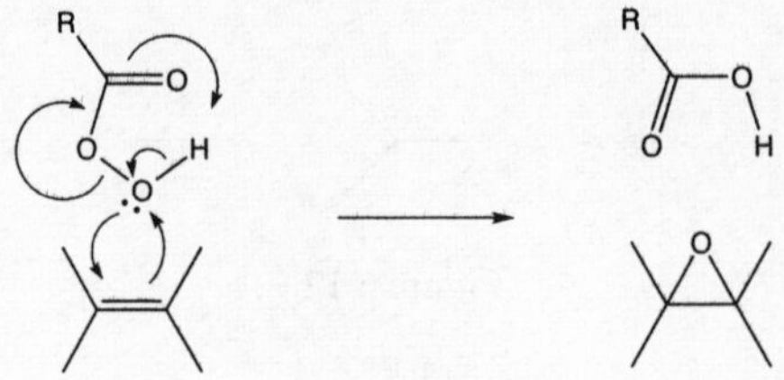

Figure 6.15

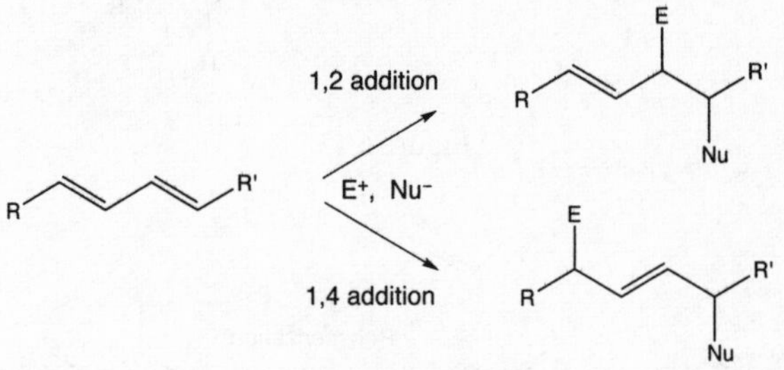

Figure 6.16

is thought to be concerted, and is a serious challenger for the distinction of having the greatest number of curly arrows in a single reaction step. It is not obvious looking at this reaction whether it is an electrophilic or a nucleophilic addition, but it proceeds at a faster rate with more substituted olefins, and can therefore be thought of as an electrophilic addition.

Conjugated olefins will also undergo many electrophilic addition reactions. These additions will often be faster than additions to non-conjugated olefins, as the conjugated olefins have higher HOMOs and are therefore more nucleophilic, at least in terms of their frontier orbital effects. The situation is more complicated for conjugated systems, as the addition may be 1,4 instead of 1,2 (Figure 6.16). Which product will form in any given reaction is not easy to predict. Either product may predominate, or mixtures may be produced, depending on the reaction and the conditions. If more than two double bonds are conjugated, there are of course even more possibilities.

Although aromatic systems are also electron-rich and can act as nucleophiles, addition reactions are rare, because addition would destroy the stabilization conferred by the aromaticity. However, aromatic systems readily undergo electrophilic substitution reactions; these are discussed below in Section 6.3.

6.1.3.2 Nucleophilic addition

As mentioned above, simple double bonds are electron rich, and are therefore not attacked by nucleophiles. However, if the double bond is conjugated to an electron-

166

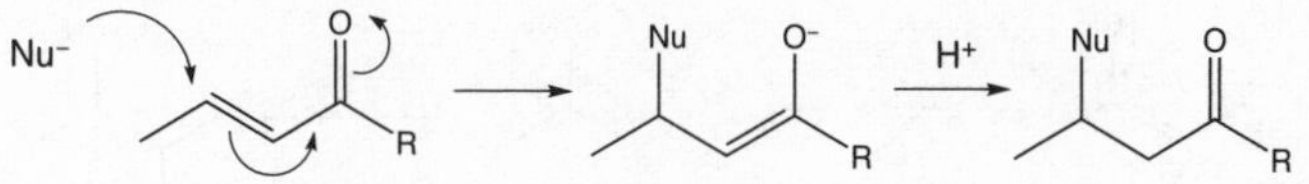

Figure 6.17

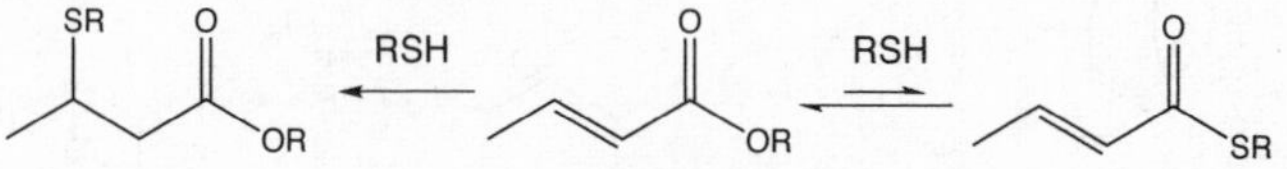

Figure 6.18

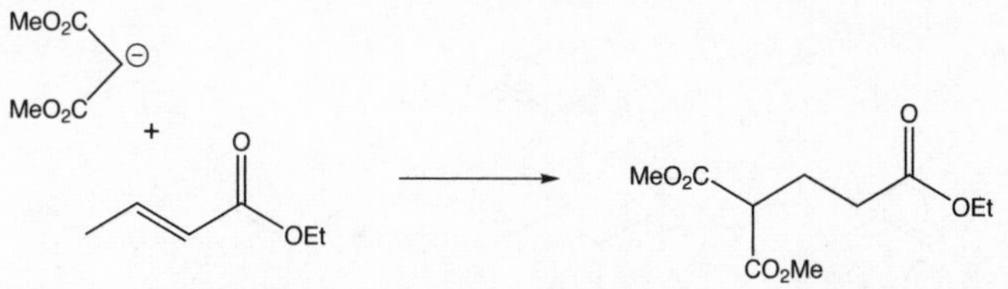

Figure 6.19

withdrawing group, then the electron distribution may change such that the double bond will undergo nucleophilic addition.

By far the most common electron-withdrawing group in this context is the carbonyl group. Olefins conjugated to carbonyl groups, more often described as α,β-unsaturated carbonyl compounds, undergo nucleophilic attack at the end of the double bond further from the carbonyl group. This results in an enolate anion, which can be protonated to complete the addition (Figure 6.17). The overall addition to α,β-unsaturated carbonyl compounds in this way is known as 1,4 addition, conjugate addition, or Michael addition.

Of course, carbonyl groups can themselves undergo addition, and so an α,β-unsaturated carbonyl compound has two alternative modes of addition: direct (1,2) addition or conjugate (1,4) addition. Although direct addition is more common, a complex interplay of factors determines which mode of addition predominates in any given reaction, as was discussed in Section 2.4.3.

The situation is complicated because these reactions may be controlled either kinetically or thermodynamically. Thiols add Michael fashion to α,β-unsaturated esters, and although it is tempting to explain this as a consequence of the soft character of sulphur, an alternative explanation is that the thioester that would be formed by direct attack is less stable than the starting material, and so any thioester formation is readily reversible and the reaction is under thermodynamic control (Figure 6.18). It is not possible to be sure which explanation is more relevant.

Anions of β-dicarbonyl compounds behave similarly (Figure 6.19). These

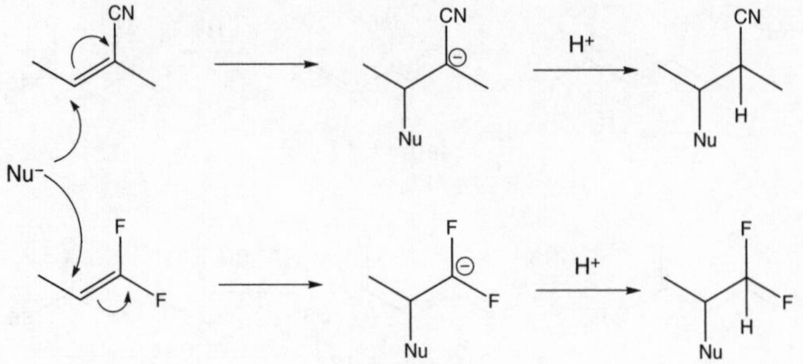

Figure 6.20

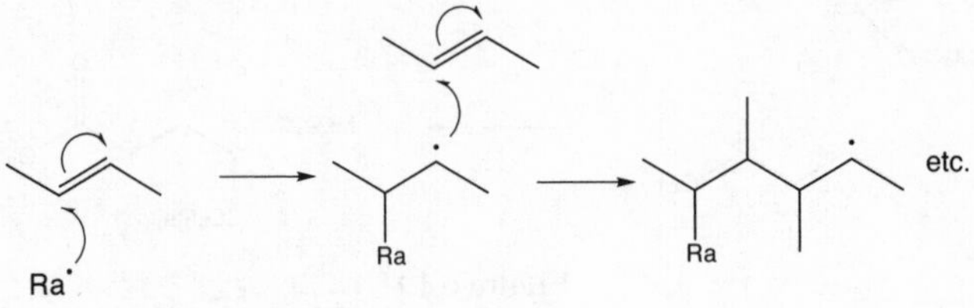

Figure 6.21

generally give conjugate addition, but again this could be for either thermodynamic or kinetic reasons.

Other electron-deficient olefins besides α,β-unsaturated carbonyl compounds can undergo nucleophilic addition, the most common of which are α,β-unsaturated nitriles and vinyl fluorides (Figure 6.20).

6.1.3.3 Free radical additions

In the laboratory, free radical addition to olefins is much less important than ionic addition, although it is of great industrial importance as one of the main mechanisms for polymerization of olefins.

The mechanism of free radical polymerization is shown in Figure 6.21. Addition of a radical to an olefin generates another radical, which may react with another olefin and so on, until some kind of termination reaction occurs (see also Section 4.3).

The most useful free radical addition to olefins in laboratory synthesis is the closure of rings containing two double bonds (Figure 6.22). Attack of a radical on the first double bond leads to another radical, which can react with the second double bond in a ring closure reaction. Another radical is thus formed, which can react with another radical in a termination reaction, or abstract an atom in a propagation reaction. The net result of the reaction is addition to both double bonds.

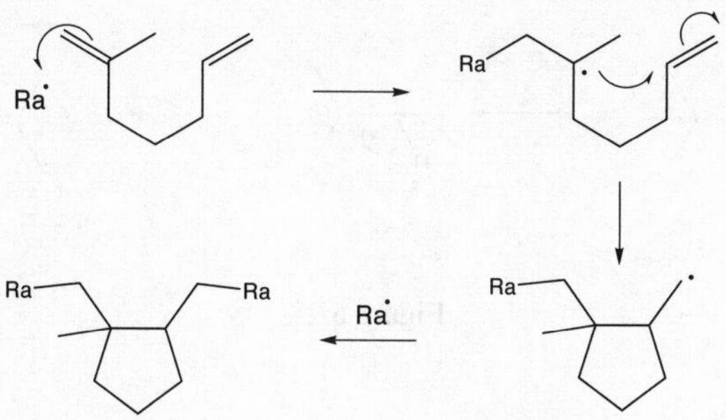

Figure 6.22

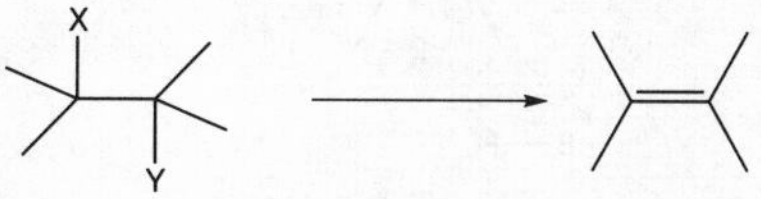

Figure 6.23

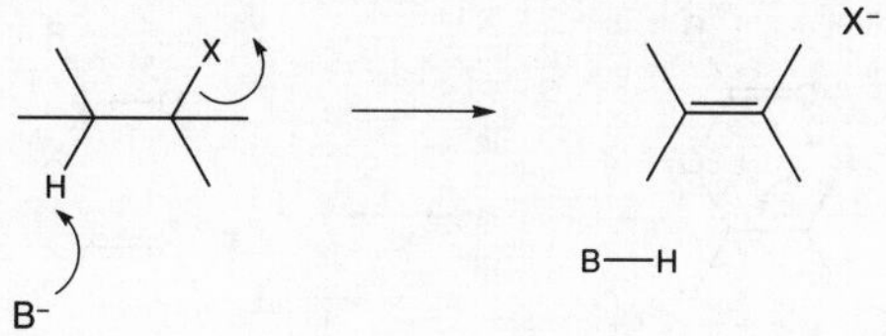

Figure 6.24

6.2 Eliminations

6.2.1 Basic principles

An elimination is the reverse of an addition. In the vast majority of eliminations, two substituents are removed from adjacent carbon atoms to leave a double bond (Figure 6.23). This is known as a β-elimination. A rarer reaction is α-elimination, which results in a highly reactive carbene (see Section 4.4.2). Long-range eliminations (i.e. in which the groups that are eliminated are separated by more than two carbon atoms) are also known. The following discussion will refer to β-eliminations unless otherwise specified. In most eliminations, the two groups that are lost are a proton and a nucleofugal leaving group, although other combinations are possible.

There are two important mechanisms for elimination, E2 and E1. In an E2 reaction, both groups are lost simultaneously, as shown in Figure 6.24. A base removes the proton, and the nucleofuge leaves spontaneously. This is called E2 because it is an Elimination and it is bimolecular (the reacting molecule and a molecule of base take part in the rate-determining step). In an E1 reaction the two bonds are broken

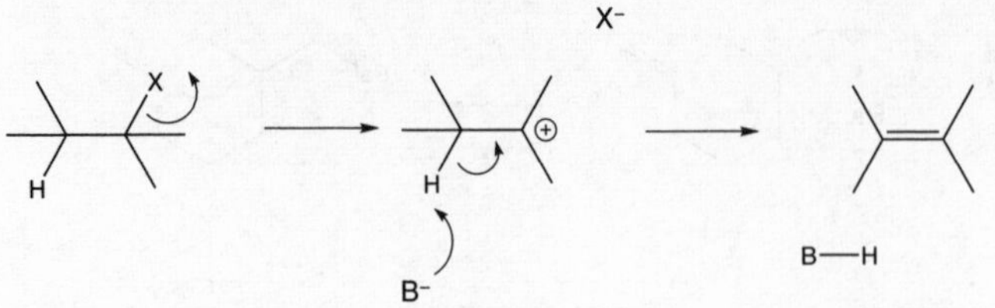

Figure 6.25

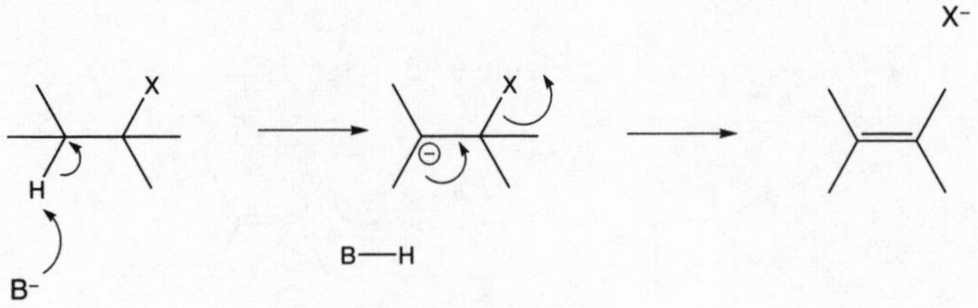

Figure 6.26

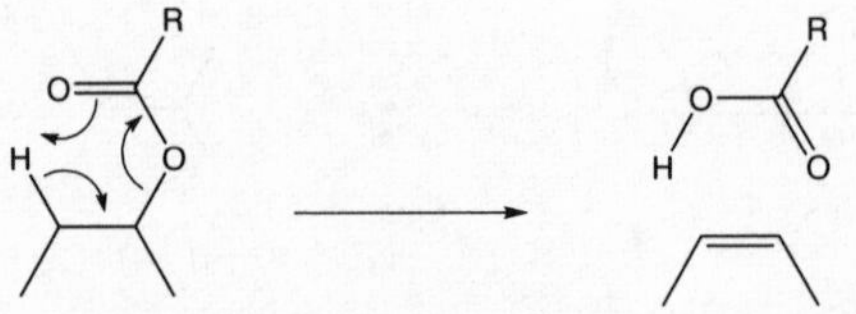

Figure 6.27

at different times. In the vast majority of E1 eliminations, the nucleofuge leaves first, and the proton is subsequently removed by a base (Figure 6.25). This is known as E1 because the rate-determining step (loss of the nucleofuge) is a unimolecular process. A much less common variation of this is that the proton is removed first (Figure 6.26). This is known as the E1cB mechanism (cB standing for conjugate base).

Although these are normally described as separate mechanisms, they are merely stages along a continuum with E1 at one end, E2 in the middle, and E1cB at the other. The bonds are not necessarily broken entirely sequentially or at exactly the same time. There may be a degree of weakening of one bond while the other is breaking. Whilst it is convenient, and usually reasonably accurate, to classify eliminations into one of these three categories, there may be reactions in which the proton is partially, but not completely, removed when the nucleofuge leaves (somewhere between E1 and E2), and others in which the nucleofuge leaves not quite at the same time as the proton (somewhere between E2 and E1cB).

A further mechanism of elimination is known as Ei (internal elimination); here the substrate acts as its own base, elimination taking place in a concerted process. An example is shown in Figure 6.27.

170

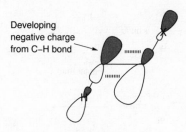

Developing
negative charge
from C–H bond

Figure 6.28

6.2.2 E2 reactions

An E2 elimination is a one-step reaction, as shown in Figure 6.24. The most common leaving groups in this reaction are halogens, but other leaving groups may also take part, for example tosylates.

An important point to remember in E2 reactions is that it is not possible to make poor leaving groups into better leaving groups by protonation, as these eliminations necessarily take place under basic conditions. For example, the hydroxyl group, which may often act as a leaving group after protonation (so that neutral water is lost), acts as the leaving group in an E2 elimination only under very specialized conditions (and such reactions may in any case be closer to an E1cB mechanism than a pure E2 mechanism).

E2 eliminations can be thought of as acid–base reactions. They are not the same as conventional acid–base reactions, because removal of a proton does not lead directly to the conjugate base, but they nonetheless have some features in common.

We can think of the substrate of the reaction as an acid, which gives up a proton to the base. Unlike in a normal acid–base reaction, the loss of the proton is assisted by the overlap of the developing negative charge with the antibonding orbital between the β-carbon and the leaving group (Figure 6.28). This migration of electrons into the antibonding orbital makes possible the loss of a proton that might not otherwise be very acidic and causes the leaving group to leave.

In a normal acid–base reaction, removal of a proton from an acid will give its conjugate base. Although this does not happen directly here, the reaction does lead to the release of a leaving group, which has features in common with a conjugate base. We can think of the molecule from which the proton and the leaving group are eliminated as being a relay stage in the formation of a conjugate base from an acid.

If we think of E2 eliminations as acid–base reactions, then it is obvious that they should be favoured by strong bases and by high acidity of the hydrogen that is lost. A stronger base increases the rate of an E2 elimination, and makes it more likely to be the dominant pathway in the face of competition from E1 eliminations or substitution reactions.

E2 eliminations also occur more readily if the hydrogen that is lost has some acidic character. Thus of the two compounds shown in Figure 6.29, the phenyl-substituted

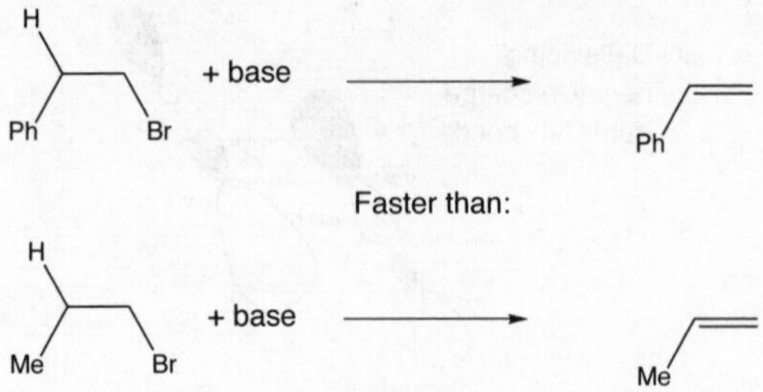

Figure 6.29

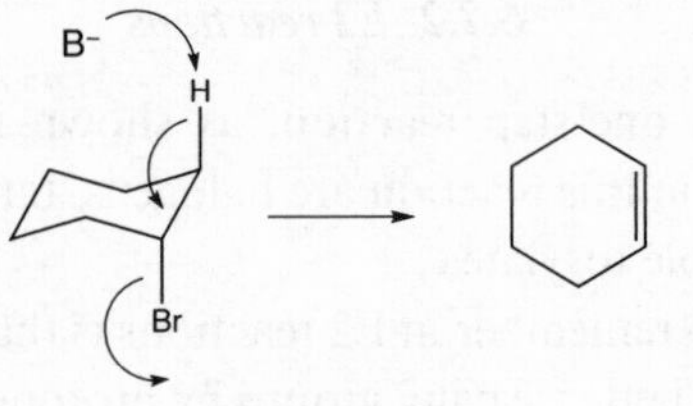

Figure 6.30

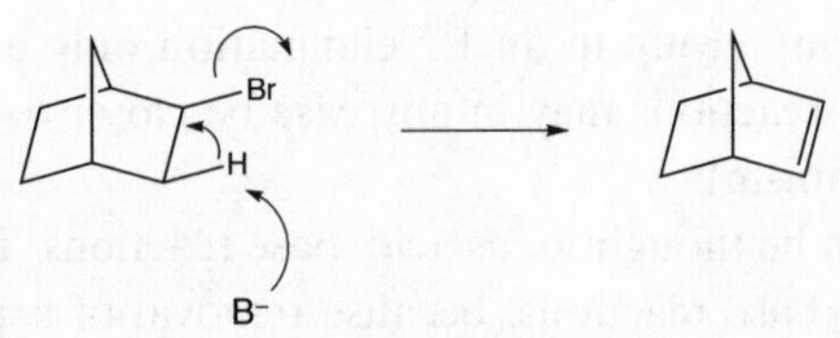

Figure 6.31

compound undergoes elimination far more quickly because the phenyl group enhances the acidity of the hydrogen.

The necessity for efficient overlap of the orbitals in E2 elimination puts a geometric constraint on the reaction. The carbon–hydrogen bond and the carbon–leaving-group bond must be in the same plane (see Section 3.3.3). The usual way for this to happen is for the two groups to be antiperiplanar (Figure 6.30), but synperiplanar eliminations (Figure 6.31) are also known.

There are two main consequences of these geometric requirements. Firstly, in cyclic systems the feasibility of an E2 elimination may depend on the ability of the molecule to adopt a conformation in which the groups are antiperiplanar, as discussed in Section 3.3.3. Secondly, in an open-chain E2 elimination, knowledge of the stereochemistry of the starting material sometimes allows us to predict the geometry of the double bond in the product, if we assume that the elimination takes place in an antiperiplanar conformation (which it almost always does). In simple compounds there may be no stereochemistry to worry about, as in Figure 6.32. However, the stereochemistry becomes relevant if there are chiral centres in the

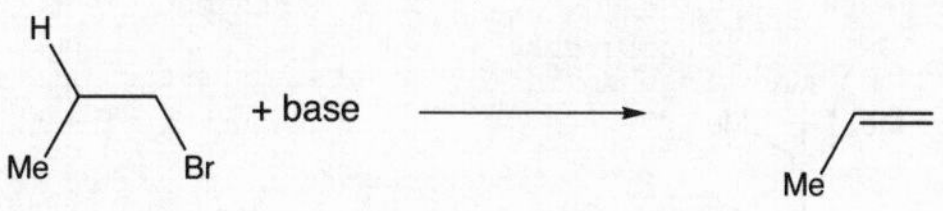

Figure 6.32

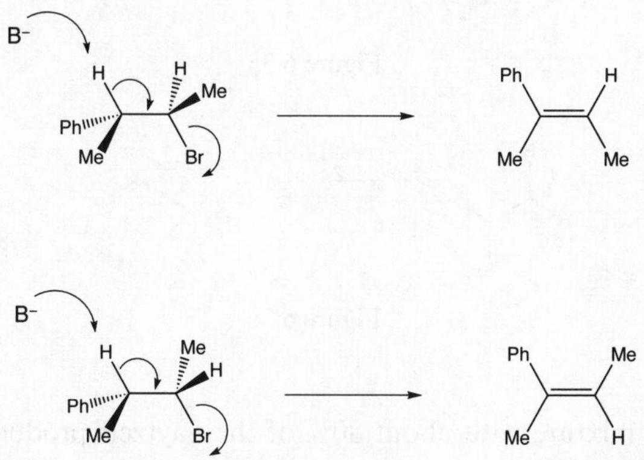

Figure 6.33

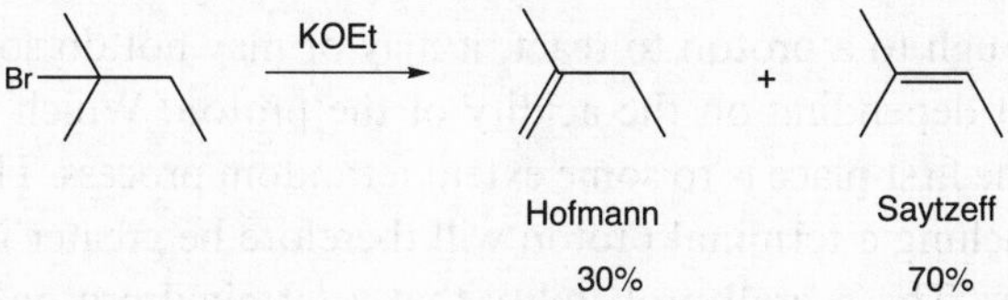

Figure 6.34

molecule. Thus the two isomers of the bromide shown in Figure 6.33 give the two isomeric olefins.

Syn eliminations occur mainly from cyclic systems in which an antiperiplanar conformation is impossible, such as the norbornyl system shown in Figure 6.31.

E2 eliminations are not usually reversible, and so control is largely kinetic. The products of E2 eliminations are therefore not necessarily the most stable olefins, but are often those that result from removal of the most acidic hydrogen. However, product stability certainly plays a part as well, since the transition state has some of the character of the product, so a transition state leading to a stable product may often be lower in energy than one leading to a less stable product.

Elimination of HBr from the bromide shown in Figure 6.34 gives a mixture of products as shown. Elimination to the more stable olefin is known as Saytzeff elimination; this is generally defined as the double bond being formed in the direction of the most highly substituted carbon. The alternative, in which the double bond is formed in the direction of the least highly substituted carbon, is known as Hofmann elimination. If the reaction were under purely thermodynamic control, we

173

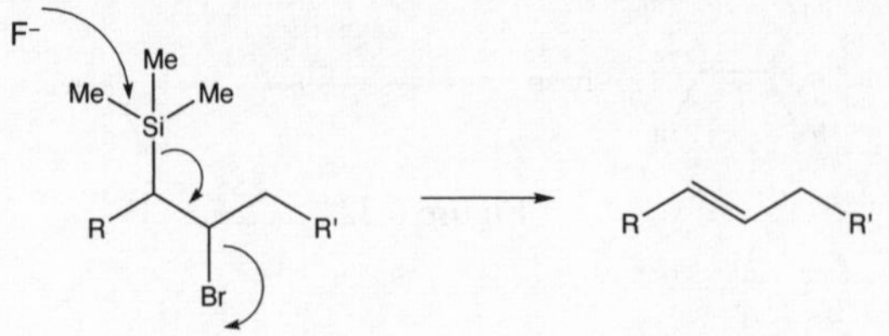

Figure 6.35

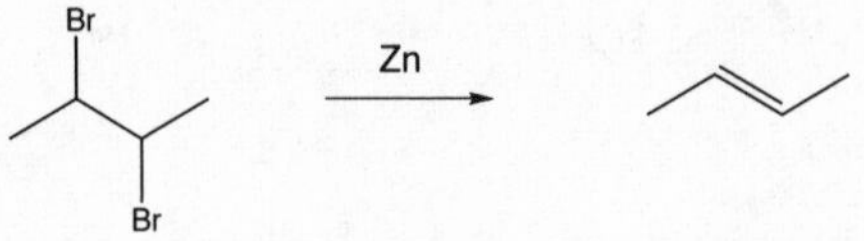

Figure 6.36

should expect a mixture with about 80% of the Saytzeff product. The Hofmann product is formed to the extent it is because the terminal hydrogens are more acidic and so more easily removed by the base. There is also a statistical effect at work here; there are more terminal protons, so they are more likely to be removed. If a base comes close enough to a proton to react, it may or may not do so, the probability that it will react depending on the acidity of the proton. Which proton the base approaches in the first place is to some extent a random process. The probability of the base approaching a terminal proton will therefore be greater if there are more terminal protons. The overall probability that a terminal proton will react is the probability that the base will approach a terminal proton multiplied by the probability that the proton will react once the base comes close enough. This is obviously increased if there are more terminal protons, since the base's probability of approaching a terminal proton is increased.

Steric effects can also be important; when the bromide in Figure 6.34 is eliminated with potassium t-butoxide as the base it gives 73% of the Hofmann product. The higher proportion of Hofmann product with t-butoxide is due to the greater importance of steric effects.

We have so far discussed E2 eliminations in which a proton is one of the leaving groups. While this is the most common occurrence, it is not the only possible reaction. Metals (including silicon) may act as the electrofugal leaving group in eliminations, and it is often impossible to prepare organometallic compounds with a leaving group in the β-position because of the readiness with which this reaction occurs. Silicon can be used to direct the position of a double bond; compounds with a trimethylsilyl group β to a nucleofuge (see Section 2.5.3) react with fluoride to give a directed elimination (Figure 6.35).

Dihalo compounds can undergo β-elimination under appropriate conditions. Treatment of 1,2-dibromides with zinc gives this reaction (Figure 6.36).

174

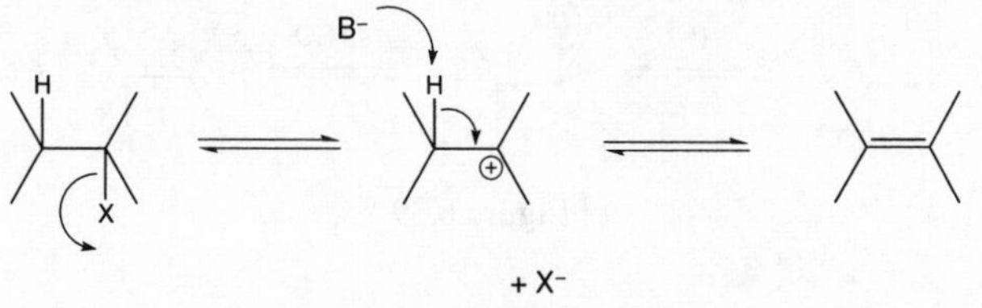

Figure 6.37

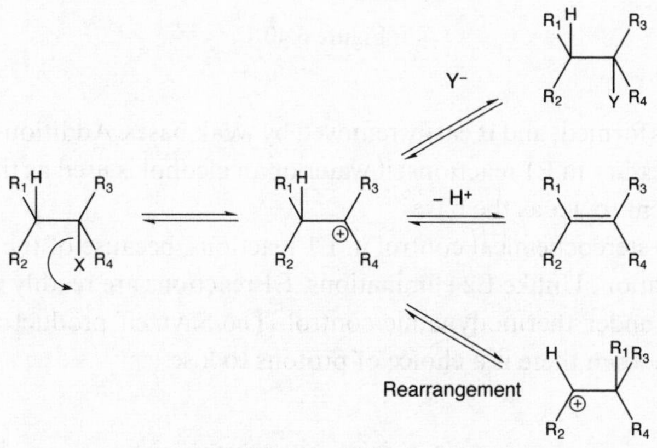

Figure 6.38

6.2.3 E1 reactions

An E1 elimination is simply the reverse of electrophilic addition to a double bond, as can be seen in Figure 6.37. The two reactions share a common intermediate, namely the carbocation. The carbocation can, of course, undergo all the usual reactions of carbocations as described in Section 4.2.3. This means that substitution and rearrangement are often side reactions of eliminations (Figure 6.38).

The rate-determining step in E1 elimination is loss of the leaving group. E1 eliminations are therefore favoured by factors that facilitate this step, namely a good leaving group, stability of the resulting carbocation, and polar solvents, which promote ionization. E1 eliminations therefore take place largely when the intermediate carbocation is tertiary or stabilized in some other way, such as by an adjacent heteroatom with a lone pair, or by conjugation with an aromatic ring or double bonds (see Section 4.2.1). The formation of simple secondary and primary carbocations is a much more energetic process and so is rarely observed.

In contrast to E2 eliminations, the E1 mechanism can be encouraged by protonation of a leaving group to increase its leaving group ability. For example, alcohols may be dehydrated by treatment with acid, so that water acts as the leaving group in the first step of the elimination (Figure 6.39). The proton is lost quickly once the

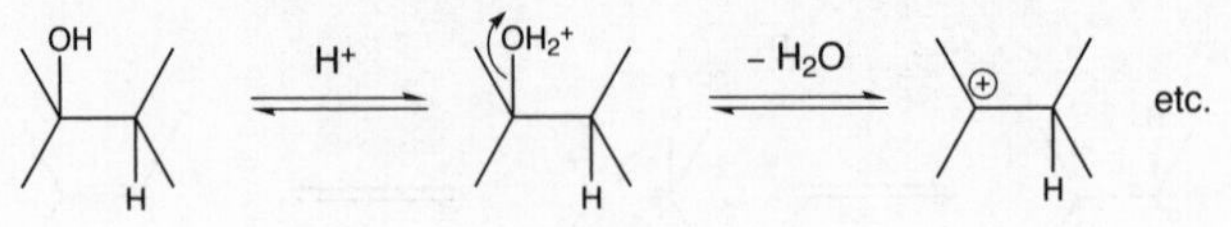

Figure 6.39

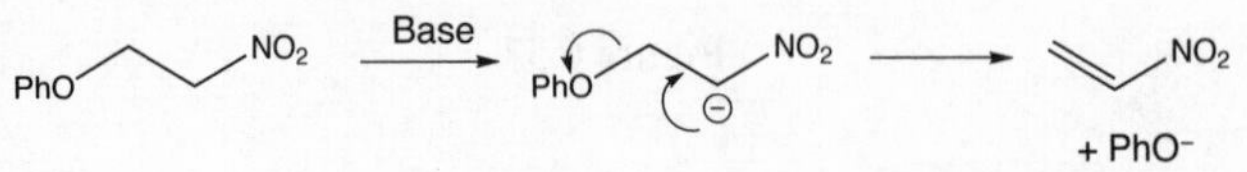

Figure 6.40

carbocation is formed, and is easily removed by weak bases. Addition of bases is not generally necessary in E1 reactions; if water or an alcohol is used as the solvent, this will be sufficient to act as the base.

There is no stereochemical control in E1 reactions, because of the planar nature of the carbocation. Unlike E2 eliminations, E1 reactions are readily reversible, and are therefore under thermodynamic control. The Saytzeff product almost always predominates when there is a choice of protons to lose.

6.2.4 Other eliminations

6.2.4.1 E1cB reactions

The E1cB mechanism is encountered much less often than the above two mechanisms. This is because when a base removes a proton as part of an elimination reaction, the resulting carbanion is usually so unstable that it breaks down at the same time as the proton is removed, leading to concerted elimination.

The E1cB mechanism is therefore favoured if the carbanion formed by loss of the proton is stabilized, and if there is a poor leaving group. An example is the base-catalysed dehydration of the nitro compound shown in Figure 6.40. The nitro group stabilizes the carbanion, and the phenoxy group is such a poor leaving group that it does not leave straight away, but must wait until the molecule has picked up some more energy by collision with another molecule. Another example of an E1cB mechanism is the dehydration of the fluorene derivative shown in Figure 6.41 (see Section 9.1). The proton at the 9-position of the fluorene system is reasonably acidic, because the resulting carbanion, being a 14 π electron system, is stabilized by aromaticity. OH^- is also a very poor leaving group, and so cannot leave until it has gained sufficient energy. This mechanism is unusual in two respects: not only is it an E1cB mechanism, but it is also a rare example of OH^- behaving as a leaving group. Hydroxyl groups normally have to be protonated if they are to behave in this way.

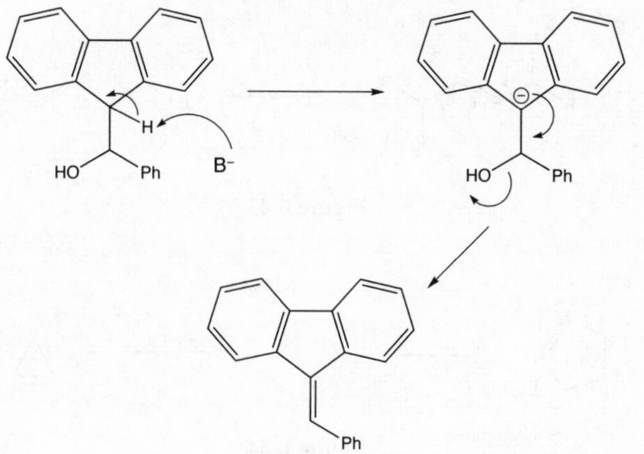

Figure 6.41

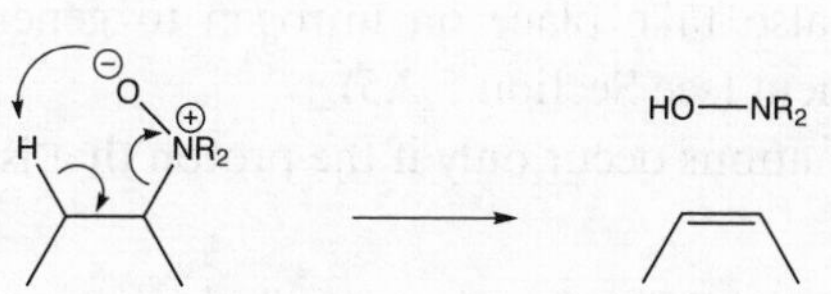

Figure 6.42

6.2.4.2 Ei reactions

Some eliminations have a concerted mechanism but are unimolecular. These are known as Ei reactions.

Two common examples of Ei reactions are the pyrolysis of esters and the Cope reaction. Esters with a β-hydrogen on the alkoxy group can undergo elimination when heated (Figure 6.27). The reaction can be run in the gas phase, since no interaction with other molecules is required. This reaction often occurs in mass spectrometry of ester compounds, giving rise to a characteristic fragmentation pattern.

In the Cope reaction, amine oxides cleave to give an olefin and a hydroxylamine (Figure 6.42). Again, no other reagents are required; the amine oxide spontaneously decomposes on heating.

A feature of both these reactions, and of other Ei reactions, is that they proceed with *syn* stereochemistry.

6.2.4.3 α-Eliminations

α-Eliminations were discussed in Section 4.4.2 as a way of generating carbenes. They are less favourable than β-eliminations, and are found only if there is no leaving group at the β-position, normally in compounds with only one carbon atom.

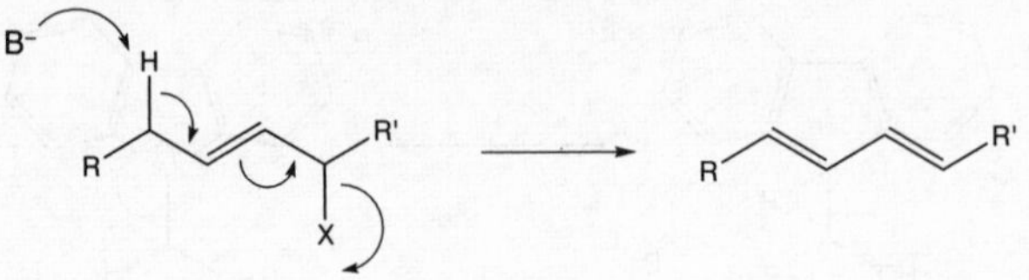

Figure 6.43

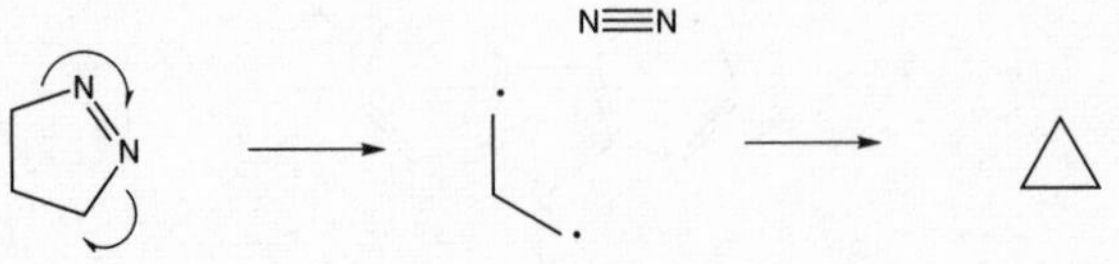

Figure 6.44

α-Eliminations may also take place on nitrogen to generate nitrenes, as in the Hofmann rearrangement (see Section 5.2.5).

In general, α-eliminations occur only if the proton that is lost is rather acidic.

6.2.4.4 Long-range eliminations

Eliminations may occur over more than two carbons. Systems with leaving groups adjacent to double bonds can undergo conjugate elimination as shown in Figure 6.43, although this reaction is rare.

6.2.4.5 Extrusion reactions

A group may be eliminated from between two other groups, joining them together. This is known as an extrusion reaction.

One of the best known extrusion reactions is the elimination of nitrogen from pyrazolines to give cyclopropanes (Figure 6.44). This takes place either on heating or photochemically. The reaction is thought to proceed by cleavage to give molecular nitrogen and a diradical, which recombines to yield the cyclopropane. Other groups that can be extruded include SO_2, CO_2, and CO.

6.3 Substitutions

6.3.1 Basic principles

In a substitution reaction, one group is replaced by another (Figure 6.45). There are a number of ways in which this can take place. The net result of addition followed by elimination, or vice versa, is substitution. Substitutions may also take place directly. Aliphatic substitutions are almost always nucleophilic, although electrophilic mechanisms are far more common in aromatic systems.

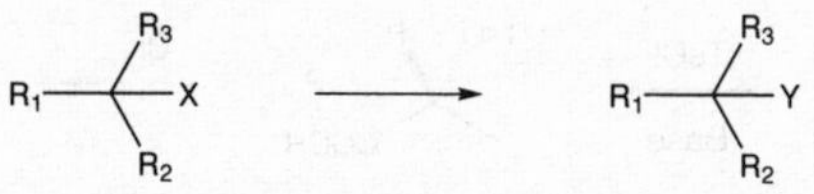

Figure 6.45

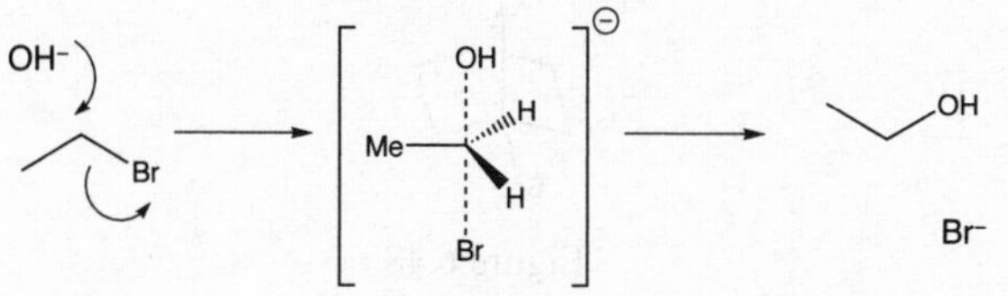

Figure 6.46

To predict what mechanism any given substitution reaction is likely to have, we need to know about the properties of the molecule undergoing the reaction. For example, aromatic rings, aliphatic carbons, and carbonyl groups all undergo substitutions by different mechanisms. The mechanism within any of these groups can also vary, depending on the precise structure of the molecule and sometimes on the conditions.

Substitutions at saturated carbon atoms can usually be classified into one of two types, S_N1 or S_N2. Which of these mechanisms operates in any particular reaction depends on a number of factors such as the structure of the molecule and the conditions of the reaction. We shall now look at each of these reactions in turn.

6.3.2 S_N2 reactions

The S_N2 reaction is the simplest type of substitution, being a direct displacement of one group by another. It is called S_N2 because it is a <u>S</u>ubstitution, it is <u>N</u>ucleophilic, and bimolecular.

The mechanism is shown in Figure 6.46 for the hydrolysis of ethyl bromide. The incoming nucleophile interacts with the antibonding orbital between the carbon and its leaving group. This antibonding orbital is the LUMO of the substrate, and interacts with the HOMO of the nucleophile. In the transition state the carbon is 5-coordinate, with the nucleophile and the leaving group in a straight line. As the interaction between the nucleophile and the antibonding orbital becomes stronger, the bond between the carbon and the leaving group is weakened (by the presence of electrons in the antibonding orbital) and eventually breaks.

Because of this mechanism, S_N2 reactions have a defined stereochemical outcome, and lead to inversion of configuration. Thus if L-lactic acid is converted to its tosylate and then treated with bromide ion, the product has the configuration shown in Figure 6.47. This inversion of configuration is sometimes known as Walden inversion and is one of the most important features of the S_N2 reaction.

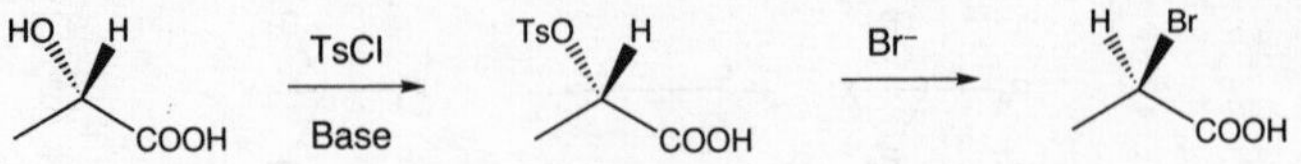

Figure 6.47

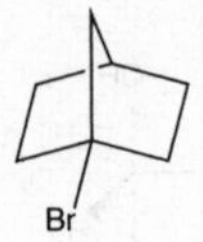

Figure 6.48

Another consequence of this mechanism is that S_N2 reactions are possible only if the back of the carbon is accessible and if the carbon is free to invert its configuration. The norbornyl bromide in Figure 6.48 is inert to S_N2 displacement because neither of these conditions is met. The reaction can take place only at sp$_3$ hybridized carbons; sp$_2$ carbons never undergo direct displacement.

The requirement for the nucleophile to attack the carbon directly means that S_N2 displacement is sensitive to steric effects, and so becomes progressively harder with increasing substitution of the carbon. S_N2 displacements are usually fast at primary carbons, take place with more difficulty at secondary carbons, and are rare at tertiary carbons. They are also sensitive to steric effects at the adjacent carbon; for example, neopentyl halides are very resistant to S_N2 displacement (see Section 3.3.2).

The rate of reaction depends on both the nucleophile and the leaving group, as both take part in the rate-determining (and only) step. The behaviour here is as we would expect; the reaction proceeds faster with good leaving groups and good nucleophiles. As with other reactions involving leaving groups, some groups may be protonated to turn them into better leaving groups, notably the hydroxyl group.

Solvent effects can be important in S_N2 reactions, which are generally fastest in polar aprotic solvents such as DMSO or acetonitrile. The polar nature of these solvents allows them to stabilize the charged transition state, while the nucleophiles are less well solvated and therefore more nucleophilic than in protic solvents.

If we consider the transition state, we can see that there is a negative charge spread over the nucleophile, the carbon, and the leaving group. The carbon therefore has some of the characteristics of a carbanion, and we would expect factors that stabilize carbanions to stabilize the transition state, and hence to increase the rate of reaction. This is indeed what we find, and so S_N2 reactions are faster when adjacent to carbonyl groups, double bonds, aromatic rings, or other groups that stabilize carbanions.

Occasionally, an S_N2 displacement will appear to take place with retention of configuration. This is almost always due to a phenomenon known as neighbouring group participation. An example is the diazotization of amino acids in the presence of an added nucleophile (see Section 5.2.1). Figure 6.49 shows the diazotization of

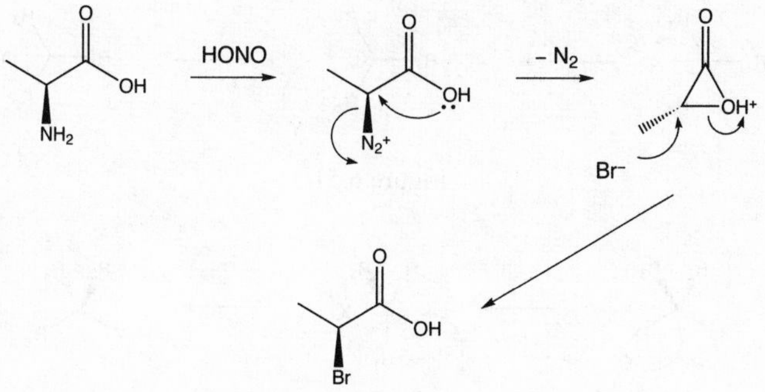

Figure 6.49

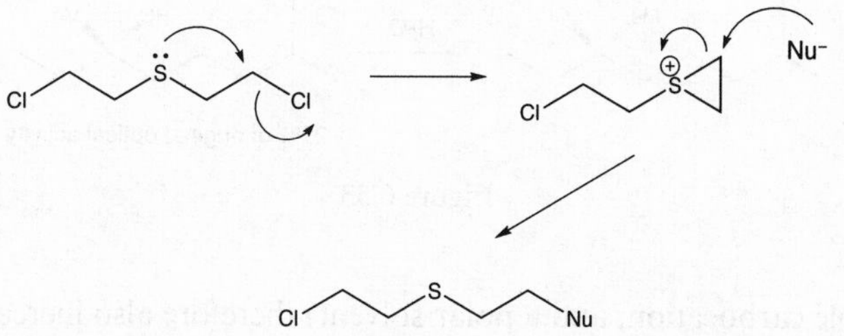

Figure 6.50

alanine in the presence of bromide ions. The product is the alkyl bromide with the same configuration as the starting material. This is due to two consecutive S_N2 displacements, each with inversion of configuration. This is one of the rare examples of an aliphatic diazotization reaction that is preparatively useful.

Neighbouring group participation can sometimes lead to considerable rate enhancements. For example, mustard gas is a very effective alkylating agent (Figure 6.50). The initial attack by sulphur on the chloride group is faster than attack by an external nucleophile would be, because the sulphur is already close. The resulting sulphonium ion is more reactive than the chloride, and so the rate of reaction is increased.

6.3.3 S_N1 reactions

An alternative mechanism in nucleophilic aliphatic substitution reactions is known as S_N1 displacement. This is shown in Figure 6.51. The initial, rate-determining step is loss of the leaving group to give a carbocation, which then combines with a nucleophile to complete the substitution.

Note that the first step of this reaction is the same as the first step of E1 elimination. Not surprisingly, E1 elimination is often a side reaction in S_N1 displacements and vice versa. Factors that increase the rate of E1 elimination (i.e. a good leaving

181

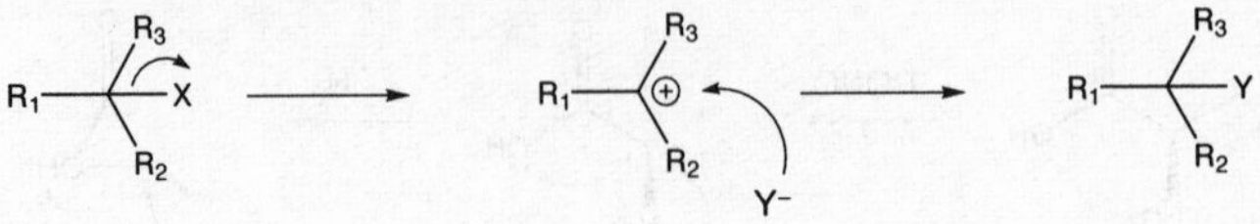

Figure 6.51

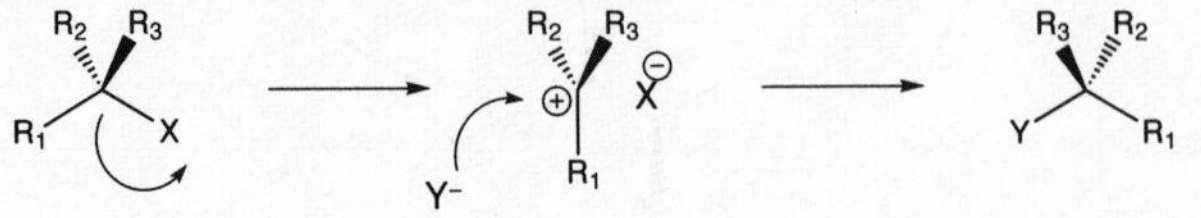

Figure 6.52

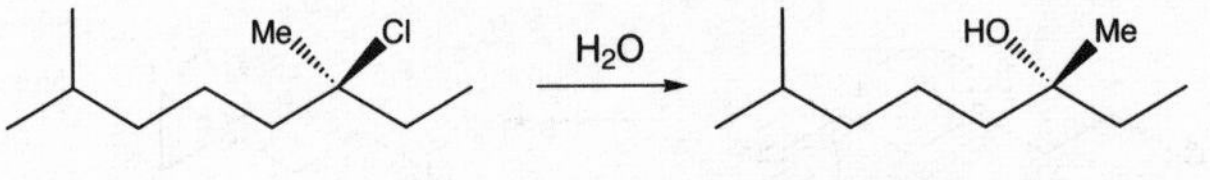

21% of original optical activity

Figure 6.53

group, a stable carbocation, and a polar solvent) therefore also increase the rate of S_N1 displacement. We will look at factors that may favour one of these modes of reaction over the other in the next section. As with E1 reactions, the leaving-group ability can be increased by protonation.

Although a nucleophile of some kind is necessary in S_N1 displacement, the nature of the nucleophile has no effect on the rate of reaction because it does not participate in the rate-determining step.

The stereochemistry of S_N1 displacements is less predictable than that of S_N2 displacements. Once a free carbocation is formed, a nucleophile can attack from either side, leading to racemization if the starting material is chiral. However, the carbocation may not always become completely free, but may exist as an intimate ion pair, in which the carbocation and the leaving group are held together within a cage of solvent molecules (Figure 6.52). The reaction will then proceed with inversion of configuration, as the nucleophile will not be able to approach from the side that the leaving group has just vacated, as it will still be blocked.

The extent of racemization or inversion depends on the reaction and the conditions. Typically, S_N1 reactions give partial racemization and partial inversion. For example, hydrolysis of the chloride shown in Figure 6.53 gave an alcohol with 21% of the optical activity of the alcohol from which the chloride was prepared.

A variation of the S_N1 mechanism is the $S_N i$ mechanism (internal nucleophilic substitution). In this, the leaving group itself gives rise to the nucleophile, and the reaction proceeds with retention of configuration. The best-known example of this mechanism is the conversion of alcohols to alkyl chlorides by treatment with thionyl chloride (Figure 6.54).

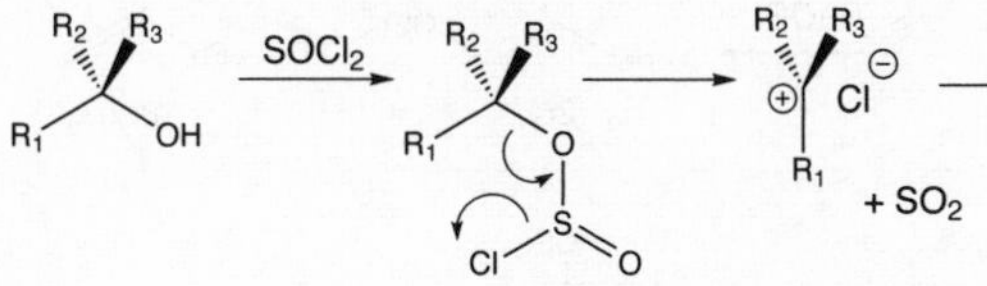

Figure 6.54

6.3.4 Substitution versus elimination, unimolecular versus bimolecular

We have now seen four mechanisms that all involve the loss of a nucleophile: E1, E2, S_N1, and S_N2. Elimination may compete with substitution leading to a mixture of products, although it is often possible to adjust the conditions of the reaction so that the desired pathway predominates. An understanding of the factors that favour one of these mechanisms over the others is therefore an indispensable tool for the organic chemist.

In general, the differences between conditions favouring the unimolecular reactions and those favouring the bimolecular reactions are more marked than the differences between conditions conducive to elimination or substitution. The unimolecular processes are favoured if the substrate is easily ionized. There are three factors at work here: the leaving group, the resultant carbocation, and the solvent. A good leaving group will promote ionization more than it increases the rate of the bimolecular reactions, so the net effect is to favour unimolecular reactions. A stable carbocation is an important prerequisite for a unimolecular process. The more stable the carbocation, the more readily it is formed, and so the more readily either an E1 or S_N1 reaction can take place. In general, E1 and S_N1 reactions take place only on tertiary substrates, or those in which the carbocation is stabilized in some other way, such as by conjugation with an aromatic ring. The solvent has an important effect by stabilizing the carbocation, and so promoting ionization. Polar solvents make ionization easier, and so favour unimolecular reactions.

The presence of a strong added base or nucleophile favours the bimolecular reactions. The base (or nucleophile) does not take part in the rate-determining step in unimolecular reactions, so the rate of these reactions is not affected by the nature or concentration of these reagents. However, bases and nucleophiles take part in the rate-determining steps of E2 and S_N2 reactions respectively, so increased strengths and concentrations of these will promote the bimolecular reactions.

In either unimolecular or bimolecular reactions, elimination will be favoured over substitution if some factor that stabilizes the developing double bond is present, such as conjugation to an aromatic ring. In general, any group that enhances the acidity of a β-hydrogen will favour elimination. An example of this effect is shown by the two bromides in Figure 6.55, which give the ratios of substitution to elimination shown under the same conditions (the proton that is eliminated in the upper compound is more acidic than the proton eliminated in the lower compound

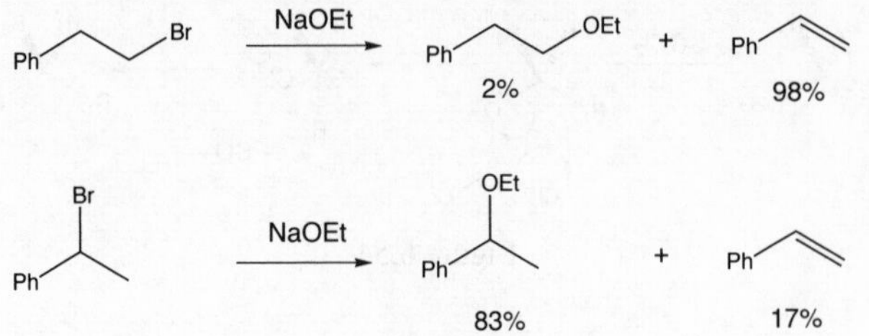

Figure 6.55

because it is next to the aromatic ring). Acidity of the β-hydrogens also favours E2 over E1, although it does not affect the rates of S_N2 and S_N1 reactions relative to each other.

In bimolecular reactions, the degree of substitution at the α-carbon has a profound effect. As discussed above, S_N2 reactions are extremely slow at tertiary carbons, so tertiary substrates give almost entirely elimination under conditions that promote bimolecular reactions, such as treatment with strong bases.

In both types of reaction, β-branching tends to favour elimination. This is partly for steric reasons: β-branching inhibits attack by a nucleophile, particularly in S_N2 reactions, and it also helps to stabilize the double bond, since more highly substituted double bonds are more stable.

Whilst there is a reasonable correlation between base strength and nucleophilicity, as we saw in Section 2.3.1 it is by no means perfect. E2 reactions can be favoured over S_N2 by the use of strong, non-nucleophilic bases, such as potassium t-butoxide or LDA. Conversely, strong, non-basic nucleophiles such as iodide will tend to encourage S_N2 reactions.

The nature of the leaving group has no effect on the E1:S_N1 ratio, as the leaving group is no longer present at the point at which the pathways diverge. However, it does affect the E2:S_N2 ratio, tosylates being more prone to substitution, positively charged groups such as NMe_3^+ to elimination, and halides somewhere between the two.

The solvent can also change the ratio of substitution to elimination. S_N1 reactions tend to be favoured over E1 reactions in nucleophilic solvents, such as water, and polar solvents promote S_N2 reactions over E2.

Finally, the proportion of elimination tends to increase with increasing temperature by both mechanisms. This is because elimination has an entropic advantage over substitution, which becomes more important at higher temperatures.

6.3.5 Carbonyl substitution

As we saw in Section 5.1.3, substitution reactions at carbonyl groups are not direct substitutions, but additions followed by eliminations. Carbonyl substitutions are of

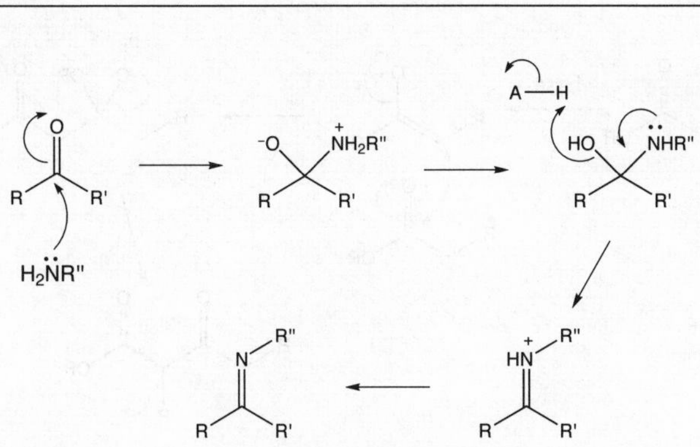

Figure 6.56

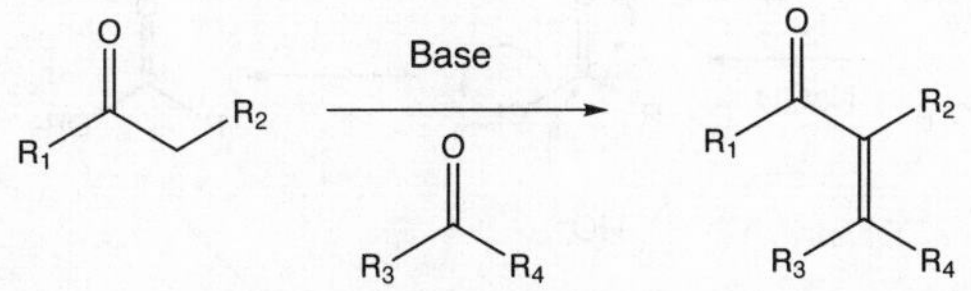

Figure 6.57

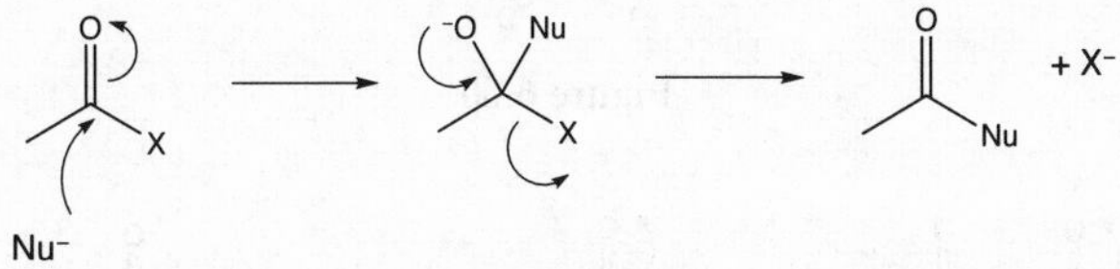

Figure 6.58

two types: those in which the oxygen of a ketone or aldehyde is replaced, and those in which the leaving group of a carboxylic acid derivative is replaced.

An example of the first sort is the reaction of an ammonia derivative with a ketone, as shown in Figure 6.56 (see Section 5.2.3). Carbon may also be the nucleophile in this reaction; any carbonyl compound with two α-hydrogens can react in this way (Figure 6.57). Aldol-type reactions (see Section 5.1.3), in which a carbonyl compound reacts with an aldehyde or ketone (this may be a self-condensation), may simply give addition to yield the alcohol, or the addition may rapidly be followed by elimination so that the net effect is substitution of the carbonyl oxygen. Which reaction predominates will depend on the nature of the reagents and on the conditions.

As discussed in Section 5.1.3, carboxylic acid derivatives can be attacked by a wide range of nucleophiles to give substitution via a tetrahedral intermediate (Figure 6.58). This is really an addition followed by an elimination, although the two are normally so closely linked that these reactions are generally thought of as substitutions. They are sometimes referred to as addition–elimination reactions.

This type of substitution reaction is of wide scope, and is the mechanism for ester

185

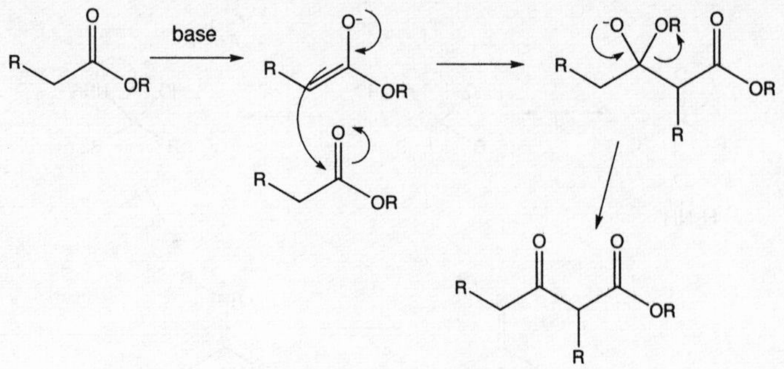

Figure 6.59

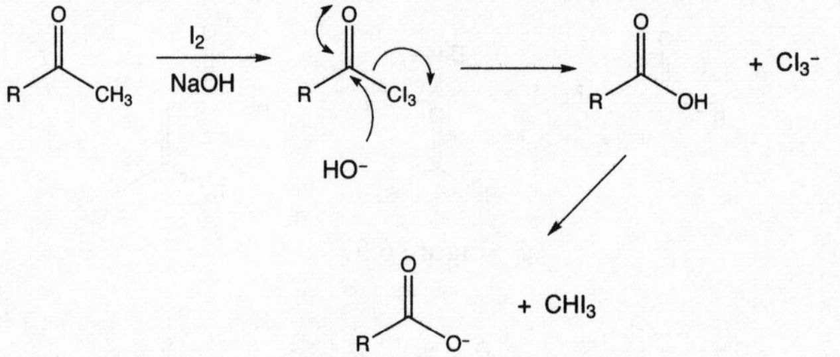

Figure 6.60

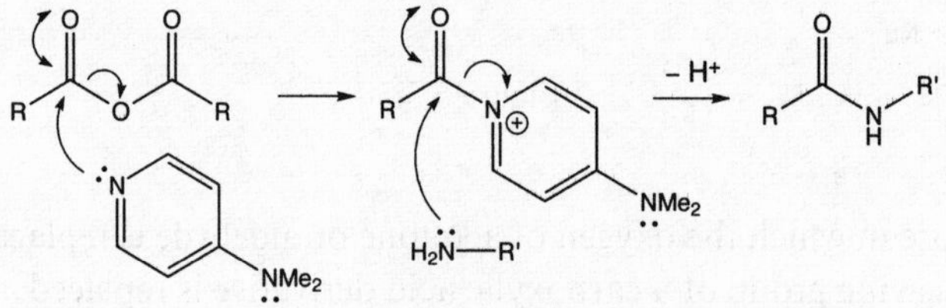

Figure 6.61

formation and hydrolysis, the Claisen ester condensation (Figure 6.59), and many other reactions, including those in which the leaving group is carbon, as in the iodo-form reaction (Figure 6.60). This reaction was used in the days before modern spectroscopic methods as a test for the $COCH_3$ group, as compounds containing this group would give yellow crystals of iodoform when treated with iodine and sodium hydroxide.

Carbonyl substitution reactions can often be catalysed by pyridine, or better still by its derivative 4-(dimethylamino)pyridine (DMAP). The mechanism of this catalysis is shown in Figure 6.61 for the formation of an amide from an amine and an anhydride. Pyridine is a good nucleophile and DMAP is still better, and conse-quently they both react well with carboxylic acid derivatives. However, the product

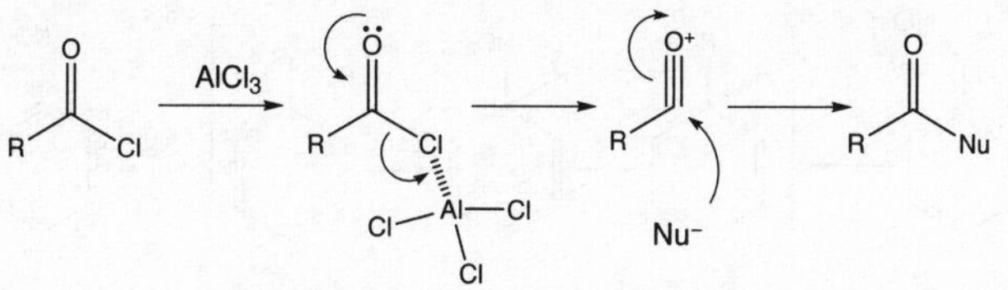

Figure 6.62

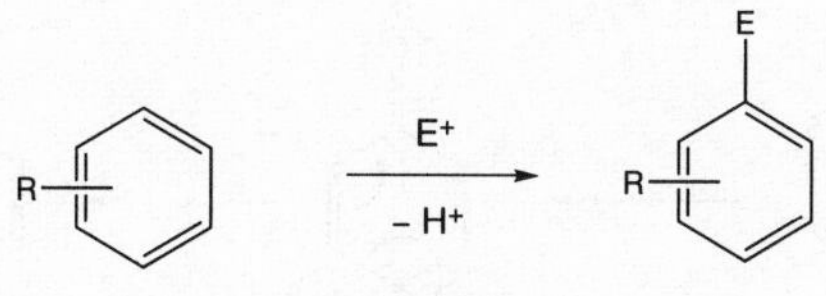

Figure 6.63

of this reaction is a positively charged pyridinium salt, which is more reactive than the starting material, and therefore reacts more rapidly with the desired nucleophile. This is similar to the use of iodide as a catalyst in S_N2 displacements (Section 5.5).

An alternative mechanism occasionally operates in carbonyl substitutions, and is analogous to the S_N1 mechanism. Here, the leaving group leaves first to yield an acylium cation which then reacts with a nucleophile. This mechanism is less common, and normally operates only if the acyl group has a particularly good leaving group, such as chloride in the presence of $AlCl_3$ as a Lewis acid catalyst, which coordinates to the chloride and makes it a better leaving group (Figure 6.62).

6.3.6 Aromatic substitution

Although other types of electrophilic substitution are known, by far the most common example is electrophilic substitution in aromatic rings. The reaction is shown in Figure 6.63; a hydrogen is replaced by an electrophile. The reason why aromatic rings undergo substitution by an electrophilic mechanism is that the ring's π electrons provide regions of high electron density above and below the ring, and these react readily with electrophiles. Electrophilic attack on aromatic rings almost always leads to substitution rather than addition, because addition would destroy the aromatic character of the ring and hence cause a considerable loss of stability.

Nucleophilic substitution of aromatic rings is also known, although it is limited to rings with strongly electron-withdrawing substituents. This will be discussed at the end of this section.

A wide variety of electrophiles can be introduced into aromatic rings by electrophilic substitution, but the mechanism is always the same. This is shown in Figure 6.64 for the nitration of benzene. The nitronium ion is formed from dehydration of nitric acid with sulphuric acid. This is a common choice of reagents for nitration of

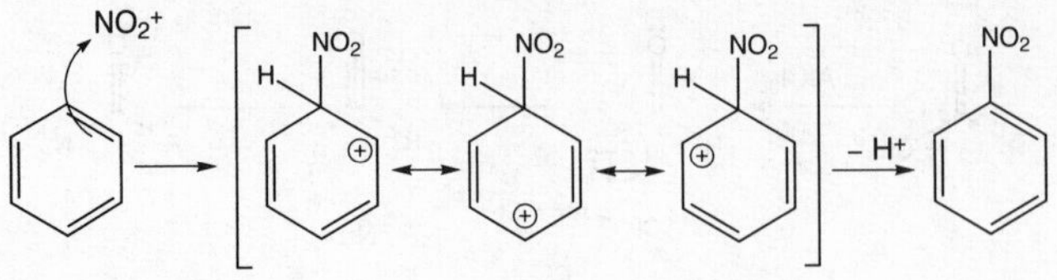

Figure 6.64

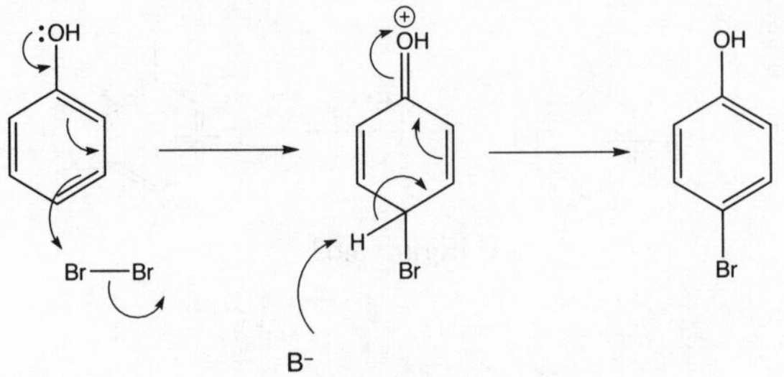

Figure 6.65

aromatic rings, but other nitrating agents can also be used, such as nitronium salts (e.g. $NO_2^+ BF_4^-$).

Attack of the electrophile on the ring leads to the delocalized carbocation, known as a Wheland intermediate. This cation then loses a hydrogen to regain its aromaticity and complete the reaction. Many electrophiles can be used in this reaction.

Unfortunately, electrophilic aromatic substitution is more complicated than this, as we seldom wish to carry out reactions on benzene itself, but rather on substituted benzenes. Substituents on the benzene ring influence substitution in two ways. Firstly, they can affect the rate of substitution, and secondly they can influence its orientation, in other words they may display a preference for *ortho*, *meta*, or *para* products.

Let us first consider the effect of substituents on the rates of reaction. There are no great surprises in store for us here. The aromatic ring behaves as a nucleophile, and so any electron-donating substituents speed up the reaction because they increase the electron density in the π electron clouds, thus making the ring more nucleophilic. Conversely, aromatic rings with electron-withdrawing substituents react more slowly than benzene.

The most powerful electron-donating substituents are oxygen and nitrogen with lone pairs available for donation into the ring. Rings with these substituents typically react considerably faster than benzene, and undergo reactions not possible for other rings. For example, bromination of aromatic rings normally requires a Lewis acid catalyst ($FeBr_3$ is commonly used) to make the bromine more electrophilic, but anilines and phenols are readily brominated without a catalyst (Figure 6.65). Aryl and alkyl substituents also speed up substitution reactions. Rates are slightly decreased by halogen substituents, rather more decreased by carbonyl groups, and strongly decreased by nitro groups.

To consider the directing effects of substituents, we first need to know whether the reactions are under thermodynamic or kinetic control. Substitution of aromatic rings is almost always essentially irreversible, so the reaction must be kinetically controlled (see Section 3.1.6). We are therefore interested not in the stability of the products, but in the energy of the transition state in the rate-determining step. The rate-determining step is the initial attack of the electrophile; loss of a proton is usually rapid.

For a rigorous treatment of the transition states, we need to consider the frontier orbital interactions. In the transition state leading to the Wheland intermediate, the main frontier orbital interaction will be between the LUMO of the electrophile and the HOMO of the aromatic ring. The HOMO of the ring will have different coefficients at different positions on the ring, and the overlap will be most efficient where the coefficients of the HOMO are largest. However, the calculations that give these coefficients are rather complex and are beyond the scope of this book. A fuller discussion of this subject can be found in textbooks on molecular orbital theory. It will suffice here just to summarize the results of these calculations.

What we find is that electron-donating groups, and groups in conjugation with the ring, such as carbon–carbon double bonds and aryl groups, have higher HOMO coefficients at the *ortho* and *para* positions than at the *meta* position. Conversely, electron-withdrawing groups (except halogens) have larger coefficients at the *meta* position. Halogens are unusual, in that they are electron-withdrawing and therefore slow the rate of reaction, but have larger coefficients at the *ortho* and *para* positions.

Fortunately, there is an alternative way of looking at the reactivity pattern which, while not as rigorous as the frontier orbital method, is much easier to understand and also correctly predicts reactivities. Since the Wheland intermediate is much higher in energy than the starting material, we can use the intermediate itself as a model for the transition state (this is justified because the Wheland intermediate is of high energy, see Section 3.1.5), and use the relative energies of the various Wheland intermediates to predict the reactivity.

Let us first consider the alternatives for electrophilic substitution of anisole. The three possible intermediates are shown in Figure 6.66. We can see from the resonance structures that in the *ortho* and *para* isomers the positive charge is stabilized by interaction with the oxygen lone pair (not all possible resonance structures are drawn for these isomers). However, it is not possible to write a resonance structure for the *meta* isomer in which the charge is stabilized in this way.

This is in agreement with the frontier orbital explanation and with experiment; anisole is indeed substituted at the *ortho* and *para* positions. The same argument can be used for other electron-donating groups, including alkyl groups, although here the stabilization comes from formation of a tertiary carbocation instead of from interaction with a lone pair. This also explains the reactivity of aryl halides;

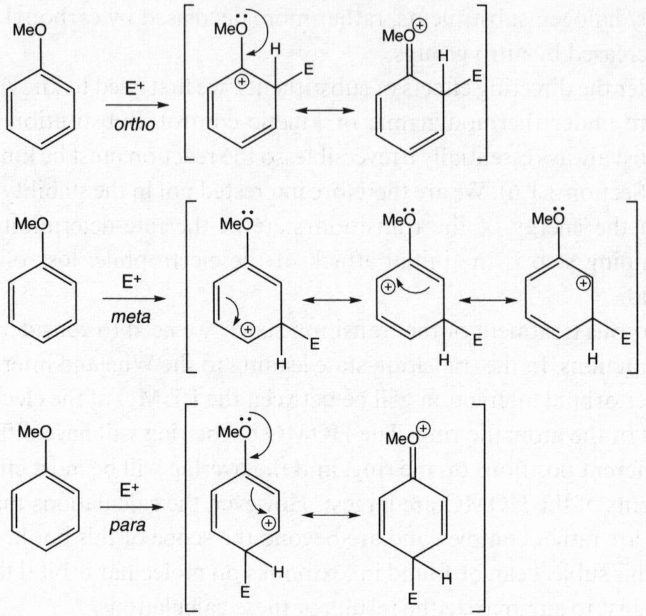

Figure 6.66

although the net effect of halogen substituents is to destabilize a positive charge, this is partly offset by the stabilization due to the halogen lone pair if substitution takes place in the *ortho* or *para* position, but not in the *meta* position, so they too are *ortho/para* directing.

Of course, none of this predicts whether the *ortho* or *para* isomer will predominate. The most important determinant of the *ortho/para* ratio is steric crowding. Bulky electrophiles and bulky substituents will tend to increase the amount of *para* product, as they make it harder for the electrophile to attack at the *ortho* position. Usually, a mixture of *ortho* and *para* isomers is produced.

Let us now consider the possible intermediates formed by attack of an electrophile on nitrobenzene (Figure 6.67). There is now no way of stabilizing the positive charge, but we can see that the *ortho* and *para* isomers are particularly destabilized, as they both have resonance structures in which the positive charge is adjacent to the positive charge on the nitrogen. Nitro groups, in common with other electron-withdrawing groups (except halogens), are therefore *meta* directing.

What happens if there is more than one substituent on the ring? Some of the time, both groups will direct to the same positions, so it is easy to predict the orientation. Some examples are shown in Figure 6.68. However, we may sometimes encounter molecules in which two groups direct to different positions, as in Figure 6.69.

As a general rule, an activating group will take precedence over a deactivating group, so, for example, *m*-nitroanisole will react as shown in Figure 6.70. If two acti-

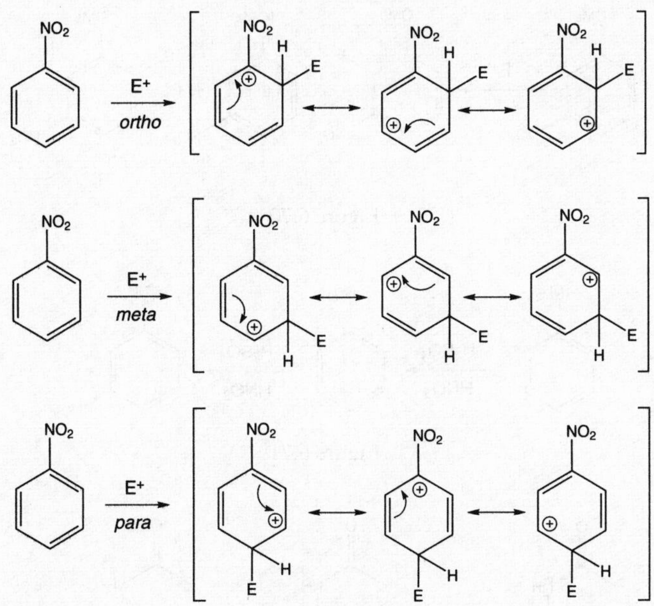

Figure 6.67

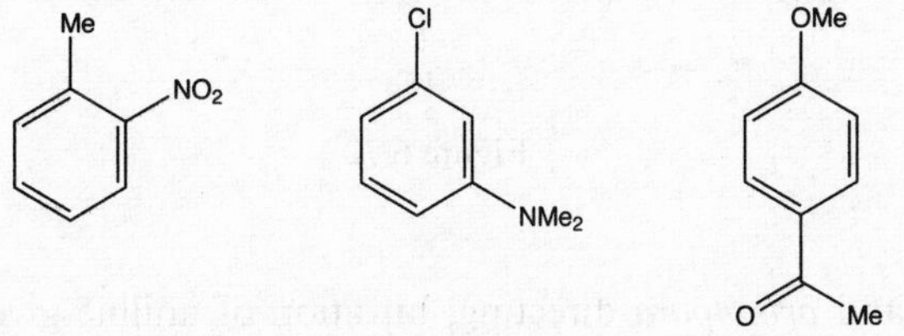

Figure 6.68

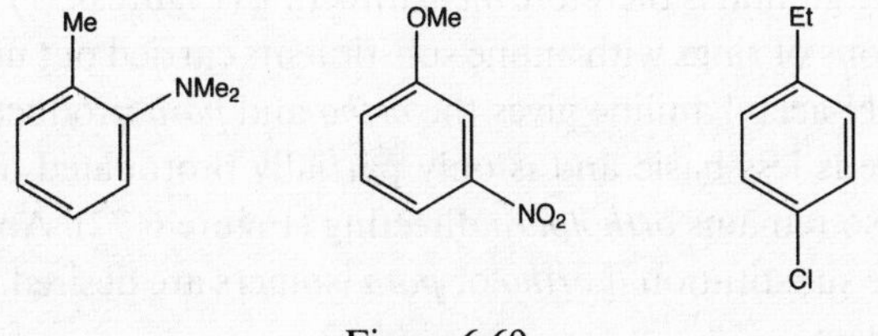

Figure 6.69

vating groups or two deactivating groups direct to different positions, it is not easy to predict which will be more important, and in fact such reactions frequently give mixtures of products.

One example is worth a special mention in discussing directing effects in aromatic rings, namely N-substituted benzenes. Although these have lone pairs and are there-

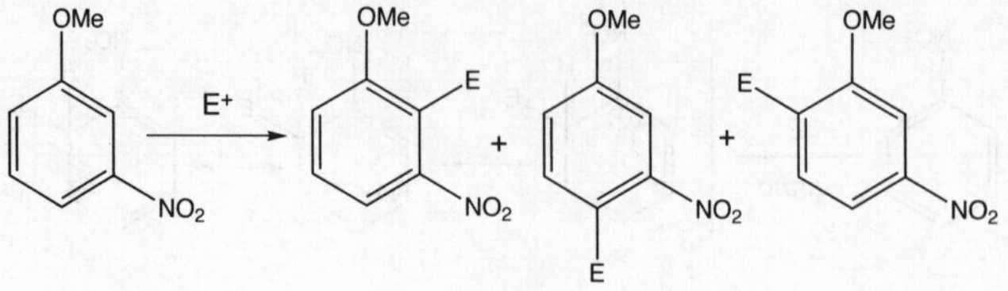

Figure 6.70

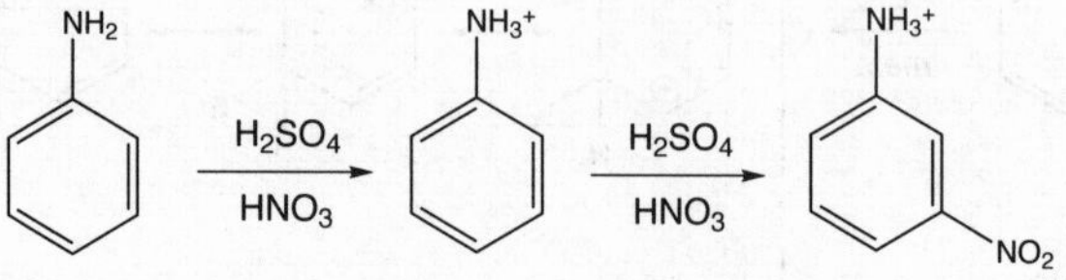

Figure 6.71

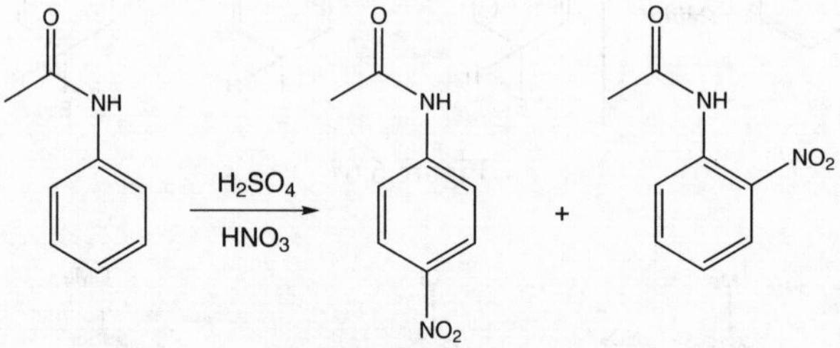

Figure 6.72

fore activating and *ortho/para* directing, nitration of aniline gives *m*-nitroaniline. The reason for this is that nitration is carried out under strongly acidic conditions, in which the substituent is not NH_2, but NH_3^+. This is of course deactivating, as it has a positive charge, and is therefore *meta* directing (Figure 6.71). The same applies to any substitutions of rings with amine substituents carried out under acidic conditions. However, N-acetyl aniline gives the *ortho* and *para* products on nitration, as the nitrogen here is less basic and is only partially protonated under the reaction conditions, and so remains *ortho/para* directing (Figure 6.72). Anilines can thus be acetylated before substitution if *ortho* or *para* isomers are desired. The acetyl group can be removed later.

As aromatic rings are nucleophiles, we might expect them to react with carbonyl compounds and alkyl halides. In fact, aromatic rings are not very strong nucleophiles, and they will only react with these compounds under rather specialized conditions. These conditions give rise to what are known as Friedel–Crafts reactions.

In Friedel–Crafts alkylations, the aromatic compound is treated with an alkyl chloride and a Lewis acid catalyst, usually $AlCl_3$ (Figure 6.73). The catalyst makes

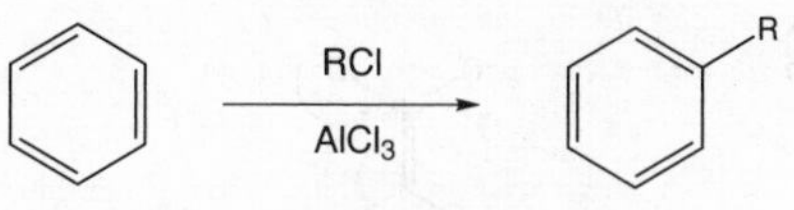

Figure 6.73

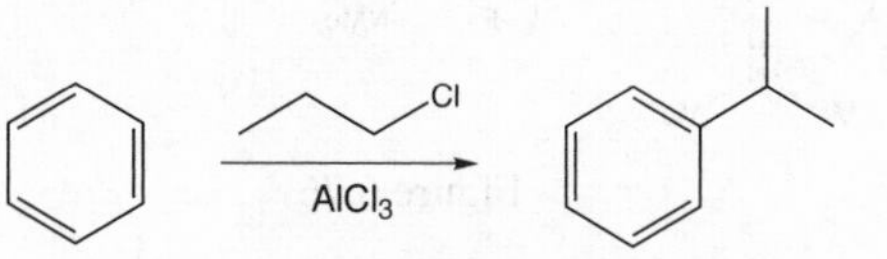

Figure 6.74

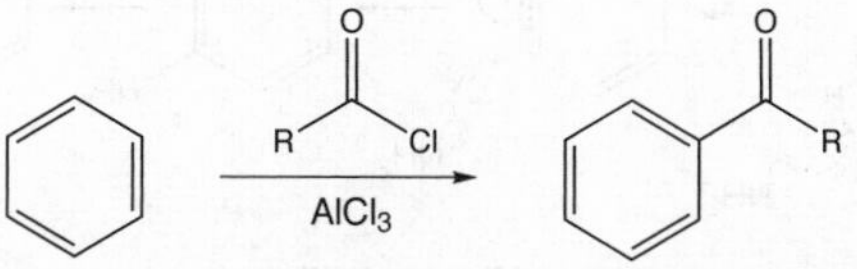

Figure 6.75

the chloride into a better leaving group and thus promotes an S_N1 reaction at the alkyl halide. However, this reaction suffers from two important drawbacks, and so is of only limited use synthetically. Firstly, carbocations are intermediates. Whilst the reaction can be successful with tertiary alkyl chlorides, primary and secondary carbocations frequently undergo rearrangements (see Section 6.4), leading to unwanted products, for example as in Figure 6.74. The other disadvantage is that the product of the reaction is alkylated, and so is more reactive than the starting material. Multiple alkylation can therefore be a problem.

A far more useful reaction synthetically is Friedel–Crafts acylation (Figure 6.75). Here, an acyl chloride is used together with a Lewis acid catalyst, again usually $AlCl_3$. The Lewis acid helps the chloride to leave, resulting in an acylium cation (see Section 6.3.5). These are normally more stable than alkyl cations, so rearrangement is less of a problem. Moreover, the product of the reaction is an acyl aromatic ring, which is deactivated, so the reaction is easy to stop after just one substitution.

As mentioned above, strongly electron-withdrawing substituents can facilitate nucleophilic substitutions. Normally nitro groups are required for this, and the reaction is much easier if the nitro group is *ortho* or *para* to the position that is substituted than if it is *meta*. The group substituted is normally a halogen, and fluorides react fastest. A typical nucleophilic aromatic substitution reaction is shown in Figure 6.76.

An alternative and somewhat esoteric mechanism of aromatic nucleophilic substitution is known as the benzyne mechanism. This, as shown in Figure 6.77, is an elimination followed by an addition. The triply bonded intermediate is very unsta-

193

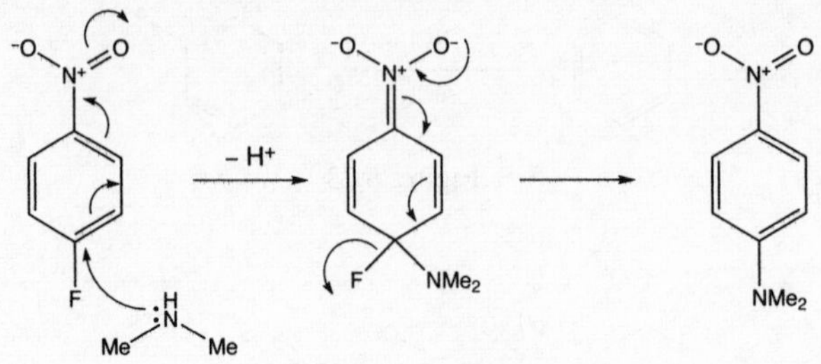

Figure 6.76

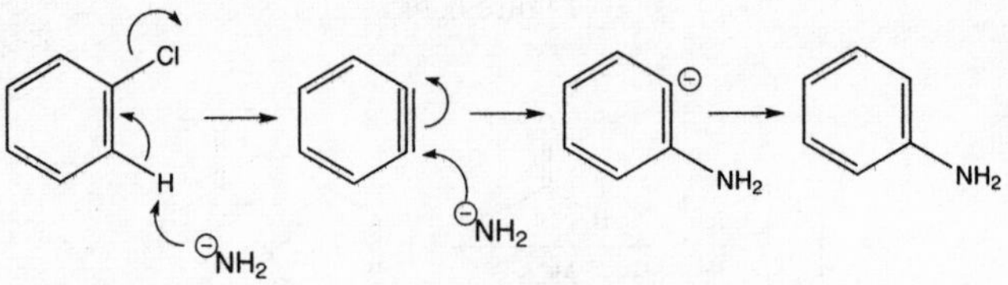

Figure 6.77

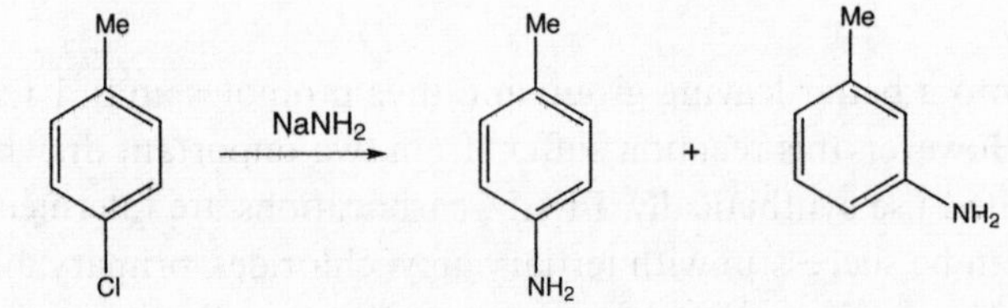

Figure 6.78

ble because the bond angles are greatly distorted from the ideal linear arrangement of a triple bond, and so it reacts rapidly with a nucleophile as shown. If there is a substituent, the nucleophile may attack at either end of the triple bond to give a mixture of products as shown in Figure 6.78. The reaction is shown here with chloride as the leaving group and NH_2^- as the base and the nucleophile. This reaction is not of wide scope, and it is this combination of groups that is by far the most likely to be encountered with this mechanism.

6.4 Rearrangements

A rearrangement is a reaction in which groups move within the same molecule, leading to a molecule isomeric with the starting material. Three types of rearrangements may be distinguished: nucleophilic rearrangements, in which the migrating group moves with its electron pair; radical rearrangements, in which the migrating group moves with one electron; and electrophilic rearrangements, in which the migrating group moves with a vacant orbital (Figure 6.79). Nucleophilic rearrange-

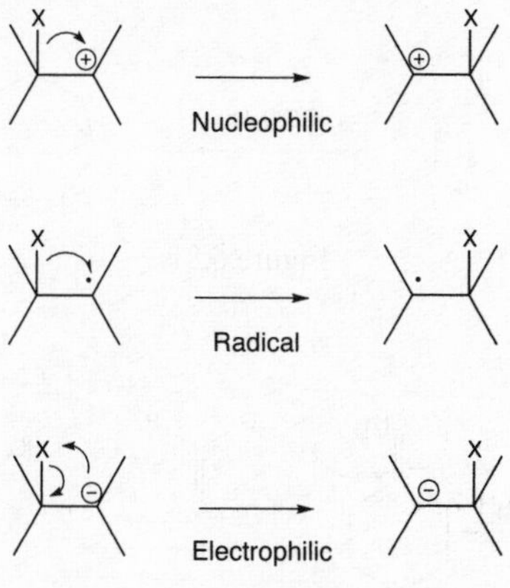

Nucleophilic

Radical

Electrophilic

Figure 6.79

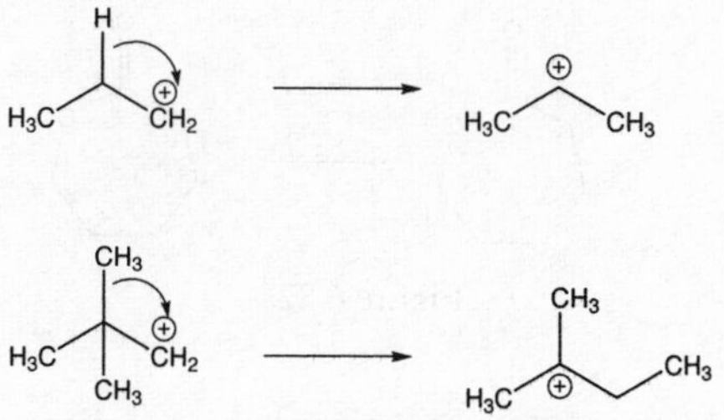

Figure 6.80

ments are by far the most common of the three; radical rearrangements are some-what esoteric, and electrophilic rearrangements are very rare indeed.

One of the most common occurrences of rearrangements is when carbocations rearrange to give more stable carbocations (see Section 4.2.3). Some examples can be seen in Figure 6.80, which shows that either hydrogens or alkyl groups can migrate. These reactions are known as Wagner–Meerwein rearrangements.

Carbocations are intermediates in E1 and S_N1 reactions, as described above, and rearrangement can sometimes be a side reaction. As already mentioned, this is par-ticularly troublesome in Friedel–Crafts alkylation of aromatic compounds (which is an S_N1 reaction from the point of view of the alkyl halide).

Of course, the products of carbocation rearrangements are themselves cations, and so can undergo the full range of carbocation reactions, such as further rearrangement, loss of a proton, or combination with a nucleophile.

There are a number of well-known reactions in which at least one step is a rearrangement. One of these is the Baeyer–Villiger rearrangement, which has the net effect of converting a ketone into an ester by inserting an oxygen atom on treatment with a peracid (Figure 6.81). This is most commonly applied to cyclic ketones, giving lactones as the products. The mechanism is shown in full in Figure 6.82, and we can

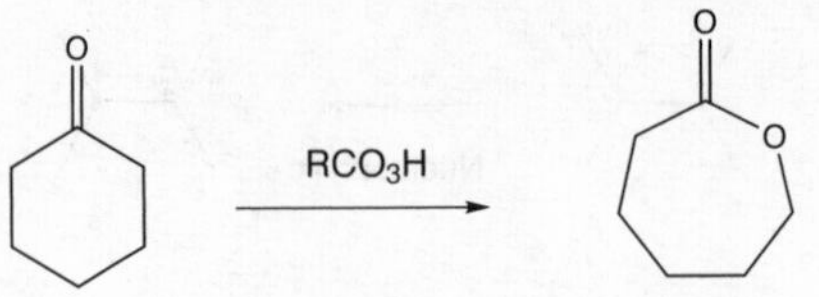

Figure 6.81

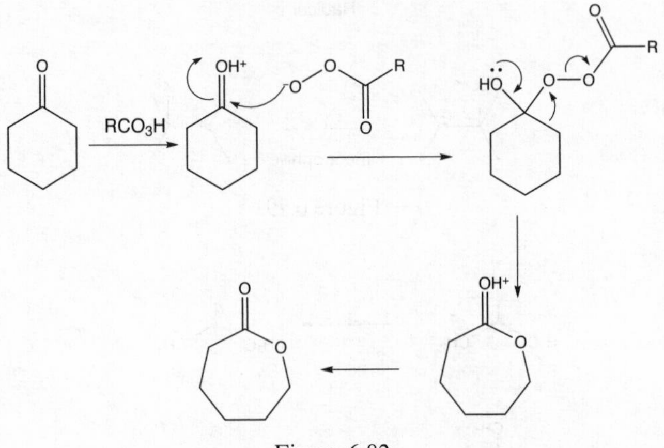

Figure 6.82

see that the key step here is a rearrangement, in which one of the alkyl groups migrates to oxygen.

If the ketone is unsymmetrical, we need to consider which of the two groups migrates. In a way, the migrating group behaves as a nucleophile, as it attacks the oxygen complete with its electron pair, and so we might expect the groups that would make the strongest nucleophiles to have the greatest migratory aptitude. This is a slightly contrived argument, because the migratory groups do not become free and are of course not nucleophiles. Nonetheless, it does give us a reasonably good guide to migratory aptitude. Thus tertiary alkyl groups migrate in preference to secondary alkyl groups, themselves having a greater migratory aptitude than primary alkyls, and methyl groups migrate least readily of all. This is as we would expect, as the electron-donating effect of the increasing alkyl substitution increases the 'nucleophilicity' of the groups. Aryl groups are generally similar in migratory ability to secondary alkyl groups, and migrate more readily if they have electron-donating substituents.

The reaction is thought to be concerted, the alkyl group migrating at the same time as the carboxylate group is expelled. The alkyl group does not become free, and if it is chiral it retains its stereochemistry.

Another rearrangement, known as the pinacol rearrangement, is shown in Figure 6.83. The driving force here is the formation of the highly stabilized oxonium cation from the carbocation.

We have already seen a series of rearrangments of nitrenes in Section 5.2.5.

196

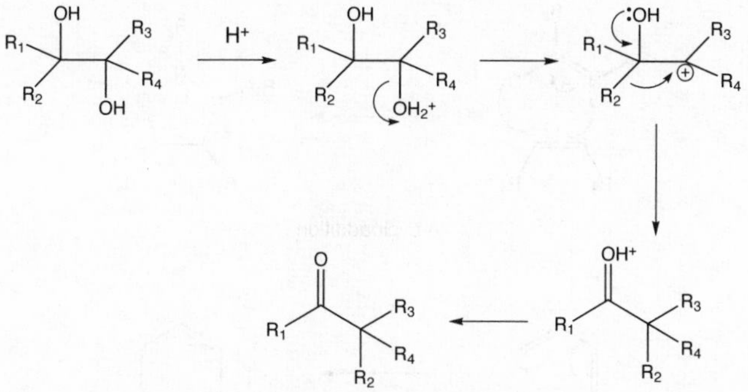

Figure 6.83

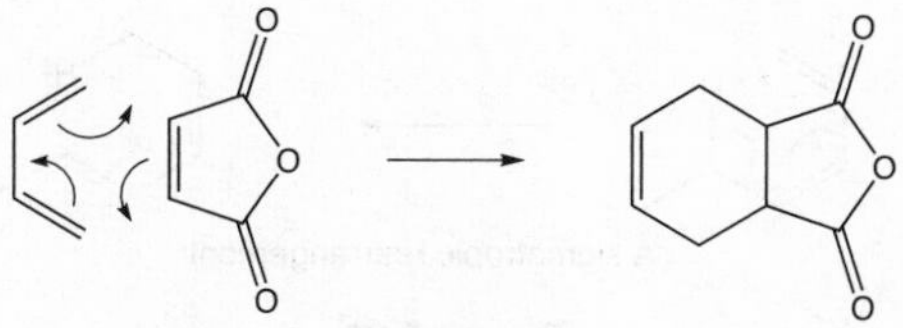

Figure 6.84

6.5 Pericyclic reactions

6.5.1 Basic principles

Although all pericyclic reactions can be classified into one of the four types of reaction described above, they have a number of features in common that set them apart from other reactions and deserve a separate discussion. As the name pericyclic implies, the electrons move in a circuit.

Figure 6.84 shows perhaps the best known of all pericyclic reactions, the Diels–Alder reaction. We can see here the main feature of pericyclic reactions, namely that concerted movement of electrons leads to the formation and breakage of bonds in a single step. Since these reactions are concerted, they do not have any intermediates, either of a polar or of a radical nature. Pericyclic reactions are therefore remarkably insensitive to the effects of changing the polarity of the solvent, and are not affected by radical initiators or inhibitors.

One of the rules for drawing curly arrows that we encountered in Section 1.3.1 is that any series of curly arrows must start at an electron source and end at an electron sink. Pericyclic reactions appear to violate this rule, because there is no obvious electron source or electron sink. We can think of any point in the chain of curly arrows as both the electron source and the electron sink, although this is rather artificial as the electrons form a circuit, which does not really have a beginning or an end.

We can draw an analogy between chains of electrons in reactions and chains of

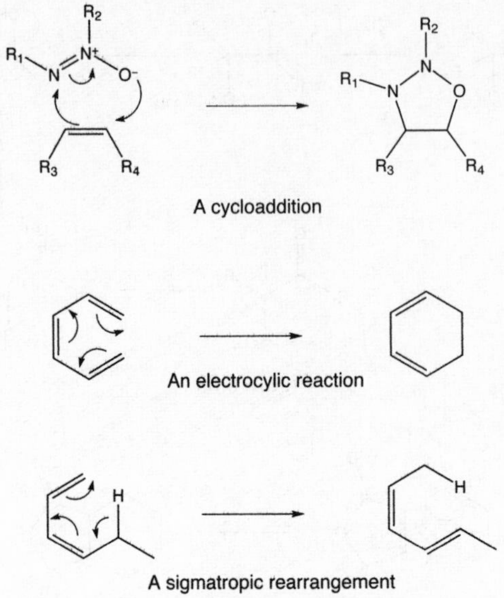

A cycloaddition

An electrocylic reaction

A sigmatropic rearrangement

Figure 6.85

people buying and selling houses. If you buy a house, the person whose house you are buying must buy another house, and so on. Similarly, when you buy the house, you must sell your old one. We thus have a chain, similar to a chain of curly arrows in a reaction. At the start of this chain is the first-time buyer, who has no house to sell. He or she is analogous to the electron source. At the end of the chain, someone sells a house without buying a new one, perhaps because the house was used as an investment rather than as a home. This person is analogous to the electron sink. Now imagine the somewhat unlikely (although nonetheless possible) situation in which there is no first-time buyer, and the person at the end of the chain buys the house of the person at the start of the chain. If all transactions are completed simultaneously, how can we say who is at the start of the chain and who is at the end of it? We cannot, because everyone is part of a circuit. This is how the electrons move in a pericyclic reaction.

We may distinguish three main types of pericyclic reactions: cycloadditions, electrocyclic reactions, and sigmatropic rearrangements. An example of each of these is shown in Figure 6.85. Each of these reactions is, in principle, reversible, which gives further possibilities.

6.5.2 The Woodward–Hoffmann rules

Pericyclic reactions cannot take place indiscriminately between any two molecules with conjugated π systems, but are subject to certain rules. These rules govern what types of π systems may react with what other π systems under what circumstances,

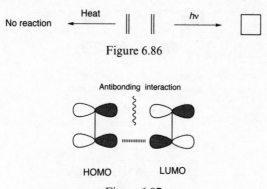

Figure 6.86

Figure 6.87

and are known as the Woodward–Hoffmann rules. As an example of allowed and disallowed reactions, let us look at cycloadditions. The Diels–Alder reaction readily takes place on heating suitable reactants. This is known as a [4+2] cycloaddition, because of the number of π electrons in the respective reactants. However, if we try to add two simple olefins in a cycloaddition by heating (a [2+2] cycloaddition), we will not normally be successful. Nonetheless, irradiation with UV light frequently brings about such cycloadditions (Figure 6.86). What is going on here?

It is tempting to think that the reason the [2+2] cycloaddition is unsuccessful is because its product would be an unstable four-membered ring, but if this were the only reason, then the reaction would not happen photochemically. Furthermore, many other reactions are disallowed, even though the products are not four-membered rings. This is where the Woodward–Hoffmann rules come in.

To understand these rules, we have to look at the molecular orbitals. Clearly, there will be an interaction between the HOMO of one reactant and the LUMO of the other. If we look at the attempted dimerization of ethylene, we already know what these frontier orbitals look like. The HOMO is the π bonding orbital, and the LUMO is the π antibonding orbital (Figure 6.87). If we try to bring these two orbitals together in the geometry that they would have to adopt to react, we see that there is an antibonding interaction. To bring the two ethylene molecules together with favourable orbital overlap would require a contorted geometry which is not possible in practice. The two molecules therefore do not react. Of course, this applies not only to ethylene itself but to any combination of olefins attempting to react in a [2+2] cycloaddition.

So why is the photochemical cyclization possible? Again, the HOMO of one molecule must react with the LUMO of the other, but UV light can promote electrons from one orbital to a higher one. In this excited state, what used to be the LUMO now becomes the HOMO, as it now contains (excited) electrons (Figure 6.88). This can interact with the LUMO of a molecule in its ground state, leading to favourable bonding interactions.

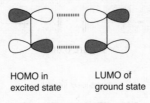

HOMO in
excited state

LUMO of
ground state

Figure 6.88

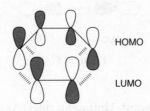

HOMO

LUMO

Figure 6.89

Let us now look at [4+2] cycloaddition. We must again consider the interaction of the HOMO of one component with the LUMO of the other. We could, in principle, consider this either way round, but in the majority of Diels–Alder reactions the important interaction is between the HOMO of the diene and the LUMO of the olefin (generally known as the dienophile), so we will look at this combination of orbitals. We saw the frontier orbitals for butadiene in Section 1.4. The HOMO (ignoring the sizes of the coefficients; we are only interested in the signs here) is as shown in Figure 6.89. We can see clearly that this is able to interact readily with the LUMO of the dienophile, so this reaction is allowed. We would have got the same answer from considering the LUMO of the diene and the HOMO of the dienophile.

The above discussion provides an explanation of why [4+2] cycloadditions are allowed thermally, but [2+2] cycloadditions are only allowed photochemically. It turns out that these arguments have general applicability to pericyclic reactions, and lead to the Woodward–Hoffmann rules. These state that the total number of electrons must be $(4n+2)$ for a thermal reaction, and $4n$ for a photochemical reaction, for processes that are suprafacial on each component. By suprafacial, we mean that the addition takes place on the same face of the respective component. The opposite of this is antarafacial. If a process is suprafacial on one component and antarafacial on the other, we need $4n$ electrons for a thermal reaction and $(4n+2)$ for a photochemical one. The geometrically impossible cycloaddition shown in Figure 6.90 is allowed by the Woodward–Hoffmann rules, because it is antarafacial on one of the components (the component whose LUMO is shown). A reaction in which both components react antarafacially has the same requirements as one in which both components react suprafacially, namely $(4n+2)$ for thermal reactions and $4n$ for photochemical ones.

200

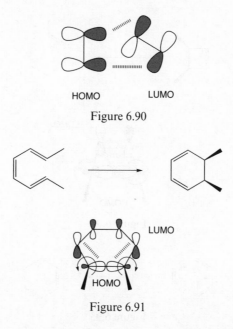

HOMO　　　LUMO

Figure 6.90

LUMO

HOMO

Figure 6.91

In most of the common cycloadditions, we are interested only in suprafacial processes, as antarafacial processes are normally ruled out for geometric reasons.
However, in cycloadditions forming large rings antarafacial processes may occasionally become feasible.

Antarafacial processes can be important in electrocyclic reactions. Let us consider the cyclization of the triene shown in Figure 6.91. This has 6 π electrons, which
is a $(4n+2)$ number, and so must be a suprafacial process. It is actually easier if we
consider the reverse reaction, which we can think of as a $[4+2]$ addition of a σ bond
to a diene. It does not matter that the reaction does not go in this direction in practice, the pathway of the true reaction must be the reverse of the fictional addition.
If we look at the frontier orbitals (the HOMO of the σ bond and the LUMO of the
diene are shown, but we could have taken these orbitals the other way round), we
see that the two ends of the molecule must rotate in opposite directions for the
orbitals to match up, which is a suprafacial process for both components. Therefore,
in the cyclization the ends of the molecule must also rotate in opposite directions.
This is known as a disrotatory cyclization and has the stereochemical consequence
shown.

In contrast, an electrocyclic reaction with 4 π electrons must take place by
an antarafacial process. This is shown in Figure 6.92, and we can see that here
the groups must rotate in the same direction, which is known as conrotatory.
This is antarafacial on the π component. Cyclobutenes therefore open with the
stereochemistry shown. This time, the reaction does go in the direction of ring
opening.

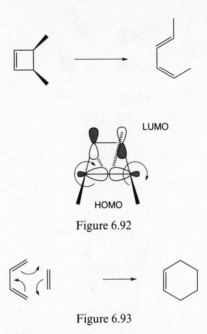

Figure 6.92

Figure 6.93

6.5.3 The Diels–Alder reaction

The most common of all [4+2] additions is the Diels–Alder reaction (Figure 6.93). The substituents on the two components are not shown here, but for most substituents the important interaction is between the HOMO of the diene and the LUMO of the dienophile. The reaction is therefore promoted by substituents that raise the energy of the diene HOMO or lower the energy of the dienophile LUMO. This can be accomplished by electron-donating substituents on the diene and electron-withdrawing substituents on the dienophile.

In fact, the reaction seldom goes at a reasonable rate unless the dienophile has some sort of electron-withdrawing group. The reaction between butadiene and ethylene itself requires high temperatures and pressures, and even then does not go to completion. A commonly used dienophile in the Diels–Alder reaction is maleic anhydride (shown in Figure 6.84). Because of its two electron-withdrawing groups, it reacts rapidly with many dienes. Azo compounds can also be used as dienophiles, diacyl azo compounds being particularly reactive (Figure 6.94).

The diene itself must be in its cisoid conformation to take part in this reaction (cisoid refers to the conformation about the central single bond, see Figure 6.95). Some dienes more readily adopt the cisoid conformation than others; this is a function of their substituents. Dienes with bulky substituents at the ends in the *cis* position will have less stable cisoid conformations (Figure 6.96), and will consequently spend less time in the cisoid form. This will have the effect of slowing the rate of reaction.

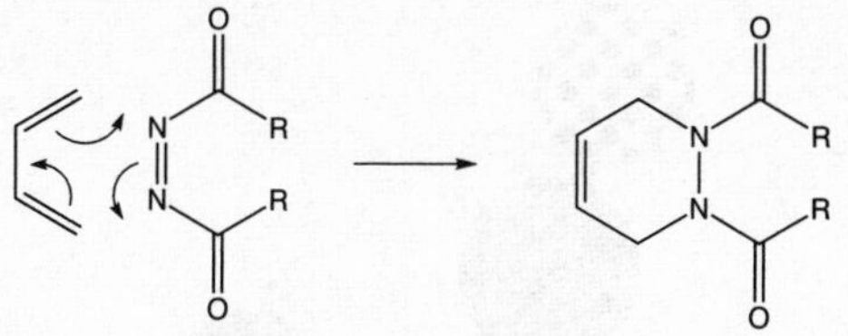

Figure 6.94

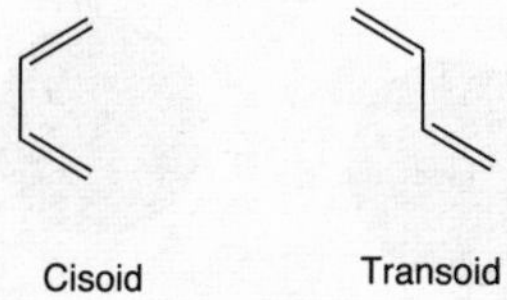

Cisoid   Transoid

Figure 6.95

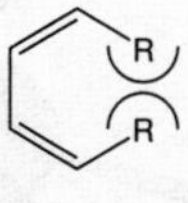

Figure 6.96

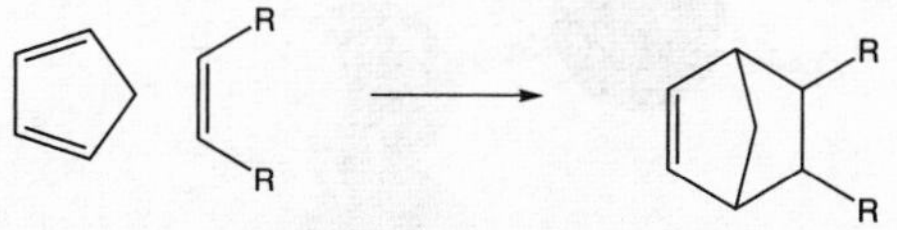

Figure 6.97

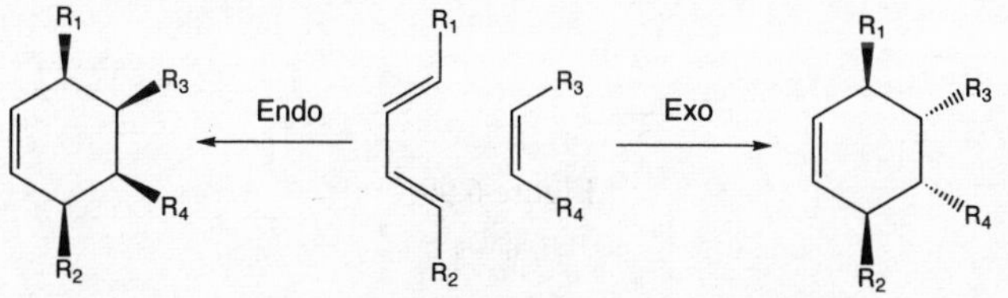

Endo   Exo

Figure 6.98

One of the best dienes in Diels–Alder reactions is cyclopentadiene, which reacts rapidly with a wide range of dienes (Figure 6.97). The reason for its high reactivity is that it is locked into the cisoid conformation. Cyclopentadiene is in fact so reactive that it forms a dimer with itself, and is stored in this form. It is regenerated before use by distillation, which produces the monomer by a retro-Diels–Alder reaction.

The Diels–Alder reaction is stereospecific, as it must be suprafacial on both of its components. The stereochemistry of the reaction is shown in Figure 6.98. There are

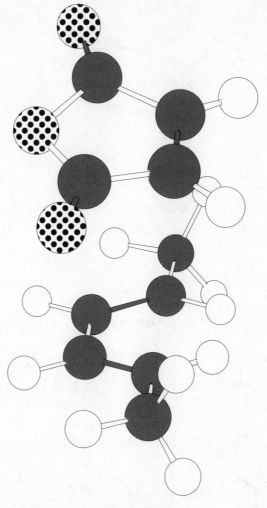

Figure 6.99

two possible stereochemical outcomes, each of which is suprafacial in both components. These are known as the endo and exo adducts, and the transition states
leading to them are shown in Figures 6.99 and 6.100 for the reaction of 2,4-hexadi-
ene with maleic anhydride. Normally, if the dienophile has carbonyl groups adjacent to the double bond, the endo adduct is formed, owing to secondary orbital
overlap in the transition state between the orbitals on the carbonyl groups in the
dienophile and the orbitals in the diene not directly involved in bonding. These
interactions do not lead to bond formation, but they do lower the energy of the
transition state.

If the starting materials are not symmetrical, we must also consider the orienta-

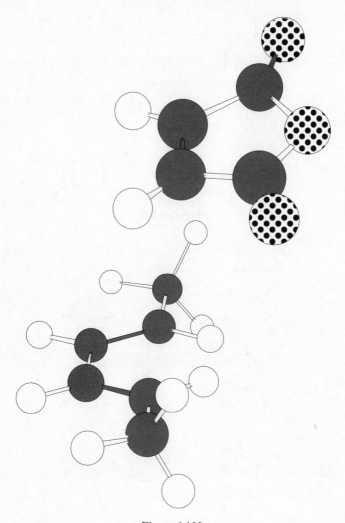

Figure 6.100

tion of the product. To explain this, we need to use rather more complex molecular orbital arguments which include a discussion of the sizes of the coefficients of the frontier orbitals on the various atoms. It will suffice here to present the conclusions of these calculations: in general, the products in which the substituents are in a 1,2 or 1,4 orientation are preferred to those with a 1,3 orientation (Figure 6.101), although there are exceptions.

Although it is normally the HOMO of the diene that reacts with the LUMO of the dienophile, if the diene contains electron-withdrawing groups and the dienophile is electron-rich, the dominant interaction may be the other way round. Such reactions are said to go with reverse electron demand.

205

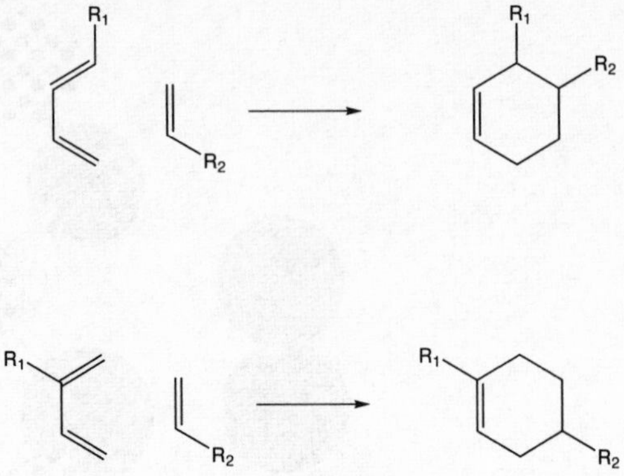

Figure 6.101

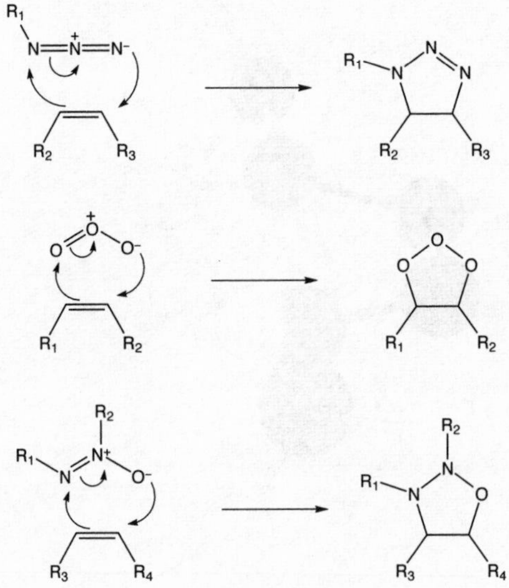

Figure 6.102

6.5.4 Other pericyclic reactions

Another common type of cycloaddition reaction is 1,3-dipolar cycloaddition. Here, a 1,3-dipole adds to a dienophile. There are many 1,3-dipoles that take part in this reaction; some common examples are azides, ozone, and azoxy compounds (Figure 6.102).

Electrocyclic reactions were discussed above (Section 6.5.2). Their most noteworthy feature is that both suprafacial and antarafacial processes are normally feasible, and the stereochemistry of the reaction can be changed depending on whether the reactions are run thermally or photochemically. In general, the reactions are reversible, although it is difficult to form cyclobutenes from butadienes; this reaction normally goes in the ring-opening direction only.

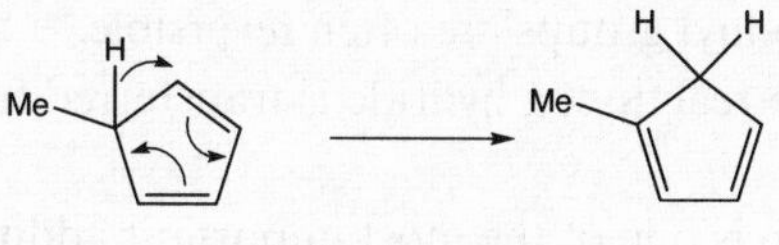

Figure 6.103

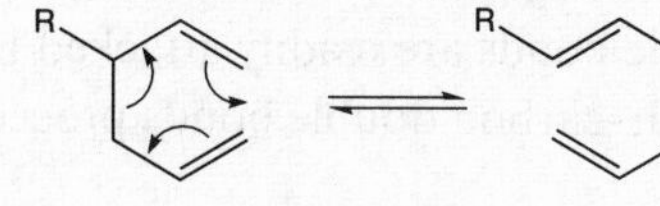

Figure 6.104

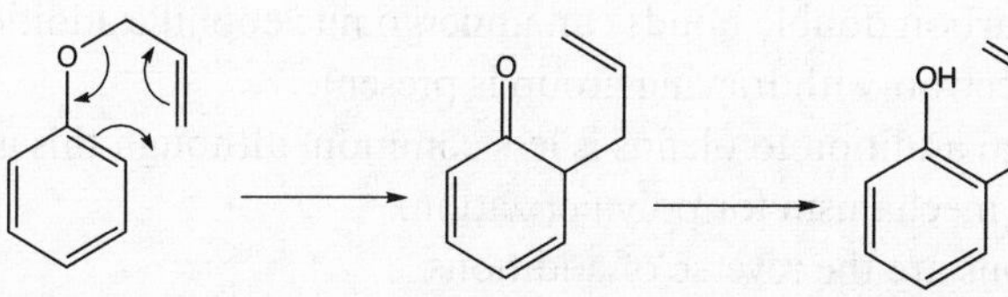

Figure 6.105

Many sigmatropic rearrangements are known. One of the simplest is the 1,5-hydrogen shift (Figure 6.103). 1,3-hydrogen shifts are rare, because they are four-electron processes, and thus would have to be antarafacial, which is difficult to achieve geometrically. Cyclopentadiene systems undergo 1,5-hydrogen shifts readily, because of the proximity of the hydrogen to the other end of the diene system.

Another sigmatropic rearrangement is the Cope rearrangement (Figure 6.104). For a symmetrical diene, the product is the same as the starting material, and so the reaction is undetectable, although unsymmetrical dienes give a detectable reaction. This will usually be an equilibrium process, leading to a roughly equal mixture, unless some factor is present that stabilizes one product over the other. An example of this is if the diene contains a hydroxyl group; the product will thus contain an enol, which tautomerizes to an aldehyde. The aldehyde cannot undergo the reverse reaction, and so the equilibrium is shifted in its favour.

The Claisen rearrangement is another example of a sigmatropic rearrangement. This takes place with allyl ethers of enols or phenols (Figure 6.105).

Summary

- Addition to multiple bonds may take place by nucleophilic, electrophilic, or radical mechanisms.
- In additions to carbonyl groups, a nucleophile adds to the carbon and an electrophile (usually a proton) adds to the oxygen.

- Additions to carbonyl groups are often reversible.
- In the Cannizzaro reaction, a hydride is transferred from one carbonyl group to another.
- The aldol reaction is one of the most important additions to a carbonyl group.
- Electrons can be the nucleophile in additions to carbonyl groups.
- Carbon–carbon double bonds are readily attacked by electrophiles.
- Bromination of carbon–carbon double bonds proceeds via a cyclic bromonium ion.
- Olefins can add nucleophiles and a proton under acidic conditions; this normally gives Markovnikov orientation.
- Carbon–carbon double bonds can undergo nucleophilic addition if a suitable electron-withdrawing group is present.
- Free radical addition to olefins is less common, although this is an important mechanism for polymerization.
- Eliminations are the reverse of additions.
- The most common mechanisms for eliminations are E2 and E1 reactions; E1cB and Ei mechanisms are also possible.
- E2 eliminations are concerted, that is both groups are lost simultaneously.
- The two leaving groups in E2 eliminations must be in the same plane; they are normally antiperiplanar, but synperiplanar eliminations are also known.
- In E1 eliminations, the nucleofugal leaving group leaves first to give a carbocation, which then loses a proton (or other electrofugal leaving group).
- E1 reactions are normally reversible and under thermodynamic control, in contrast to E2 reactions which are normally kinetically controlled.
- E1cB eliminations are favoured by stable carbanions and poor leaving groups.
- Substitution reactions may take place directly or they may be the net result of an addition and an elimination.
- The simplest type of substitution is the S_N2 reaction.
- In the S_N2 mechanism one group replaces another at a saturated carbon with inversion of configuration, via a five-coordinate intermediate.
- In the S_N1 mechanism a leaving group leaves to give a carbocation, which then recombines with a nucleophile.
- Substitution and elimination reactions are often in competition with each other.
- Carbonyl substitution is the result of an addition followed by an elimination; direct substitutions never take place at carbonyl groups.
- Carbonyl substitution reactions can often be catalysed by pyridine.
- In aromatic substitution, the aromatic ring first reacts with an electrophile and then loses a proton.

- The substituents on the ring affect the reactivity in aromatic substitution reactions: electron-withdrawing substituents decrease the rate while electron-donating substituents increase it.
- Substituents also affect the orientation.
- Aromatic substitution can also take place nucleophilically or by the benzyne mechanism.
- The most common type of rearrangement is Wagner–Meerwein rearrangement of carbocations.
- In pericyclic reactions the electrons move in a closed circuit.
- Pericyclic reactions are governed by the Woodward–Hoffmann rules.
- The Diels–Alder reaction is a cycloaddition between a diene and a dienophile.
- Other pericyclic reactions include electrocyclic reactions and sigmatropic rearrangements.

Problems

1. When propene is treated with bromide in the presence of chloride ions, a mixed halide is formed. What is its structure, and why does this particular compound form?

2. But-2-ene can be converted to 2-bromobut-2-ene by the following two transformations. What is the structure of the intermediate? Draw mechanisms for these steps and explain the stereochemistry.

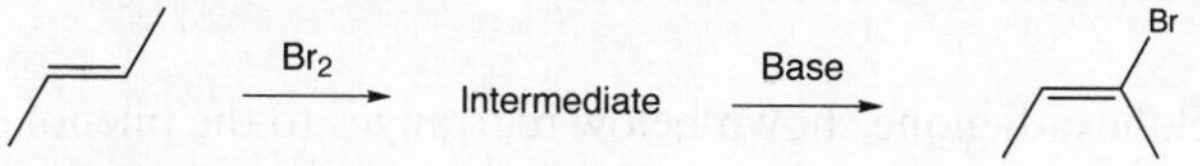

3. What would you expect to be the products of the following reactions of carbonyl compounds? Write mechanisms for the reactions.

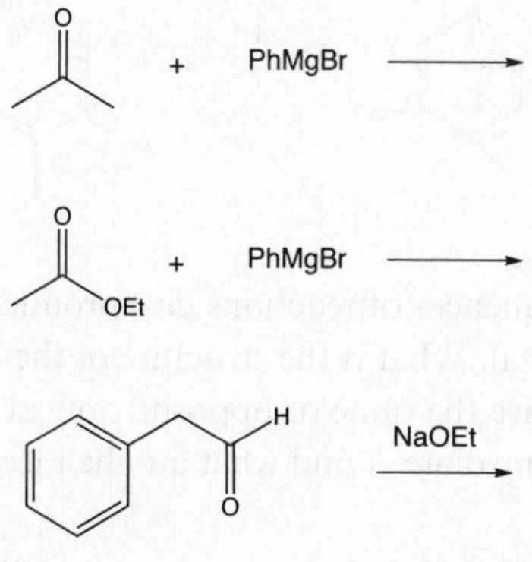

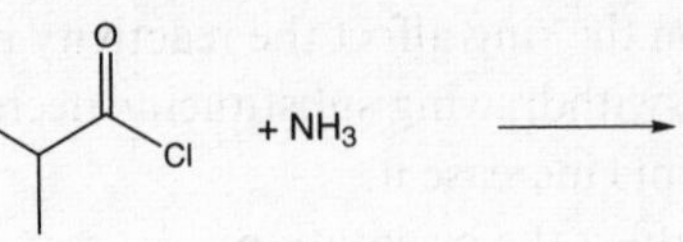

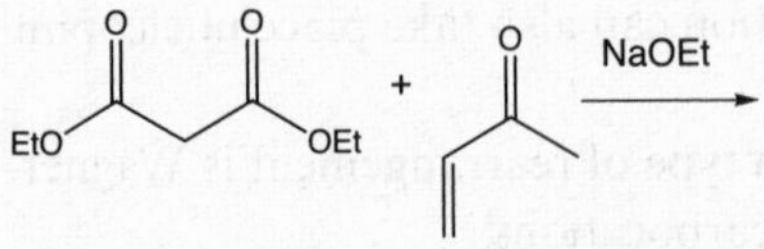

4. Enol ethers are readily hydrolysed in acid. Draw a mechanism for this reaction.
5. When the amino acid threonine is treated with nitrous acid in the presence of chloride ions, two isomeric products are formed in addition to the expected product. What is the mechanism of their formation, and what would you expect their stereochemistry to be?

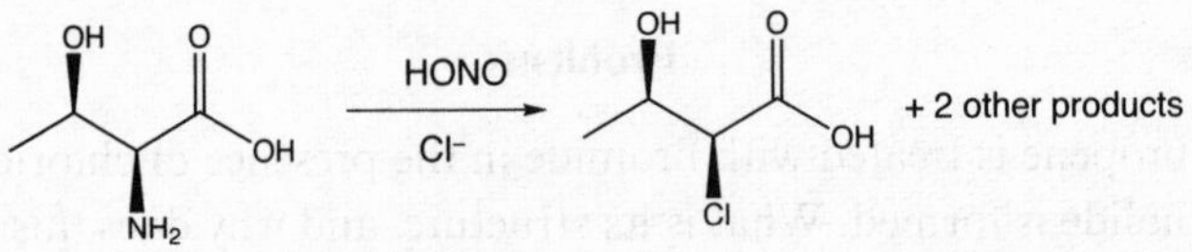

6. In aliphatic nucleophilic substitution reactions, fluoride is the most difficult of the halides to displace. However, in aromatic nucleophilic substitutions, it is the easiest to displace. Why?
7. What is the driving force for the Baeyer–Villiger rearrangement (Section 6.4)?
8. The cyclohexadienone shown below rearranges to the phenol on treatment with acid (this is known as the dienone–phenol rearrangement). What is the mechanism of this reaction?

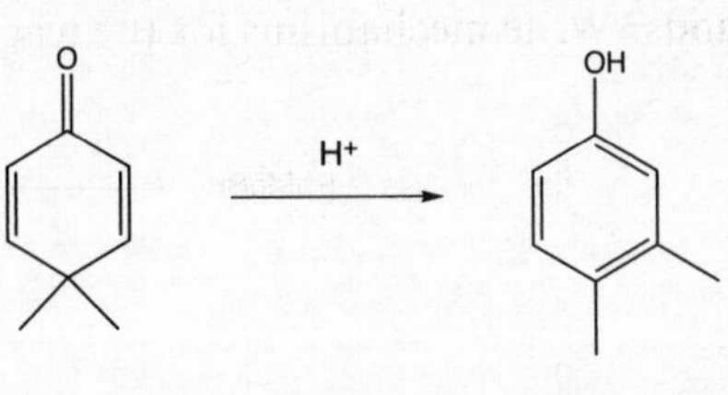

9. The following sequences of reactions give products (B and C) that are chemically identical. What is the structure of these products? Would you expect them to have the same or opposite optical rotations? What is the structure of intermediate A and what are the mechanisms of the individual steps?

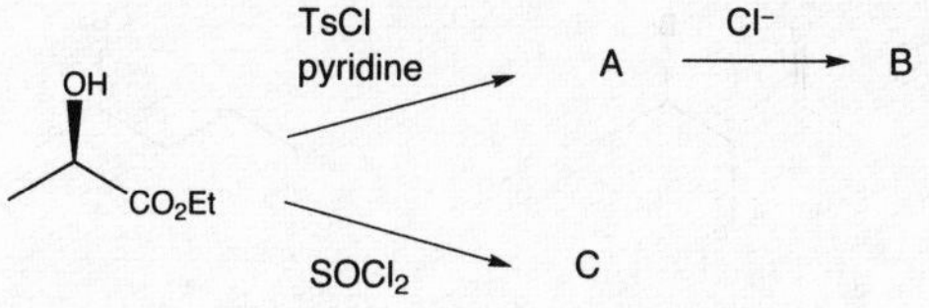

OH
CO₂Et

TsCl
pyridine → A $\xrightarrow{Cl^-}$ B

SOCl₂ → C

10. What would you expect to be the products of the following reactions?

a

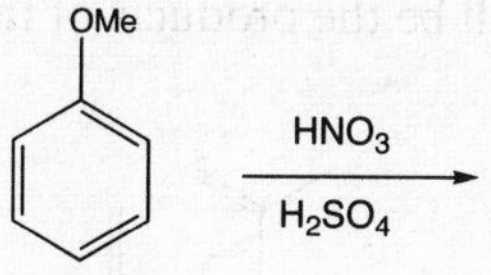

OMe

$\xrightarrow[H_2SO_4]{HNO_3}$

b

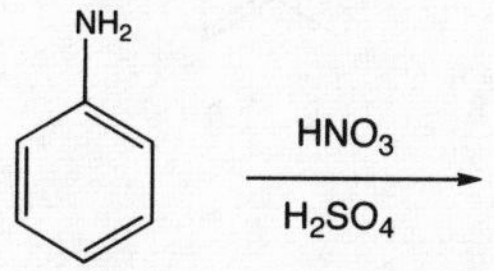

NH₂

$\xrightarrow[H_2SO_4]{HNO_3}$

c

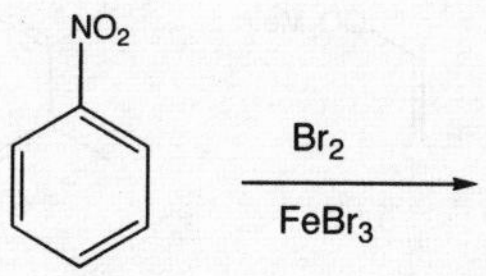

NO₂

$\xrightarrow[FeBr_3]{Br_2}$

d

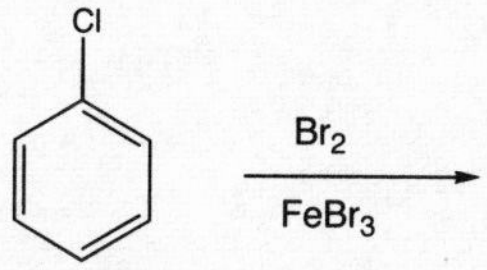

Cl

$\xrightarrow[FeBr_3]{Br_2}$

e

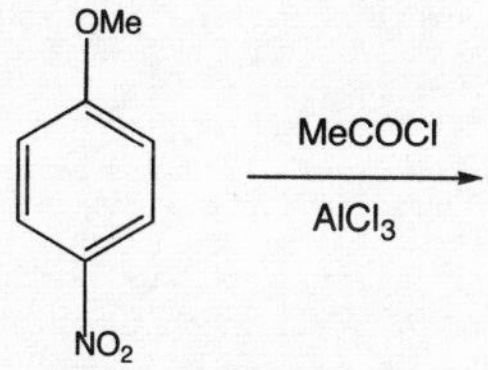

OMe

NO₂

$\xrightarrow[AlCl_3]{MeCOCl}$

f

Br

$\xrightarrow[AlCl_3]{MeCOCl}$

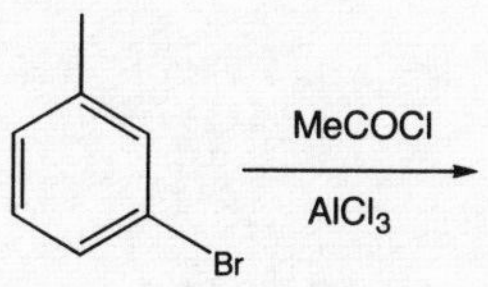

11. Which of the following compounds would you expect to give a greater ratio of elimination to substitution on treatment with sodium ethoxide? What would the effect on these ratios be of using potassium t-butoxide instead of sodium ethoxide?

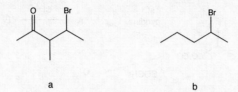

a   b

12. Arrange the following pairs of reagents in order of their rate of reaction with each other. What will be the products of these reactions?

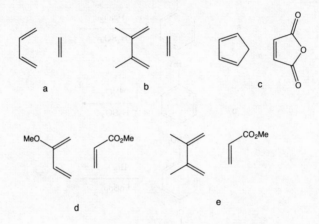

7

Techniques for investigating mechanisms

So far in this book, mechanisms of reactions have been presented as established truths. Although not all mechanisms are known with certainty, the mechanisms described here are generally thought to be accurate. Nonetheless, we should not forget that mechanisms of reactions are not handed down on tablets of stone, but are discovered by experiment. In this chapter we will look at some of the experimental techniques used for establishing mechanisms.

7.1 Basic principles

To specify the mechanism of a reaction, ideally we should know the precise positions of all atoms and the energy of all intermediates at every stage of the mechanism. In reality this ideal is seldom achieved, and we must usually be content with the gross features of a mechanism. Since we cannot look directly at atoms on the timescales of chemical reactions, we can never conclusively prove a mechanism to be correct. We can often prove mechanisms to be incorrect, and if we have ruled out all plausible alternatives to a particular idea, then this is usually seen as good evidence for it. This is not the same thing as proof, of course, since the true mechanism may be one that we have not considered.

To establish a mechanism with reasonable certainty, we normally have to carry out many experiments, as the result of any one experiment, even if it rules out some possible alternatives, will often be consistent with more than one mechanism. Quantum mechanical calculations have also been used to test the plausibility of reaction mechanisms. As computing power continues to increase, this is becoming more important. The calculations themselves are, however, extremely complicated, and are best left to theoretical chemists. We will restrict ourselves here to experimental methods.

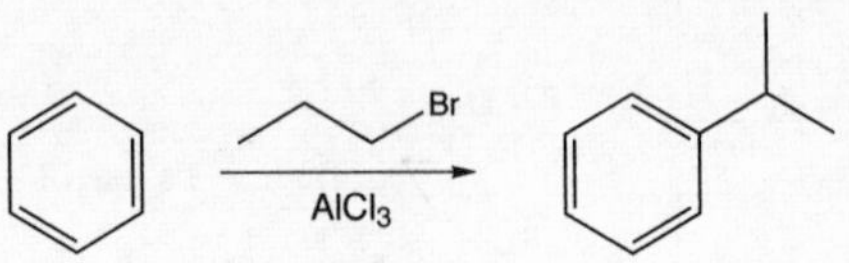

Figure 7.1

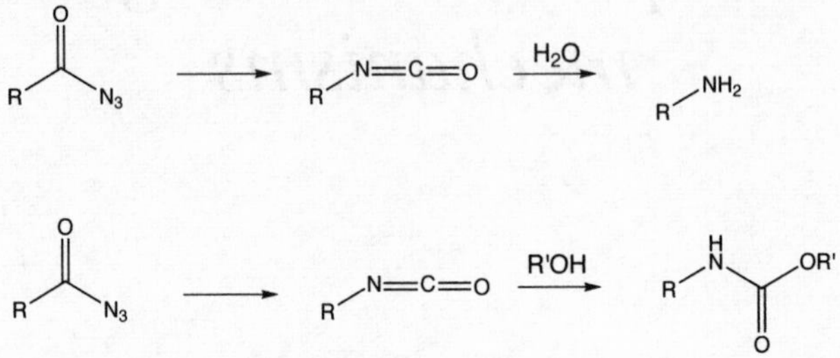

Figure 7.2

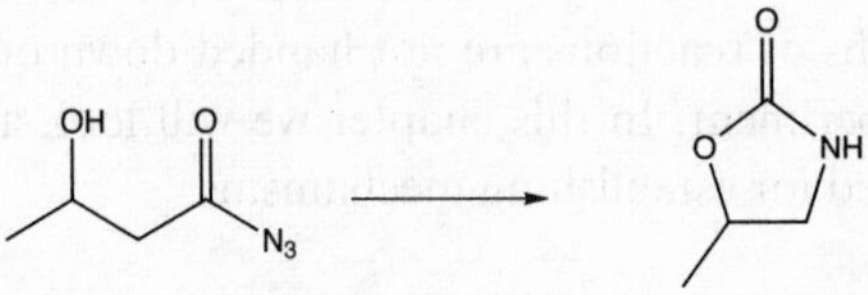

Figure 7.3

7.2 Clues from products

The first thing to consider is the nature of the products of a reaction. It is of course essential to be absolutely certain what the product of the reaction actually is. This may sound obvious, but some reactions may give unexpected products. For example, in the Friedel–Crafts reaction of 1-bromopropane with benzene (Section 6.3.6), the major product is isopropylbenzene, not n-propylbenzene (Figure 7.1). This rearrangement is an important piece of evidence for the intermediacy of the carbocation, since the 1-propyl cation is known to rearrange easily to the 2-propyl cation.

We may also find that some reactions give different products depending on the conditions. An example is the Curtius rearrangement (Section 5.2.5). When carried out in water, the product is an amine, but if an alcohol is used as a solvent, the corresponding carbamate is formed (Figure 7.2). A cyclic carbamate can even be formed if the substrate contains a hydroxyl group (Figure 7.3). The formation of all these products can be explained by the presence of an isocyanate, as this will react readily to give the various products (after decarboxylation if the nucleophile is water).

214

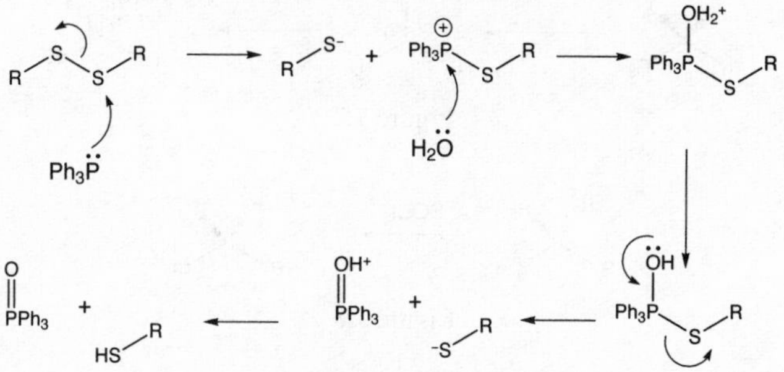

Figure 7.4

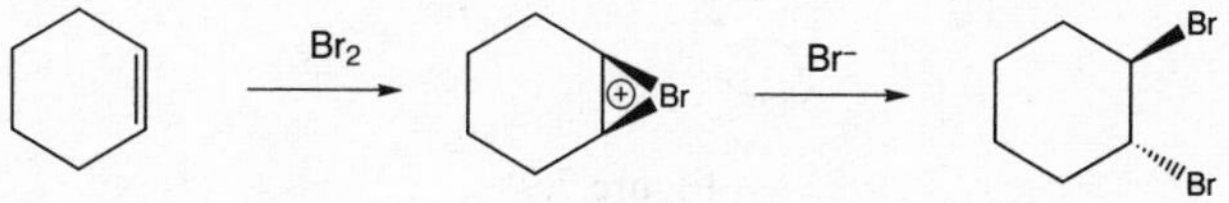

Figure 7.5

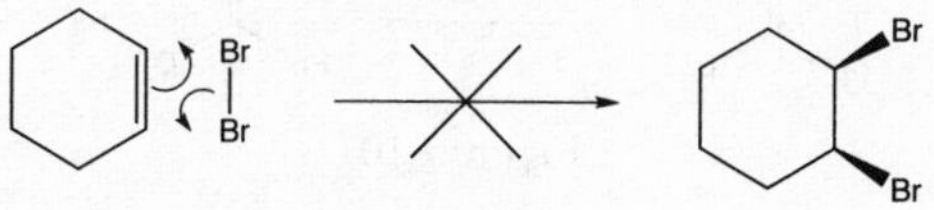

Figure 7.6

Not only are the desired products of interest, but the side products of a reaction can also provide important clues. For example, disulphides can be cleaved to thiols by treatment with triphenylphosphine in a solvent that contains water. An important clue to the mechanism of this reaction is that the triphenylphosphine is converted into triphenylphosphine oxide. This shows that the triphenylphosphine acts as a reducing agent, and allows us to postulate the mechanism shown in Figure 7.4 with increased confidence.

The stereochemistry of products can provide important clues to the mechanism. We saw in Section 6.1.3 that addition of bromine to olefins proceeds through a bromonium ion. An important piece of evidence for this mechanism is the addition of the two bromine atoms to opposite sides of the double bond, as can be seen in the bromination of cyclohexene (Figure 7.5). This rules out a mechanism in which the bromine adds directly to the bond in a pericyclic mechanism (Figure 7.6), as this would give *syn* addition.

If optically active compounds are used, we can often derive useful information from the stereochemistry of the product. If the product of the reaction is racemic, then this shows that it must proceed through a planar intermediate. The classic example of this is S_N1 substitution. The racemization observed in this reaction is a strong piece of evidence for the intermediate carbocation (Figure 7.7). (Remember,

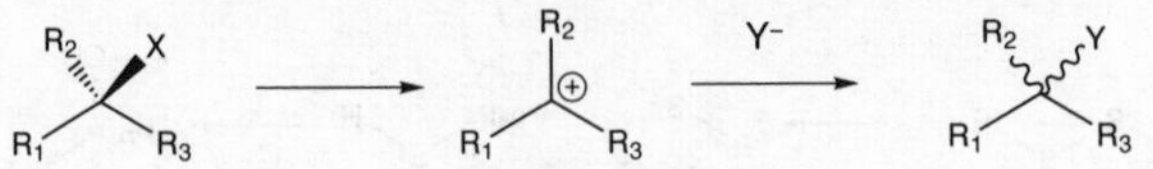

Figure 7.7

Figure 7.8

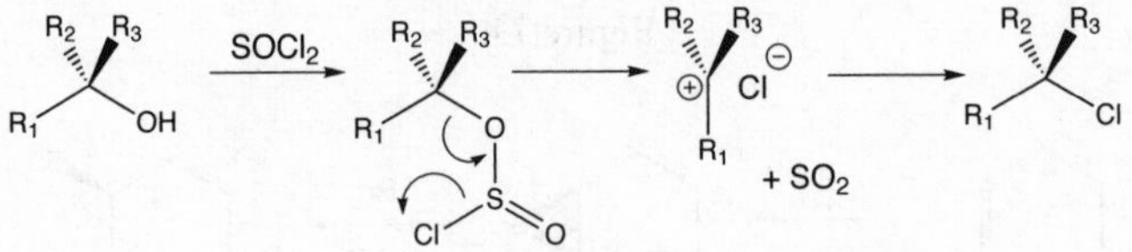

Figure 7.9

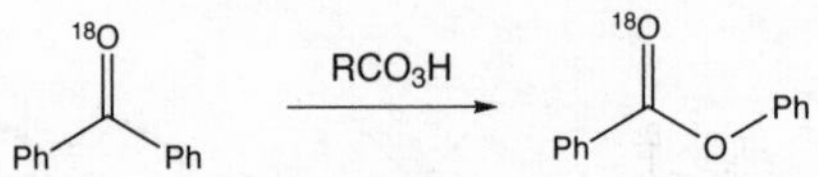

Figure 7.10

however, that S_N1 reactions sometimes proceed with partial inversion, because the carbocation intermediate is part of an intimate ion pair, as discussed in Section 6.3.3.)

Clearly, a mechanism that proceeds with inversion of configuration at a chiral centre is not the same as one proceeding with retention. Retention of configuration in the reaction of alcohols with thionyl chloride (Figure 7.8) shows us that this reaction is not a normal S_N2 inversion, and leads to the postulation of the S_Ni mechanism (Figure 7.9).

Much information about the mechanisms of reactions can be obtained from judicious use of isotopic labelling. An experiment that helped to elucidate the mechanism of the Baeyer–Villiger rearrangement (see Section 6.4) used benzophenone labelled with ^{18}O. The product of the reaction was labelled with ^{18}O entirely in the carbonyl position, with none of the label in the alkoxy oxygen (Figure 7.10). This ruled out a carboxylate intermediate, in which the label could become scrambled between the two positions, and strongly suggested the concerted mechanism described in Section 6.4.

Another example of isotopic labelling is the demonstration of the stereochemistry of the elimination of HBr from the norbornyl bromide shown in Figure 7.11. Without labelling, it is impossible to tell whether the elimination is *syn* or *anti*. However, deuterium labelling shows that this is a *syn* elimination; if the elimination

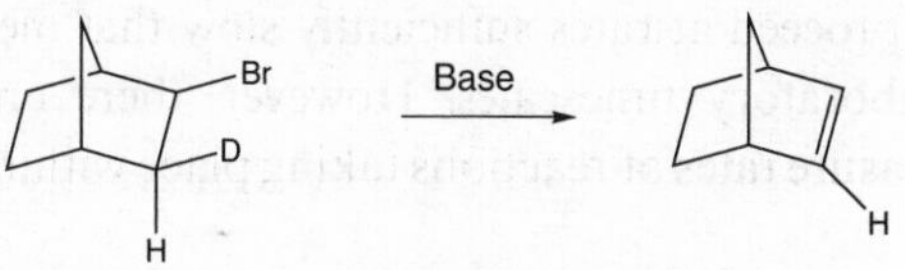

Figure 7.11

were *anti*, the product would contain deuterium. Recall from Section 6.2 that eliminations are usually *anti*, the reaction here is different because the rigidity of the molecule prevents the bromine adopting an antiperiplanar conformation to a hydrogen, but allows a synperiplanar conformation.

A wide range of isotopes, both radioactive and stable, may be used in mechanistic investigations. Some of the more commonly used radioisotopes are ^3H (tritium, T), ^{14}C, ^{32}P, ^{35}S, and ^{131}I. Stable isotopes that may be used for labelling include ^2H (deuterium, D), ^{13}C, ^{15}N, ^{17}O, and ^{18}O.

The presence of radioisotopes may easily be detected by measuring their radioactivity. This is a sensitive method, requiring only minute amounts of sample, but suffers from the disadvantage of revealing nothing about the position of the radioactive label within a molecule.

All the stable isotopes may be detected by mass spectrometry. Again, this is a sensitive method, but also does not give definitive information about the location of the label. However, the fragmentation patterns in the spectrum may give useful clues.

Some isotopes can be detected by NMR, such as deuterium, ^{13}C, ^{15}N, and ^{17}O (but not ^{18}O). This method is less sensitive, but has the advantage of telling us the position of the label within the molecule.

7.3 Kinetics

We can often obtain useful information by measuring the rate of a reaction and how it changes under different conditions. We can divide kinetic investigations into two types: firstly, those which discover the rate law for a reaction, and secondly, those which measure changes in the rate when various conditions are changed, such as the structure of the starting material or the nature of the solvent.

Rates can be measured in practice by a variety of methods. Either the rate of disappearance of the starting material or the rate of appearance of the product can be measured. One of the simplest ways to do this is spectroscopically. If either the product or the starting material has a characteristic UV absorption, this may be useful, as it can be continuously monitored to follow the course of the reaction. Reactions can also be run in NMR tubes and their NMR spectra recorded at various times as the reaction proceeds. A more laborious way of measuring rates is to quench the reaction at various times (for example by the addition of base if the reaction is acid-catalysed) and to analyse the reaction mixture. All these methods require reactions to proceed at rates sufficiently slow that measurements can take place on normal laboratory timescales. However, there are also more hi-tech methods that can measure rates of reactions taking place within milliseconds or even less.

Rate laws for reactions can be determined by plotting the concentrations of the various reagents against time. The more simple rate equations will give straight lines if plotted with appropriate transformation of the variables (for example logarithmic), the form of which will depend on the nature of the equation. For example, if the rate equation is of the form $r=k$ [A], or $-\mathrm{d}[\mathrm{A}]/\mathrm{d}t=k$ [A], then integration of the equation gives $\ln [\mathrm{A}]=-kt+C$. If a plot of ln [A] against time gives a straight line, this shows that the rate equation for this reaction is $r=k$ [A].

In general, the rate of a reaction will be proportional to the concentrations of the species that participate directly in the rate-determining step (RDS). If the rate of a reaction is measured at varying concentrations of the reagents used, a rate law will be obtained which should tell us about the species participating in the RDS. For example, if the rate of reaction is proportional to the concentration of reactant A times the concentration of reactant B, or $r=k$ [A][B], then it is likely that one molecule of A reacts with one molecule of B in the RDS. An example is an S_N2 displacement reaction, in which the rate is proportional to the concentration of the substrate multiplied by the concentration of the nucleophile, reflecting the participation of both of these in the RDS (here the only step). This is said to be an example of second-order kinetics, since the rate is proportional to two concentrations multiplied together. A reaction with a rate proportional to the square of the concentration of one reagent would also be said to have second-order kinetics. A reaction with first-order kinetics is proportional to only one concentration, for example the concentration of the substrate in an S_N1 reaction. Here, the substrate is the only molecule to appear in the rate equation, because it is the only molecule taking part in the RDS.

The order of a reaction is determined experimentally, and is the sum of the powers of the concentrations in the rate equation. First-order and second-order reactions are by far the most common, but third-order reactions, such as $r=k$ [A][B]^2, are sometimes encountered. It is even possible to have reactions with non-integer orders.

We should not confuse the order of a reaction with molecularity. Molecularity is the number of molecules taking part in a step of a reaction. This is a fundamental property of the reaction step. Reactions as a whole do not have molecularities unless they proceed in a single step; the term refers only to individual steps. Normally, however, the molecularity of the RDS will be the same as the order of the reaction.

Sadly, there are a number of factors that can complicate the interpretation of rate equations. A common pitfall is that the solvent may participate in the RDS, but does not appear in the rate equation because it is present in such large excess that its concentration is effectively constant. A reaction with a bimolecular RDS would

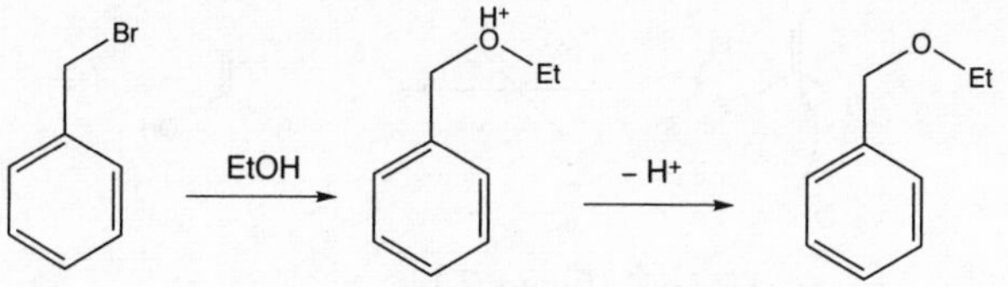

Figure 7.12

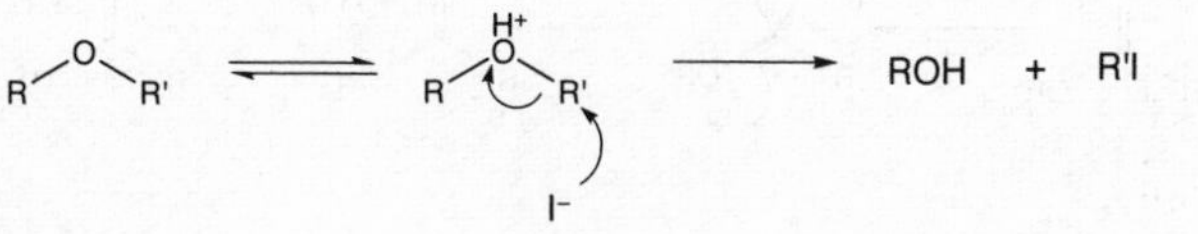

Figure 7.13

thus appear to have first-order kinetics. This is known as a pseudo first-order reaction. An example is an S_N2 displacement in which the solvent acts as the nucleophile, such as the ethanolysis of benzyl bromide (Figure 7.12). We cannot measure changes in the concentration of ethanol, because it is the solvent. The rate equation will therefore be $r=k$ [BnBr]. The reaction will thus appear to have first-order kinetics, but its RDS has a molecularity of 2.

Most organic reactions proceeding in more than one step will have one step that is clearly slower than any of the others and is therefore rate-determining. However, this is not a universal law, and it is possible for reactions to have two steps proceeding at roughly the same speed. If this happens, the rate equation is more complicated, and it is no longer possible to use it to draw simple conclusions about molecularity. Fortunately, such reactions are rare in standard organic chemistry, although they are much more common among reactions catalysed by enzymes.

Often, the RDS will not be the first step of a reaction, and the rate is then proportional to the concentration of an intermediate (and possibly other species as well). We cannot easily measure the concentration of the intermediate, so its concentration will appear in the rate equation indirectly. This also complicates the rate equation.

Sometimes there is a rapid equilibrium between the starting material and the intermediate. We can then obtain the concentration of the intermediate from the concentration of the starting material and any other species taking part in the equilibrium. An example of this is the cleavage of an ether by HI. The RDS is the S_N2 reaction between the iodide and the protonated ether (Figure 7.13). The protonated ether will be in equilibrium with the unprotonated form, and the concentration will be proportional to the concentration of the ether multiplied by the H^+ concentration. The overall rate equation will therefore be $r=k$ [ROR'][H^+][I^-]. This appears to be a third-order reaction, although the RDS is only bimolecular.

Although a rate law can give useful information about a reaction, it does not tell us the mechanism. A given rate law will sometimes be compatible with more than

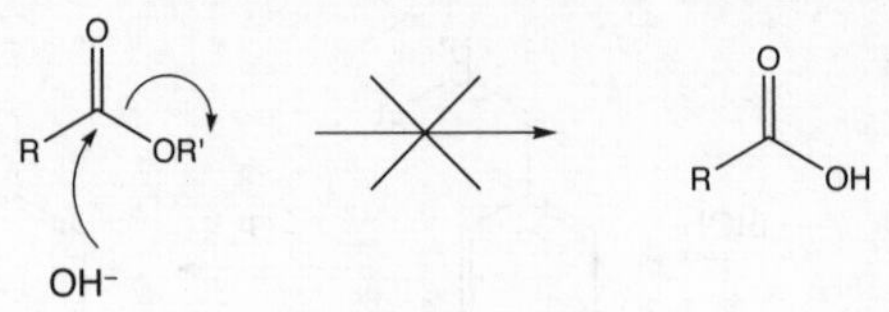

Figure 7.14

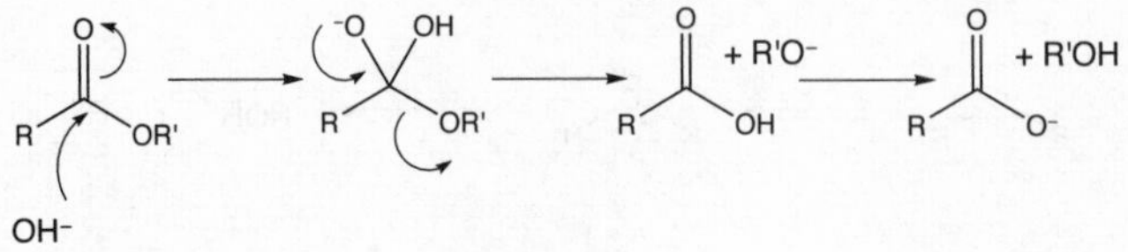

Figure 7.15

one mechanism. For example, basic hydrolysis of an ester has the rate law $r=k$ [ester][OH$^-$], which is the same as the rate law for an S_N2 displacement. This would therefore be compatible with direct displacement of the alkoxy group by hydroxide, as shown in Figure 7.14. However, we know from other evidence that displacement reactions at acyl carbons do not take place by this direct mechanism, but by the tetrahedral mechanism shown in Figure 7.15. The first step of this is rate-determining, and as this is a bimolecular process, the reaction overall obeys simple second-order kinetics, even though there are other steps in the reaction.

As stated above, knowing about rates of reaction is useful not only because of the information supplied by rate laws, but also because of how the rate is affected by various changes in the reaction. One such change that is often of interest is isotopic labelling of the reactants.

Compounds that differ only in their isotopic composition are often described as being chemically identical, but this is not strictly true. Quantum mechanical calculations show that a carbon–deuterium bond has a lower zero-point energy than a corresponding carbon–hydrogen bond, this being the residual vibrational energy of the bond when it has the minimum vibrational energy allowed by the laws of quantum mechanics (which cannot be exactly zero). Because of this difference in the zero-point energies, the carbon–deuterium bond is slightly stronger than the carbon–hydrogen bond. This is a general finding, applicable to bonds between hydrogen/deuterium and other atoms, or in fact to any pair of isotopes; thus a $C-^{18}O$ bond is stronger than a $C-^{16}O$ bond.

This difference in bond strengths means that if a $C-H$ bond is broken in the RDS of a reaction, the reaction will be slower if the starting material contains a $C-D$ bond at this position instead. This is known as a primary kinetic isotope effect. An example is E2 elimination, in which there is a primary kinetic isotope effect for the β-hydrogens with $k_H/k_D=3-5$, or in other words the reaction is 3–5 times faster when the reaction is carried out with hydrogen than when it is carried out with deuterium.

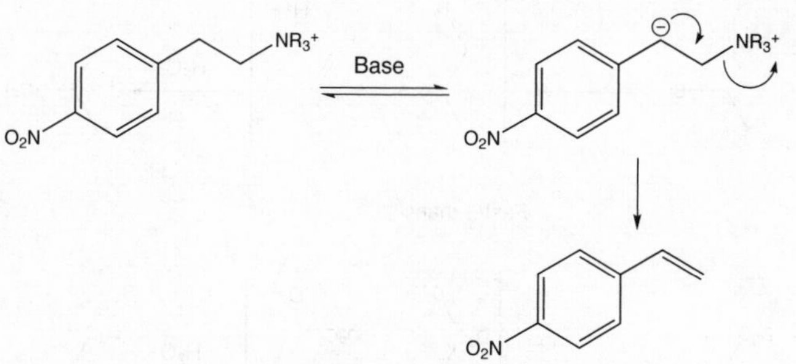

Figure 7.16

Primary kinetic isotope effects typically range from 1 (no isotope effect) to about 7 or 8, depending on the extent of C−H bond breaking in the RDS.

Inverse isotope effects are also known, in which k_H/k_D is less than 1. An example is in the base-catalysed elimination of the amine constituent in the quaternary ammonium compound shown in Figure 7.16, where $k_H/k_D=0.13$ for the water used as the solvent. This is consistent with an E1cB mechanism. When the carbanion intermediate is formed, it may eliminate the amine or it may revert to the starting material. If the reaction is carried out in D_2O, the return to the starting material will be slower, as an O−D bond must be broken instead of an O−H bond. This increases the concentration of the intermediate, which in turn increases the overall rate of reaction.

Primary kinetic isotope effects may be measured for isotopes other than hydrogen and deuterium, but more sensitive measurements are required because the differences are much smaller. A C^{12}/C^{13} isotope effect, for example, will typically be of the order of 1.02–1.10.

Isotope effects may be observed even when the bond to the hydrogen showing the effect is not itself broken during the reaction. These are known as secondary kinetic isotope effects. They are generally smaller than primary isotope effects, and are more difficult to interpret, so they are less useful in providing information about a reaction. In general, knowledge of a reaction mechanism is necessary to rationalize a secondary isotope effect, rather than the isotope effect providing useful information about the reaction mechanism. An example of a secondary isotope effect is found in the S_N1 hydrolysis of isopropyl bromide, which is marginally faster for the protonated compound. This is thought to be because of the less effective hyperconjugation in the transition state for the deuterated compound (Figure 7.17).

7.4 Intermediates

For reactions that proceed in more than one step, isolation of the intermediates may provide invaluable evidence about the course of the reaction. For example, a proposed intermediate in the Hofmann rearrangement (see Section 5.2.5) is an

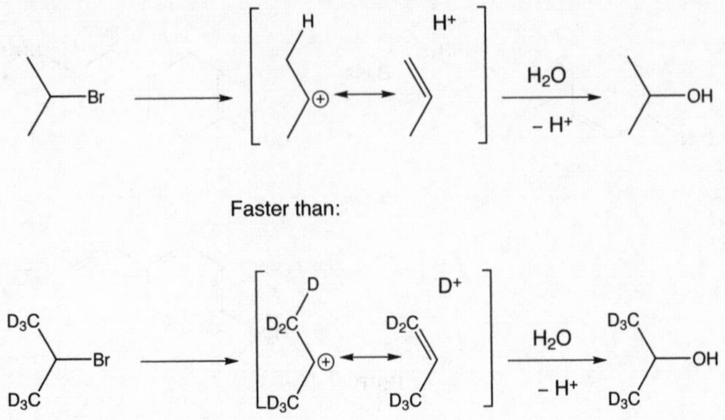

Figure 7.17

isocyanate. Whilst there is much indirect evidence for this, the best piece of evidence is that the isocyanate has been isolated.

Needless to say, as with any other experimental technique for investigating mechanisms, there are a number of pitfalls. Just because a compound that looks like an intermediate can be isolated from a reaction mixture, it does not prove that it really is an intermediate; it may be a side product. Conversely, if a proposed intermediate cannot be isolated from a reaction, this does not mean that it is not present; it may be that it is very short-lived.

An important test to show that a compound isolated from a reaction mixture is a true intermediate is to show that it can be converted to the normal products under the conditions of the reaction. Returning to our example of the isocyanate in the Hofmann rearrangement, this isocyanate does indeed give the expected amine when treated with water under the conditions of the reaction, thus strengthening the evidence that it is an intermediate. Of course, it still does not prove that it is the only intermediate; the reaction could proceed by more than one pathway, with the isocyanate on a minor route to the product. Nonetheless, if a suspected intermediate is shown not to be converted to the products under the conditions of the reaction, this provides unequivocal proof that it is not an intermediate. Once again, we see that reaction mechanisms are far easier to disprove than to prove.

The failure to isolate a postulated intermediate does not necessarily mean that we cannot obtain some evidence of its existence. One piece of evidence for the mechanism of the Fischer indole synthesis (Figure 7.18) is that the intermediate containing the aniline and imine groups has been detected in reaction mixtures by ^{13}C and ^{15}N spectroscopy. Another example is the Beckmann rearrangement (Figure 7.19), in which the nitrilium intermediate has been detected by UV spectroscopy.

A rather less direct form of evidence for intermediates can be provided by trapping experiments. Here, a reagent that would be expected to react rapidly with the postulated intermediate is added to the reaction. One important piece of evidence for the benzyne mechanism (Figure 7.20) is that addition of cyclopentadiene to the

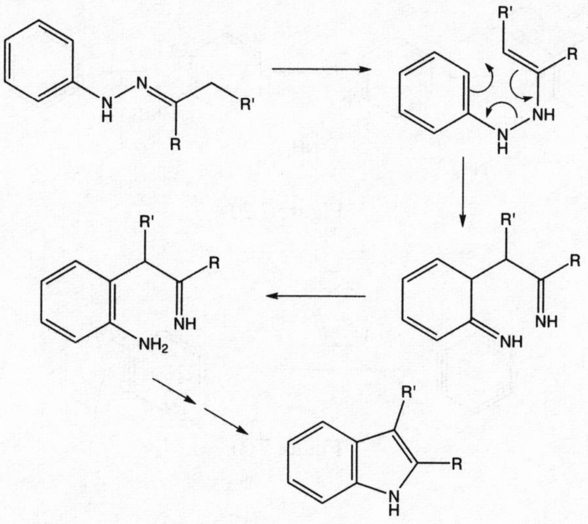

Figure 7.18

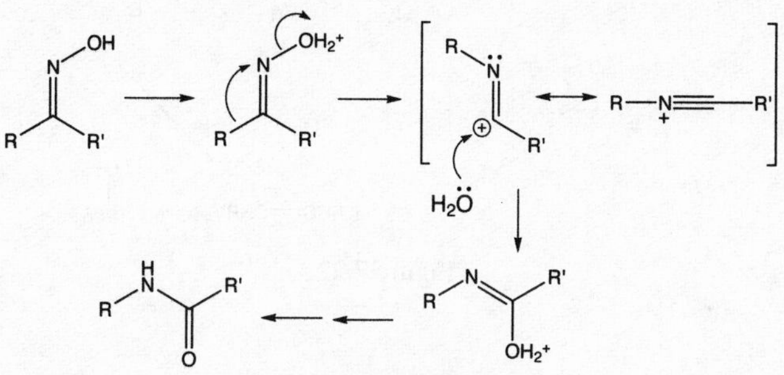

Figure 7.19

reaction medium gives the product shown in Figure 7.21, which presumably arises from a Diels–Alder reaction between cyclopentadiene and the benzyne.

We can also learn about intermediates by synthesizing them by a separate route to see if they can be converted to the expected products. Again, this does not prove that the postulated intermediates are true intermediates, but it can rule out compounds that are not. A piece of evidence that supports the presence of the betaine intermediate in the Wittig reaction is that the betaine may be prepared by reaction of triphenylphosphine with an epoxide. This then forms the expected olefin (Figure 7.22).

7.5 Other methods

There are many ways of investigating mechanisms in addition to those described above. One common technique is to carry out a series of reactions with different but

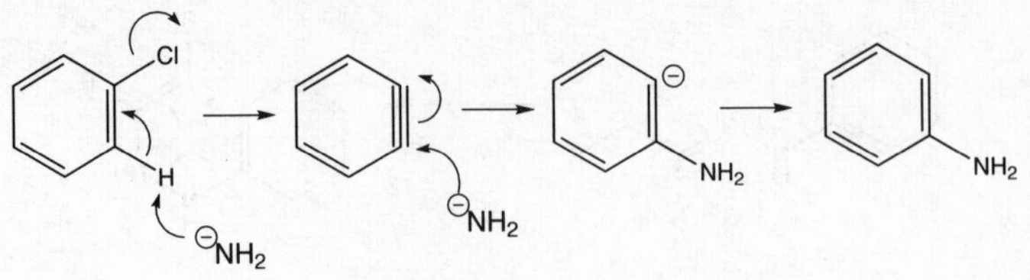

Figure 7.20

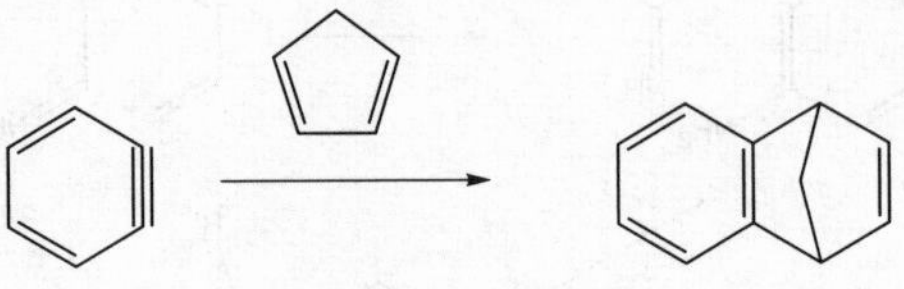

Figure 7.21

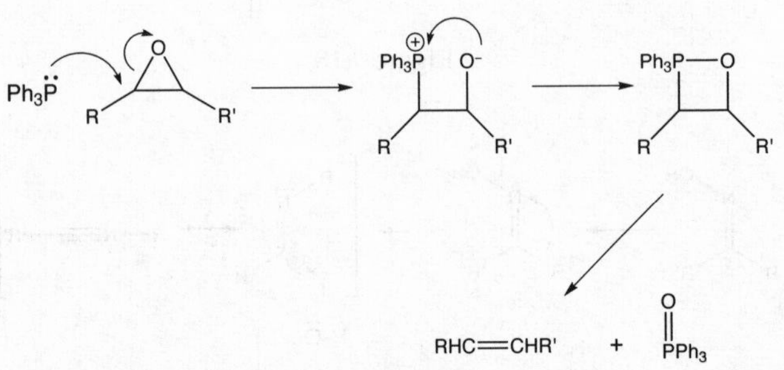

Figure 7.22

related starting materials to find out how the rate or the outcome of a reaction varies with the structure of the starting material.

Clearly we must be careful here, because even small structural changes may lead to different mechanisms coming into play. In Section 3.3.1 we saw one way in which the structure of the starting material may be varied with minimal effects on the reaction mechanism. This is where we change the substituents on an aromatic compound at a site remote from the reaction centre. The Hammett plots obtained in this way can provide useful information about the electronic demands of the transition state. As described in more detail in Section 3.3.1, a positive value for ρ means that there is a build-up of negative charge in the transition state for the RDS, whereas a negative ρ means that the transition state has some positively charged character.

We may look at changes in the structure of aliphatic molecules as well, but obviously we must be much more careful in altering substituents close to the reaction centre (which are of course the most interesting ones to vary) as they may change the mechanism of the reaction. We will look at some examples of changing structures of starting materials and the pitfalls that may accompany this in Chapter 9.

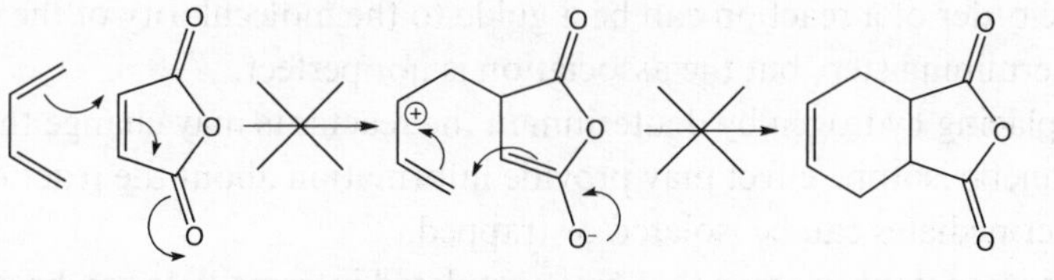

Figure 7.23

For example, the Favorskii reaction can go by different mechanisms, depending on the structure of the starting material.

Changing the conditions of a reaction can have a marked effect on the rate, and this can often give us useful information about the reaction. For example, if a reaction has polar intermediates, it is likely to be faster in polar solvents. Part of the evidence for the formation of carbocations in S_N1 reactions is that the reactions are faster in solvents that promote ionization, such as water. In contrast, the Diels–Alder reaction is unusually insensitive to the effects of changing the solvent. This is one of the main pieces of evidence that led to the proposal of a concerted pathway with no polar intermediates. Were the reaction faster in polar solvents, then it would be more likely to proceed by an ionic intermediate (Figure 7.23).

By changing the conditions, it is usually fairly easy to discover whether a reaction goes by a free radical pathway. If addition of a radical inhibitor, such as hydroquinone, slows a reaction, then this is evidence that free radicals are among the intermediates. Similarly, free radical reactions are normally faster in the presence of a radical initiator such as AIBN (see Section 4.3). Once again, radical reactions are usually insensitive to changes in the polarity of the solvent.

It is often useful to know whether a reaction is catalysed by acid or base, and this is of course straightforward to discover by measuring the rate of reaction at different pHs.

Many ingenious methods have been used in the investigation of mechanisms. Careful observation is the essence of scientific discovery, and any fact about a reaction may provide information about its mechanism if the experimenter is sufficiently alert. We will see in Chapter 9 how some mechanisms were discovered in practice.

Summary

- We can never prove a reaction mechanism to be correct, we can only rule out alternative mechanisms.
- Before speculating about the mechanism of a reaction, we must always be sure what the product is.
- Side products of reactions can often give useful clues to the mechanism.
- Isotopic labelling can be used to track the fate of individual atoms.
- The rate law of a reaction can often give us useful information.

- The order of a reaction can be a guide to the molecularity of the rate-determining step, but the association is not perfect.
- Replacing hydrogen by deuterium in the reactants may change the rate; such a kinetic isotope effect may provide information about the reaction.
- Intermediates can be isolated or trapped.
- It is important to show that any postulated intermediate can be converted to products under the normal conditions of the reaction.
- Postulated intermediates can be synthesized by a different route to see if they give the same products as the original reaction.
- Changing the structure of starting materials or the reaction conditions may give useful information, but the results should be interpreted with caution.
- Any experimental observation may provide a useful clue to an alert observer.

Problems

1. We know that it is generally the acyl–oxygen bond that is cleaved in ester hydrolysis rather than the alkyl–oxygen bond. One of the main pieces of experimental evidence for this came from an experiment that used ^{18}O. How might such an experiment be carried out? What result would you expect, and how would this be different if the alkyl–oxygen bond were cleaved?

2. When the alkyl group of an ester is tertiary, hydrolysis under acidic conditions takes place with alkyl–oxygen cleavage. Could this be distinguished from the above mechanism kinetically? Apart from ^{18}O labelling, what other evidence might be used to verify this mechanism?

3. Treatment of methyl α-bromopropionate with hydrazine gives a mixture of products as shown below. A possible mechanism for the reduction is that the hydrazine function is somehow oxidized to an azo function, which then undergoes an elimination reaction to give the enolate, and hence the reduced product. Suggest a way of discovering whether this mechanism is possible.

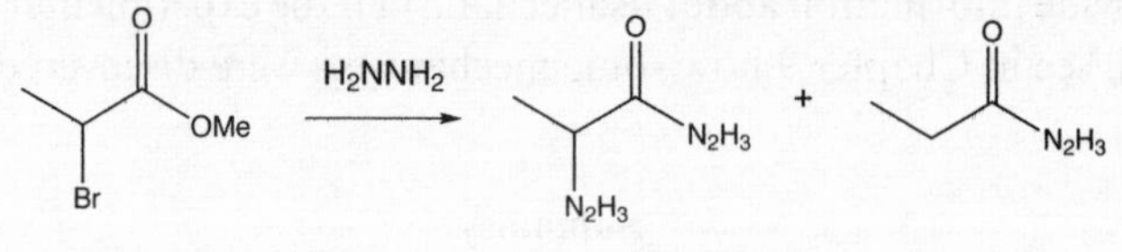

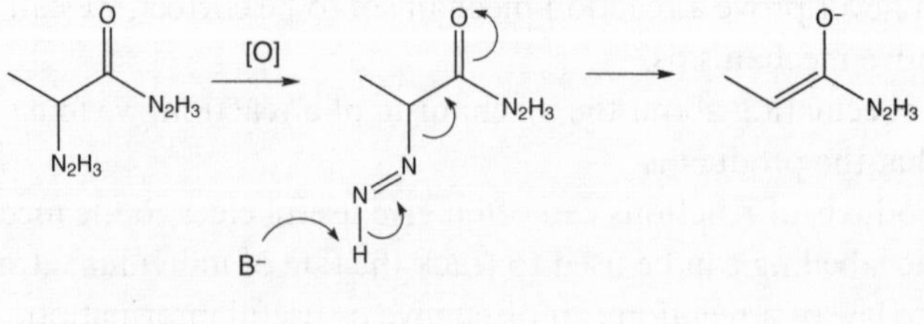

4. When the reaction in Problem 3 was carried out in the presence of azobenzene, the azobenzene was rapidly reduced to hydrazobenzene. What does this tell us about a likely side product of the reaction?

5. When the tosylate shown below undergoes displacement in acetic acid, the product has the stereochemistry shown, and is a racemic mixture. Is this consistent with a straightforward S_N2 displacement of the tosylate group? The reaction is considerably faster if there is a *p*-methoxy group on the aromatic ring. What does this tell us about a likely mechanism?

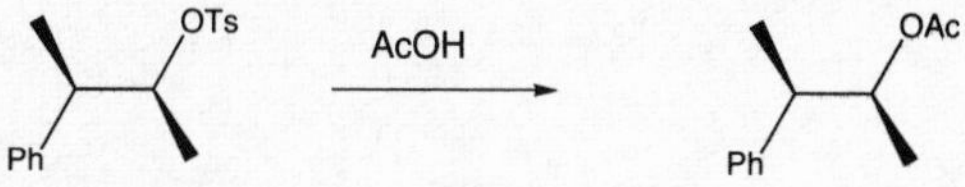

6. When acetone is brominated, the rate of reaction is independent of the concentration of bromine. Deuteriated acetone is brominated more slowly than normal acetone. What do these facts tell us about the mechanism of the reaction?

8

How to suggest mechanisms

A common question in undergraduate examinations is 'suggest a mechanism for the following reaction', followed by some unfamiliar and frequently rather daunting reaction. The purpose of this chapter is to guide you through the thought processes required for this task.

Suggesting mechanisms for an unknown reaction may seem scary at first, but there are in fact a number of questions you can ask yourself about any reaction that should lead to the mechanism. In this chapter we will look at the clues that can help you to suggest a plausible mechanism, and we shall study some examples of how these clues can be used in practice.

8.1 Introduction

Being able to suggest sensible mechanisms for reactions is useful not only in exams, but frequently in 'real-life' chemistry as well. A chemist may carry out a reaction with the intention of effecting a particular transformation only to find that something completely unexpected happens. This unexpected reaction may prove to be synthetically useful in its own right and the chemist may wish to develop it further, but in any case he or she will probably still wish to carry out the transformation originally intended. A knowledge of the mechanism of this new reaction will frequently be of invaluable assistance for either of these purposes.

We saw in the last chapter that we can never be certain of a reaction mechanism, although some reactions can be specified with far more precision than others. This should be borne in mind when attempting to suggest a mechanism for a reaction on paper. It is often possible to write more than one plausible mechanism that is consistent with the data available. To distinguish between these possibilities further information may be required; this may be obtained by reference to relevant literature or from further experiments. This is not possible in an examination, of course, although experiments can always be suggested, together with how their results might be interpreted. Obviously, if more data are available, the mechanism can be specified more precisely.

8.2 Types of clues

8.2.1 Basic principles

We saw in the last chapter how conclusions may be drawn about mechanisms from experimental evidence. In suggesting a mechanism for any reaction, therefore, we should start by being clear about what evidence is available. Any experimental fact about the reaction may provide useful clues. We should be interested in not only the starting material and products, but also in any side products, the solvent, whether any catalyst is required (acid or base for example), the temperature, etc.

Many reactions take place in a number of steps, and so to write a mechanism for the reaction we in fact need to write several mechanisms. When looking at the starting material for a reaction, it is not always at all obvious what the first step on the pathway to the product might be. However, we can use a number of principles to help us suggest plausible steps. Although a reaction sequence may not be readily apparent to start with, the product of one step of a reaction might suggest the next step, and so on, until a logical sequence is formed leading to the product. Sometimes, we may get ideas by working backwards from the product; we may be able to make an educated guess about what the final intermediate might be by considering likely immediate precursors to the product. Of course, we will not always be so lucky, and it may be that the products of a suggested step have no sensible reaction pathways open to them; we may then have to go back a step or two and look at other possibilities. Let us now look at some of the clues that may guide us towards sensible pathways.

8.2.2 Consideration of the carbon skeleton

The most important steps of a reaction are generally those in which carbon–carbon bonds are formed or broken, and so normally the first thing to look at is whether the carbon skeleton of the molecule has changed during the reaction. If it has not, then it will normally be a relatively straightforward task to come up with a mechanism.

We should start by counting the number of carbon atoms in the starting material and in the product. This may sound like a rather elementary step that is beneath sophisticated organic chemists, but the importance of counting carbon atoms should never be underestimated. It is all too easy to lose count of the number of atoms, and so miss an interesting rearrangement or some other important step in a reaction sequence.

If the carbon skeletons of the starting material and product are reasonably similar it should be simple to account for all the carbons. However, if a considerable change

229

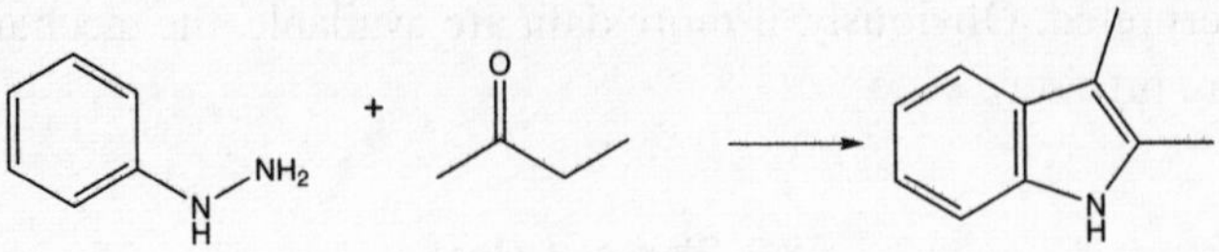

Figure 8.1

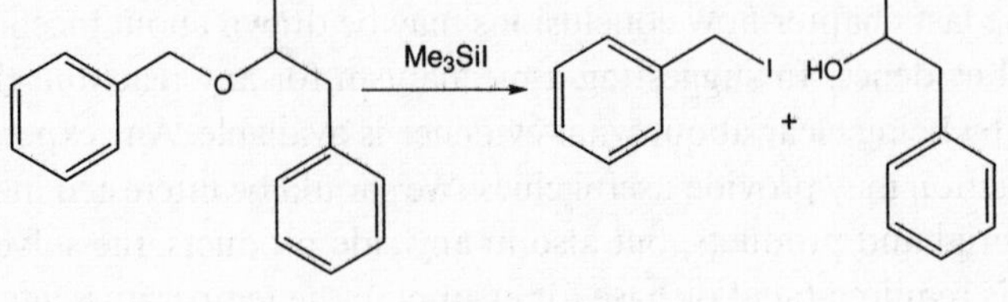

Figure 8.2

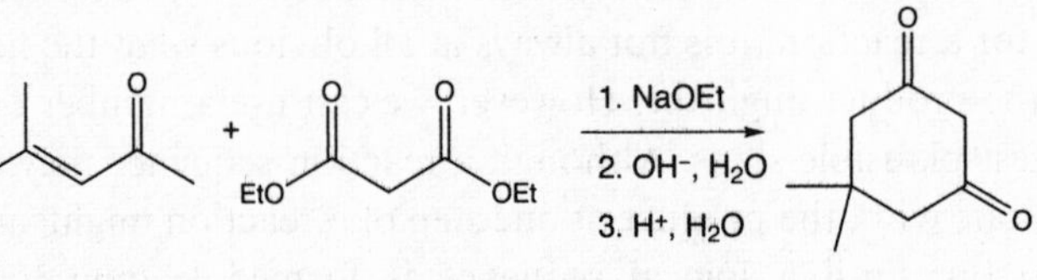

Figure 8.3

in the carbon skeleton has taken place in the reaction, it is often useful to write down a numbering system for the carbon skeleton of the product and to try to relate these numbers to the carbons of the starting materials (or vice versa). This can often help to show where all the carbons come from and to make sure they are all accounted for. We will look at an example of this in Section 8.3.5.

If the product and starting material have equal numbers of carbon atoms and their arrangements are the same, then it is usually safe to conclude that only a simple functional group interconversion has taken place, with no carbon–carbon bonds being made or broken.

Even if the starting material and products have different numbers of carbons, then there still may be no change in the carbon framework, provided that all the carbons are accounted for and portions of the carbon framework joined by heteroatoms remain intact. Compare the two reactions in Figures 8.1 and 8.2. In the first of these (the Fischer indole synthesis, see Section 5.2.4) we can see that the carbon skeleton has changed in going from starting material to product, and so we must expect to find a step in the reaction in which a carbon–carbon bond is made. In the second, however, we can see that all that has changed is that the ether group has been cleaved, and both portions of the carbon framework remain intact.

In Figure 8.3 we see that the product has one fewer carbon atom than the total in the carbon skeletons of the starting materials (disregarding the ethoxy groups, which are obviously lost). We should certainly expect to find some carbon–carbon bond forming steps here, but we must also lose a carbon atom at some stage. One of the most common ways in which a single carbon atom is lost is by decarboxylation, and this is what happens here, as we shall see in Section 8.3.5.

8.2.3 Ionic, radical, or pericyclic?

Reactions may be divided into ionic, radical, and pericyclic. It is always a good starting point if we know in advance which of these our mechanism is likely to be. Of course, if a reaction proceeds in many steps, it is possible for different steps to be of different types.

The overwhelming majority of organic reactions are ionic and it is usually sensible to assume that an unknown reaction has an ionic mechanism unless there is evidence to the contrary. We saw in Section 7.5 some of the experimental evidence that may suggest a radical mechanism. If a reaction is run in the presence of a known radical initiator, such as AIBN, then this is one such piece of evidence. Halogens give radicals on exposure to UV light, so the presence of halogens and UV may suggest a radical mechanism. Iodine is more likely to take part in radical mechanisms than chlorine or bromine.

It is perhaps less easy to describe the clues that should alert one to a pericyclic reaction, although these are not too difficult to recognize with practice. The main types of pericyclic reaction were discussed in Section 6.5. Familiarity with these will make it less likely that a pericyclic step in a reaction will be overlooked.

Sometimes it is not possible to say whether a reaction is ionic, radical, or even pericyclic, and it may be possible to draw plausible mechanisms from each of these categories. Further experiments may help clarify which mechanism is operating, but it may sometimes be difficult to distinguish between them even when much experimental evidence is available.

8.2.4 Identifying nucleophiles and electrophiles

If we assume that a reaction, or at least a step of a reaction, is ionic, then the reaction will consist of a nucleophile reacting with an electrophile (or possibly an acid reacting with a base, which is no more than a variation on this theme). Identifying species that are likely to react as nucleophiles and electrophiles is useful in predicting what the first step of the reaction might be.

Let us take the reaction we encountered above (Figure 8.2), in which an ether is cleaved in the presence of trimethylsilyl (TMS) iodide. We have already established that it is unlikely that any carbon–carbon bonds will be made or broken in the reaction, and the next step is to look at likely electrophiles and nucleophiles.

TMS iodide is likely to react as an electrophile. Even if we were not familiar with the chemistry of this compound, we could predict this by analogy with alkyl iodides

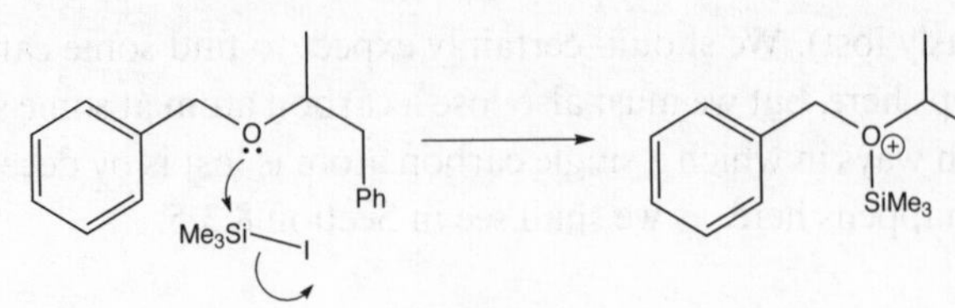

Figure 8.4

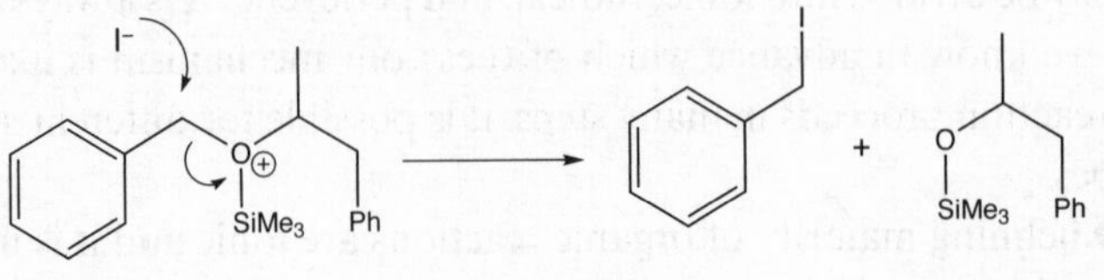

Figure 8.5

(silicon is directly below carbon in the periodic table). We need a nucleophile to react with this; since the reaction takes place at the ether group, it would make sense for the ether oxygen to act as the nucleophile. We saw in Section 5.1.2 that this is one of the few reactions we can expect from ethers. A plausible suggestion for the first step of this reaction would therefore be displacement of the iodide at the silicon by the ether oxygen, as shown in Figure 8.4. A further clue that should point us towards this mechanism is that it forms a silicon–oxygen bond. We know that silicon has an affinity with oxygen, and a knowledge of natural affinities such as this can often help us in suggesting mechanisms.

This clearly does not give us the complete mechanism leading to the final product of the reaction, but this should not worry us at this stage. What is important is that we have suggested a plausible first step, in which a compound known to act as an electrophile reacts with a group known to be nucleophilic, and furthermore in which a strong bond is formed (silicon–oxygen). If we can suggest sensible steps, we are making progress. There is no need to be able to see the entire mechanism straight away, as we will get there in the end by considering each step separately.

We now need to think about what happens next, and again we can look at the electrophiles and nucleophiles present. The first step liberated iodide ion, which is an excellent nucleophile. We also have an oxonium ion; the positively charged oxygen is now a good leaving group for a displacement reaction at one of the attached carbons, so here is our electrophile. We can now postulate the next step as being attack of the iodide on the benzylic carbon (Figure 8.5). We choose the benzylic carbon for two reasons: firstly, because benzylic positions are more reactive in S_N2 displacements than secondary alkyl groups, and secondly, and perhaps more importantly, because attack on this position is required to give the products of the reaction as shown. We should never forget that the products can give useful clues to the reaction mechanism.

Hydrolysis of the TMS group on acid work-up completes the reaction.

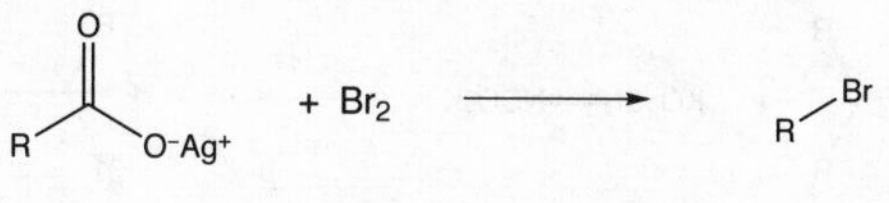

Figure 8.6

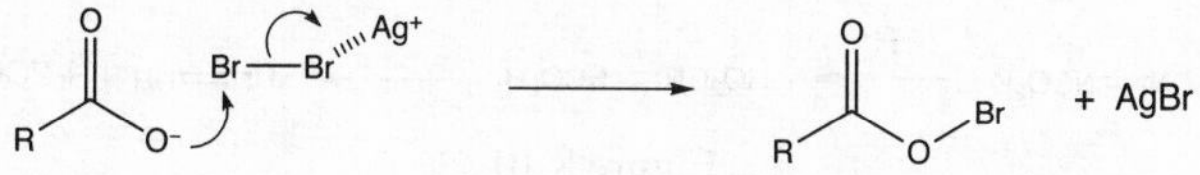

Figure 8.7

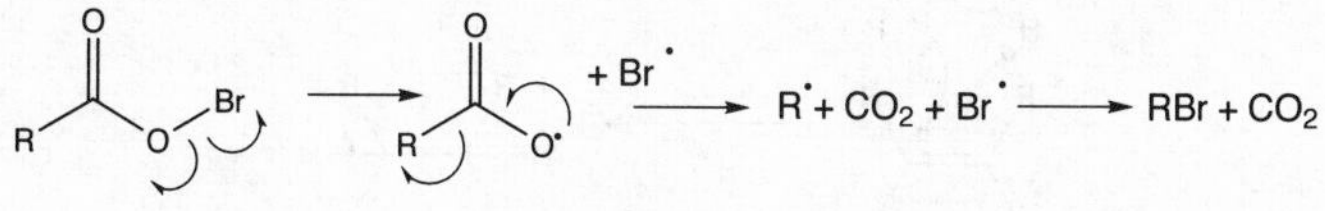

Figure 8.8

8.2.5 Common reaction patterns

In the above example we made use of the known affinity between silicon and oxygen. There are many examples of likely reaction patterns that can give clues to reactions. If we see combinations of elements or functional groups that seem likely to react together, then we should make use of this information.

As a further example of this, let us consider the reaction shown in Figure 8.6, which is known as the Hunsdiecker reaction. The first thing we should notice here is that we have a silver salt and we have bromine. It would be astonishing if we were not to have silver bromide formed in one step of the reaction. Silver forms stable insoluble salts with halides; this is just the sort of combination we should be looking for. Formation of silver bromide from the starting materials would lead to the acyl hypohalite, as shown in Figure 8.7. A plausible mechanism for this step would be attack on the bromine by the carboxyl oxygen, the silver acting as a Lewis acid, although the actual mechanism of this step is not known with certainty. The remainder of the reaction proceeds by a radical process (Figure 8.8), which we might have predicted because of the weak oxygen–bromine bond, which is easily cleaved homolytically. A carbon atom count will alert us to the need for decarboxylation.

8.2.6 Thermodynamic driving force

A similar idea to looking for known affinities between atoms is to look for the thermodynamic force that drives a reaction. In the Hunsdiecker reaction the driving force is largely due to the formation of the stable silver bromide, but other thermodynamic clues may give us information about a reaction mechanism.

For example, when potassium azodicarboxylate is treated with acid in the

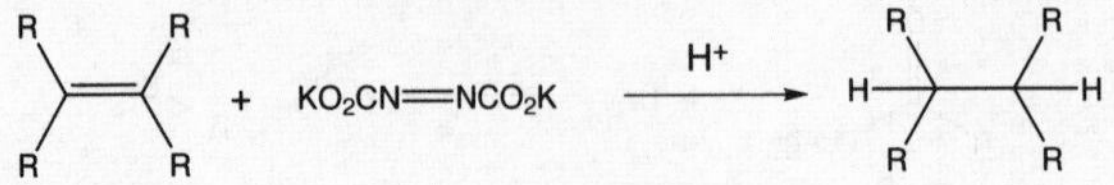

Figure 8.9

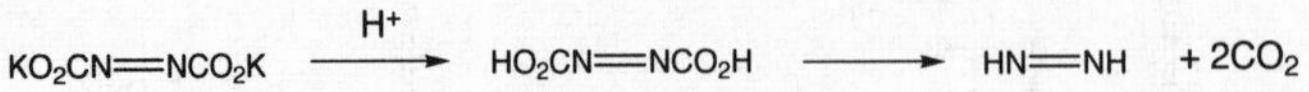

Figure 8.10

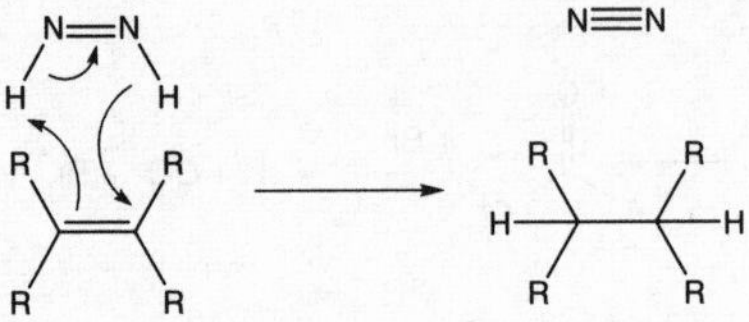

Figure 8.11

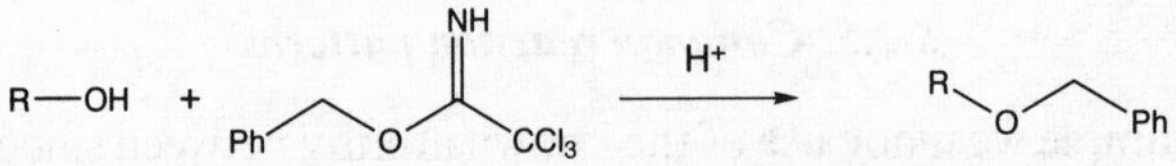

Figure 8.12

presence of an olefin, the olefin is reduced (Figure 8.9). A clue to the mechanism here comes from the nitrogens. The molecule we start out with bears enough resemblance to molecular nitrogen that we might imagine its formation during the reaction. This would provide a powerful thermodynamic driving force for the reaction, and so is likely to happen. Bearing this in mind, it becomes easier to suggest a mechanism, as we are now looking for a way of forming nitrogen, as well as reducing the double bond. We must therefore first get rid of the carboxylate groups, which is easily done, as carboxylate groups attached to nitrogen decarboxylate on treatment with acid (Figure 8.10). The resulting diimide can transfer its hydrogens directly to the double bond to give the expected product and molecular nitrogen. This is most likely to be by a pericyclic mechanism (Figure 8.11).

8.2.7 Conversion of starting materials into reactive species

Sometimes there may not be obvious nucleophiles or electrophiles in the reaction mixture as given, but they may be formed by the action of acids or bases. If a reaction is carried out in acid, we should always be alert to the possibility that the first step may be a protonation, or a deprotonation if a base is present.

Treatment of alcohols with benzyl trichloroacetimidate in the presence of an acid catalyst (Figure 8.12) is a common procedure for their protection as benzyl ethers

234

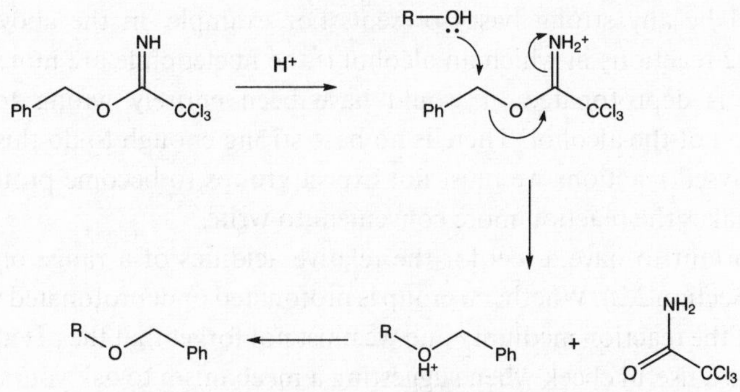

Figure 8.13

(which can subsequently be cleaved with trimethylsilyl iodide, as described above). We can see easily enough that no carbon–carbon bonds have been formed or broken, and so the reaction is just a functional group interconversion. When we look for the available nucleophiles and electrophiles, we see that the alcohol may be a nucleophile, but there is no obvious electrophile. However, since the reaction takes place under acidic conditions, it is reasonable to expect that some group may be protonated, which could make it a better electrophile. One possibility is that the alcohol could be protonated. This would give us an electrophile, as water could then be displaced from the alcohol in either an S_N1 or S_N2 reaction, depending on the structure of the alcohol. This does not help us, however, as the only nucleophile to react with this protonated alcohol is the unprotonated alcohol, and we need to involve the benzyl trichloroacetimidate in the reaction.

You will often find that seemingly good ideas in suggesting reaction mechanisms turn out to be unproductive, as we have just found for the idea of protonating the alcohol. When first getting the feel for suggesting mechanisms, it is often a good idea to write down several ideas for a first step, if they all seem plausible. It will soon become apparent that some of these ideas do not lead anywhere, and then attention can turn to the remaining ideas. As your experience and confidence grows, it will be easier to see at an early stage which ideas are likely to be unproductive.

If instead we protonate the nitrogen of the benzyl trichloroacetimidate (a more sensible protonation anyway, as nitrogens are normally more basic than oxygens), we now have a suitable electrophile. The protonated trichloroacetimidate group is now a reasonable leaving group, and we can write an S_N2 displacement at the benzylic carbon (Figure 8.13).

8.2.8 The importance of pH

In any reaction with acid or base catalysis, it is important always to keep in mind the nature of the catalysis. If a reaction is catalysed by acid, we must remember that there cannot be any strong bases present. For example, in the above reaction, although S_N2 reactions in which an alcohol is the nucleophile are more efficient if the alcohol is deprotonated, it would have been entirely wrong to postulate deprotonation of the alcohol. There is no base strong enough to do this. Similarly, in base-catalysed reactions we must not expect groups to become protonated just because it makes the reaction more convenient to write.

It is important to have a feel for the relative acidities of a range of functional groups (see Section 2.2). Whether a group is protonated or deprotonated will depend on the pH of the reaction medium,[1] and we must not forget that the pH applies to all reagents. It is a useful check when suggesting a mechanism to ask yourself whether the reaction takes place under acidic, basic, or neutral conditions, and to make sure that all the reagents and intermediates are compatible with these conditions.

8.3 Worked examples

8.3.1 General principles

We shall now look at some reactions with apparently complex mechanisms and examine the thought processes needed to come up with plausible mechanisms for them. If you have assimilated all the material in the book so far, you should now have a good understanding of reaction mechanisms, which should, in theory, allow you to predict the mechanisms of these reactions with ease. However, if you still feel that your grasp of reaction mechanisms is less than perfect, do not worry. As with any other skill, applying your knowledge of reaction mechanisms to real-life examples takes practice. These examples should give you some practice, as will the problems at the end of the chapter. The next chapter will also help you to discover how a knowledge of reaction mechanisms is used in real life.

You will get more out of the following examples if you try to stay one step ahead of the text, in other words if you suggest some ideas of your own before reading each stage in the thought processes described. If you find this difficult, you may find it helpful to look back again at the last section. Do not be too disheartened if at first you find it difficult to know where to start. Thinking of mechanisms for reactions is a skill that develops with familiarity, and as you become more experienced in organic chemistry your confidence and expertise will grow.

8.3.2 The formation of benzoic anhydride from benzoyl chloride

When benzoyl chloride is dissolved in acetone and treated with pyridine and a small amount of water, benzoic anhydride is formed (Figure 8.14). How are we to go about suggesting a mechanism for this reaction?

---

[1] Strictly, pH refers only to aqueous media, but we use the term loosely here to mean the general level of acidity or basicity in whatever solvent we are using.

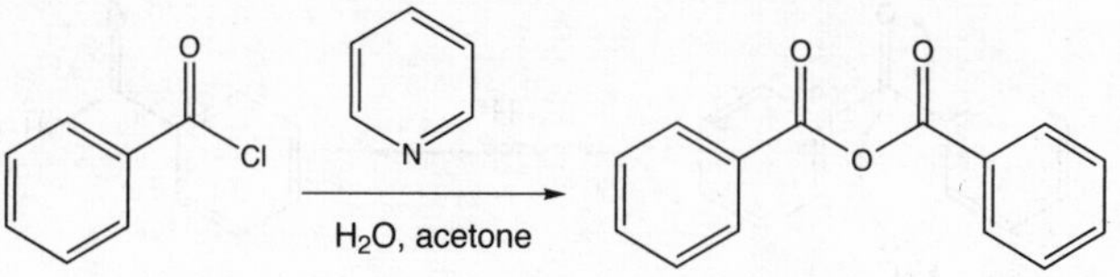

Figure 8.14

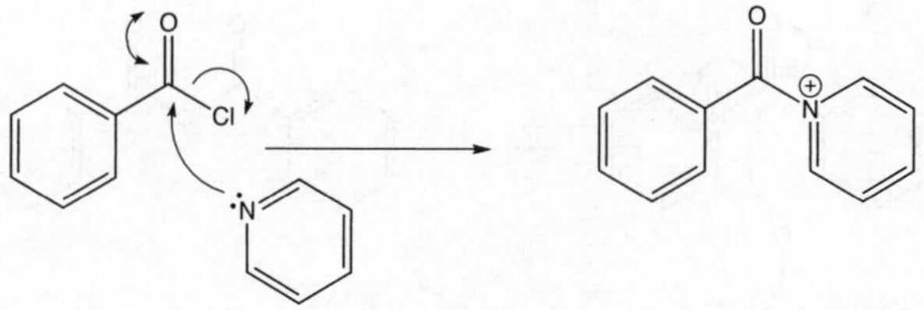

Figure 8.15

The first thing we should notice is that there has been no change in the carbon skeleton. The product has fourteen carbons, while the starting material has only seven, but if we look at the product carefully, we will see that it is composed of two halves, each of which has the same carbon skeleton as the starting material. We are therefore looking only for functional group transformations.

There is nothing about this reaction suggestive of a radical or pericyclic mechanism, so we are probably safe in assuming that it is ionic.

Our next step should be to look for obvious nucleophiles and electrophiles. We do indeed have a good electrophile here, namely benzoyl chloride. We should perhaps be aware that the solvent, acetone, could also act as an electrophile, although if we look at the structure of the product we see no part that appears to have come from acetone, so we can probably ignore the solvent this time.

We have two molecules that can act as nucleophiles, pyridine and water. A sensible first step would be a reaction between benzoyl chloride and either pyridine or water. We should always be on the lookout for familiar reaction patterns and we will recall from Section 6.3.5 that pyridine is an effective catalyst for carbonyl substitution reactions. We can therefore postulate that the first step is a reaction between pyridine and benzoyl chloride (Figure 8.15).

The product of this step is the pyridinium salt, and we are aware that this is a reactive species that will react rapidly with any nucleophile present. We still have another nucleophile available, namely water. The next step is therefore likely to be hydrolysis of the pyridinium salt (Figure 8.16).

This provides benzoic acid, and the product we are looking for is benzoic anhydride. We should never get so carried away in predicting sensible reaction steps that we lose sight of the product. We can see here that we can now obtain our product by the reaction of benzoic acid with benzoyl chloride.

237

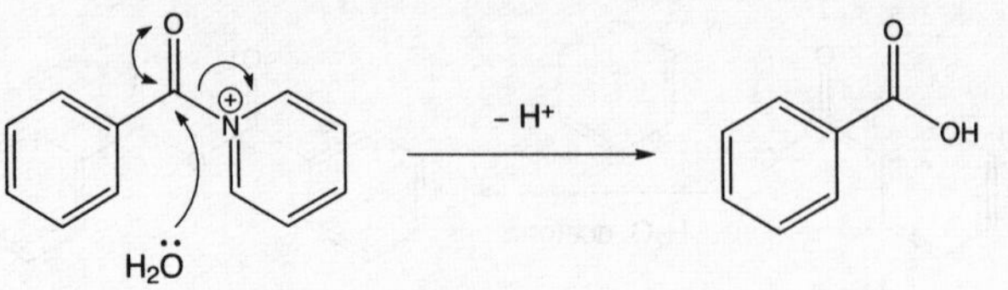

Figure 8.16

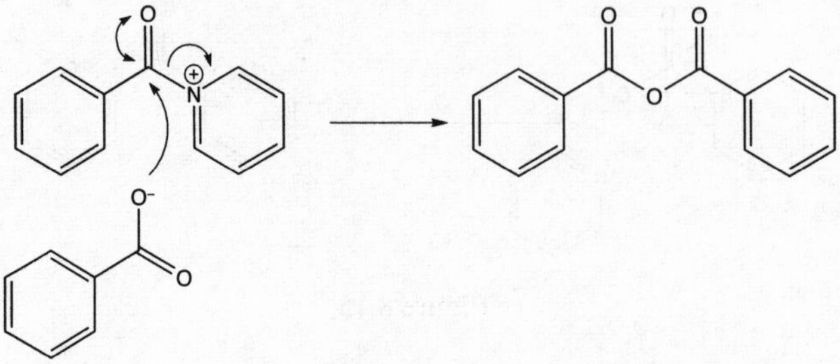

Figure 8.17

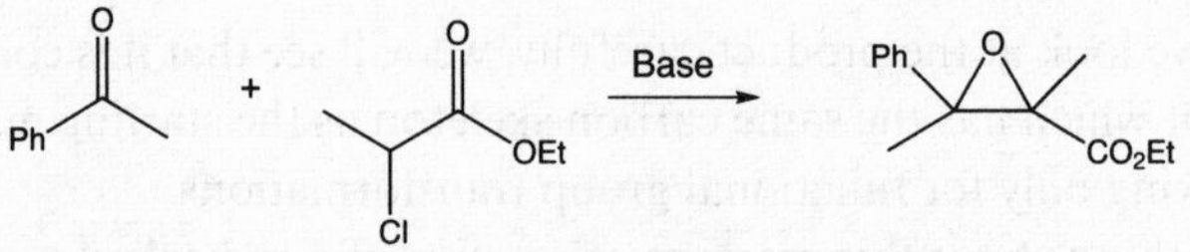

Figure 8.18

Benzoic acid is in itself a rather poor nucleophile, but remember that pyridine is present in the reaction mixture and may act as a base. In this way, the benzoate anion can be formed, which is a better anion than the acid. Again, we must expect the reaction to be catalysed by pyridine, and so the final step is the reaction of the benzoate anion with the benzoyl pyridinium salt (Figure 8.17).

We might reasonably ask why the benzoyl chloride is not hydrolysed completely, giving benzoic acid as the sole product of the reaction. The reason for this is the amount of water present; there is only enough to hydrolyse half the benzoyl chloride and the rest must react with the benzoic acid formed by hydrolysis of the first half. If an excess of water were used, the product would indeed be benzoic acid.

8.3.3 The Darzens condensation

In the Darzens condensation, an aldehyde or ketone reacts with an α-haloester in the presence of a base to give an α,β-epoxyester. An example of a Darzens condensation is shown in Figure 8.18.

Here we see that there has been a change in the carbon skeleton. At some stage in the reaction a carbon–carbon bond must be formed between the two reactants. We

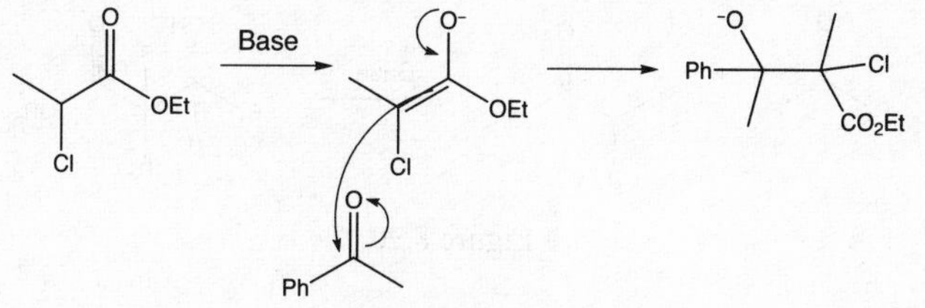

Figure 8.19

should therefore keep our eyes open for any reaction pattern which is likely to lead to the formation of a carbon–carbon bond.

It may not be immediately obvious how the epoxy group is formed, but we need not worry about this for the time being. Provided we look at the likely nucleophiles and electrophiles, include a suitable carbon–carbon bond forming step, and take the reaction one step at a time, it should be easier to see how this group is formed as we write more of the mechanism.

Since a base has been added, it is a safe bet that the reaction is ionic, at least to start with. The base is added for a reason, and we should therefore expect a deprotonation at some stage during the reaction.

We should now look at the nucleophiles and electrophiles present. The reaction mixture seems rich in electrophiles: the ketone group is electrophilic, and the ester can act as an electrophile at two sites, either at the ester group itself or at the α-carbon, which could react by nucleophilic displacement of chloride. There is no obvious nucleophile, and this is where we should expect the base to play a part. Either the ketone or the ester could be deprotonated, resulting in an anion that would be an excellent nucleophile. We should therefore consider which starting material is more likely to be deprotonated.

There are two factors that argue for deprotonation of the ester. Firstly, the chlorine atom at the α-position enhances the acidity (see Section 5.5), thus making deprotonation easier here. Secondly, if we look carefully at the carbon skeleton of the product, we can see that the carbon–carbon bond has formed between the α-carbon of the ester and the carbonyl carbon of the ketone. Deprotonation of the ketone would lead to a carbon–carbon bond between the α-carbon of the ketone and either the carbonyl or α-carbon of the ester. In other reactions, it may not always be clear which of two possible deprotonations (or indeed any other process) will occur. If this is so, then it is usually a sensible policy to write down two reaction schemes, one starting with each possibility. It will normally become clear later on which scheme is the correct one.

Having thus formed our nucleophile, we can now write our carbon–carbon bond forming step, with the ketone as the electrophile (Figure 8.19). Not only is this a reaction between the most likely nucleophile and a suitable electrophile, but we should also recognize this as a standard aldol-type reaction, a familiar pattern of reactivity.

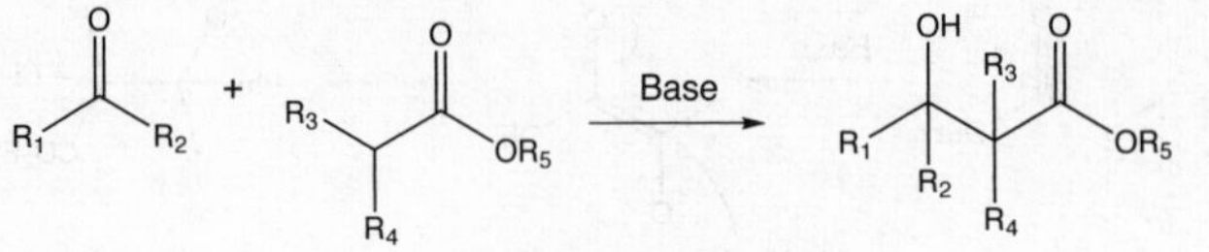

Figure 8.20

Figure 8.21

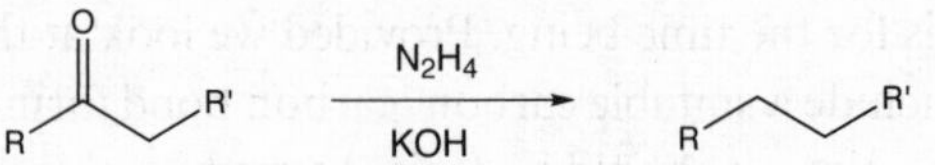

Figure 8.22

This gives us the desired carbon skeleton, but we still have to account for the formation of the epoxy group. This will now be obvious, provided we have not tried to be too clever and write down too many processes occurring in a single step. Even though we recognize this as an aldol-type reaction and we know that these proceed as shown in Figure 8.20, we must not assume that the reaction proceeds in this manner (since this is obviously something more than a simple aldol reaction) and write down too much at once. In an unfamiliar reaction, it is always sensible to write down each step individually, including such simple processes as protonation and deprotonation. If we proceed in such a way here, we will see that an oxygen anion is adjacent to a chloride, and it is now not difficult to see how the epoxide is formed (Figure 8.21).

8.3.4 The Wolff–Kishner reduction

The Wolff–Kishner reduction is a method for reducing a carbonyl group to a methylene group. The reaction is carried out by treating an aldehyde or ketone with hydrazine and a base at high temperature, either in a refluxing protic solvent with a high boiling point, such as diethylene glycol, or in ethanol in a sealed tube (Figure 8.22).

It is obvious here that there is no change in the carbon skeleton. Again, the presence of a base suggests that the mechanism is ionic. Looking at the reactants, we see a good nucleophile (hydrazine) and a good electrophile (the ketone). The hydroxide can also act as a nucleophile, but this leads only to hydration of the carbonyl group – an unremarkable reaction that is not likely to lead to reduction.

Let us then write a reaction between hydrazine and the carbonyl group as a first

240

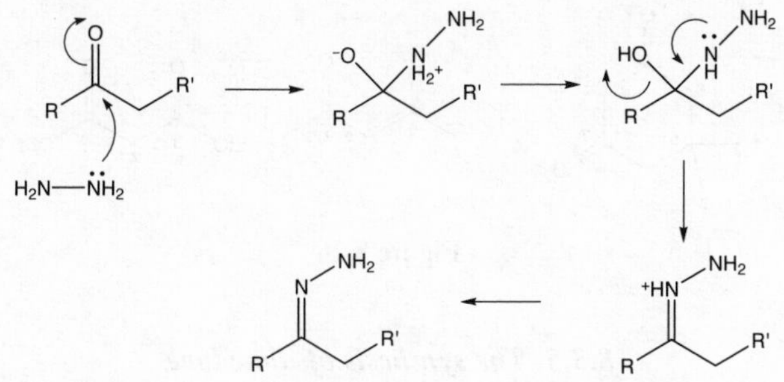

Figure 8.23

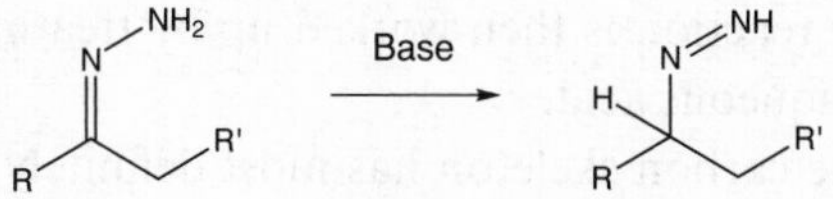

Figure 8.24

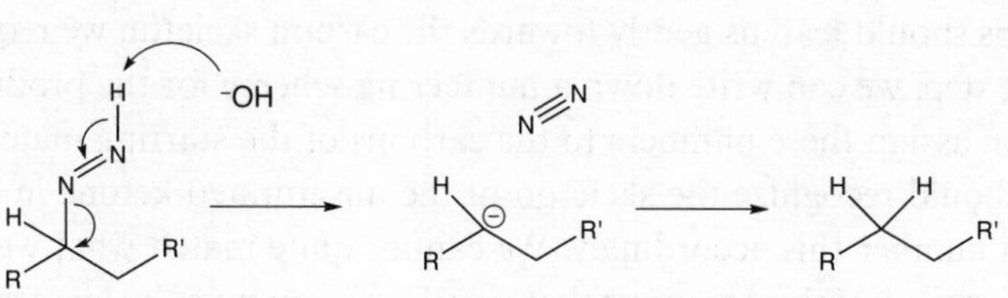

Figure 8.25

step. As a matter of habit, we should write each step of this individually (Figure 8.23), although no obvious pathways to reduction present themselves here from the intermediate steps.

It may not be clear at this stage how this hydrazone function is reduced to a methylene group, but a clue is the nitrogen–nitrogen bond. Given that our original molecule is reduced, something else must be oxidized. A likely candidate is the hydrazine function; oxidation of these groups to molecular nitrogen is a very favourable process. It would therefore make sense to increase the bond order between the nitrogens as a step towards molecular nitrogen. This can be easily done by base-catalysed tautomerization (Figure 8.24). We are now almost at the stage of having molecular nitrogen, and this may be obtained by a base-catalysed elimination from the azo group (Figure 8.25). We would not normally expect a simple carbanion to act as a leaving group like this, but there is a considerable driving force given to this step by the formation of molecular nitrogen. Also, we should remember that the reaction is carried out at a high temperature, which facilitates steps with high activation energies.

The carbanion thus produced is quickly protonated by the solvent, which completes the reaction.

241

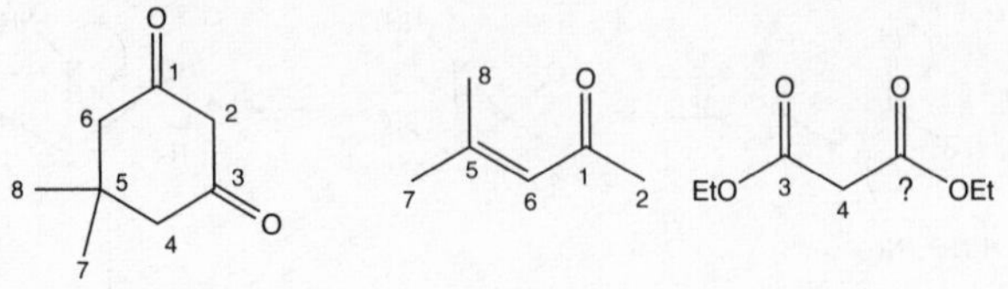

Figure 8.26

8.3.5 *The synthesis of dimedone*

For our last example, let us look at the synthesis of dimedone (Figure 8.3). The unsaturated ketone and diethyl malonate are mixed in ethanol in the presence of sodium ethoxide. The reaction is then worked up by treatment first with aqueous alkali, and then with aqueous acid.

We see here that the carbon skeleton has most definitely been changed, almost unrecognizably so at a first glance. Nonetheless, this should not frighten us: a methodical consideration of likely nucleophiles and electrophiles and of their reaction patterns should lead us gently towards the carbon skeleton we require.

As a first step, we can write down a numbering scheme for the product and then see if we can assign these numbers to the carbons of the starting materials (Figure 8.26). We should recognize the skeleton of the unsaturated ketone in the product, and we can number this accordingly. We cannot quite match what we have left to diethyl malonate, but this shows us that a carbon atom must be lost from the skeleton somewhere.

Just as in the Darzens reaction, we find an abundance of electrophiles (the ester groups of diethyl malonate and the two electrophilic sites of the ketone), but no nucleophile. The function of the base must therefore be to provide us with a nucleophile. We know that the methylene group of diethyl malonate is reasonably acidic, because of the combined effect of the two adjacent carbonyl groups. We can therefore expect the anion of diethyl malonate to be our nucleophile.

We must now decide which is the likely site of reaction on the other molecule. This can react either at the carbonyl group, or in a Michael reaction at the β-position. We should remember that malonyl-type anions generally attack Michael fashion, as they are soft nucleophiles and the β-position of α,β-unsaturated ketones is softer than the carbonyl position. However, even if we had not remembered that, we could write both forms of attack and see that attack at the β-position leads to something resembling the desired product, whereas attack at the carbonyl position leads to something else entirely (Figure 8.27). We could also look at the numbering of our carbon atoms as mentioned above, which would also suggest that the central carbon of diethyl malonate should bond to the β-position of the ketone.

Although attack at the β-position gives something resembling the product, it is certainly not the product itself. We can see that the product is cyclic, and so it would make sense for a cyclization reaction to take place next. There are three possible

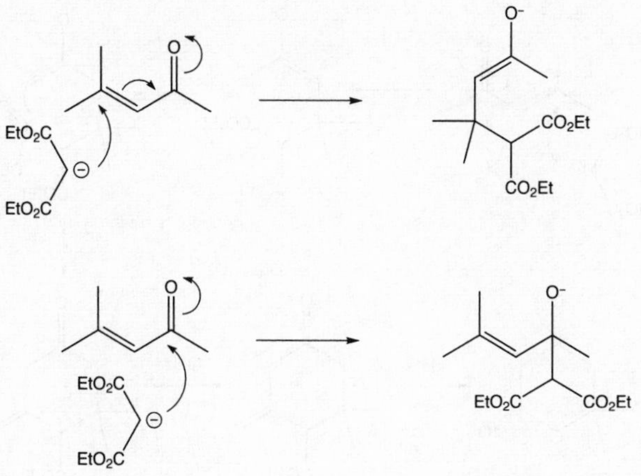

Figure 8.27

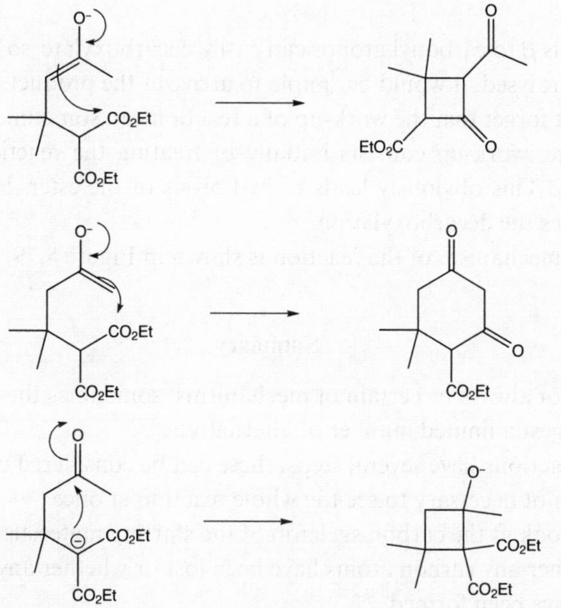

Figure 8.28

ways in which this molecule can cyclize, depending on where the anion is formed (Figure 8.28). We can see that two of these lead to four-membered rings, whereas the third leads to a six-membered ring. As our product is six-membered, it is not too difficult to choose between these options.

Cyclization in this way brings us closer to our product. All we need to do now is suggest a way of removing the ethoxycarbonyl group. We should be aware that

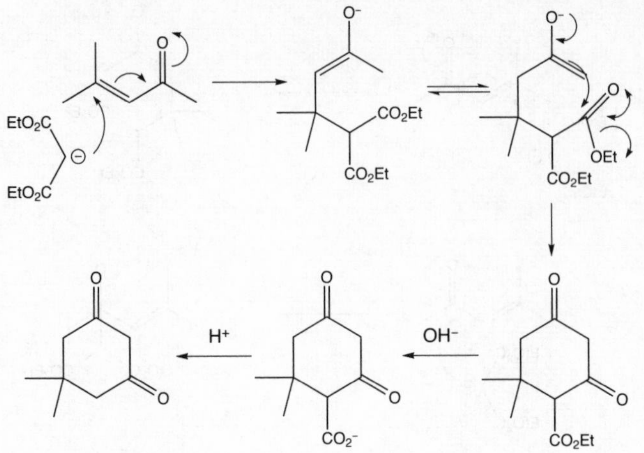

Figure 8.29

carboxylic acids β to carbonyl groups can easily decarboxylate, so if the ester group were to be hydrolysed, it would be simple to arrive at the product.

We must not forget that the work-up of a reaction can sometimes play an important part. Here, work-up consists initially of treating the reaction mixture with aqueous alkali. This obviously leads to hydrolysis of the ester. Finally, treatment with acid causes the decarboxylation.

The overall mechanism of the reaction is shown in Figure 8.29.

Summary

- We cannot always be certain of mechanisms; sometimes the best we can do is to suggest a limited number of alternatives.
- Many reactions have several steps; these can be considered one at a time, and it is not necessary to see the whole reaction at once.
- Always look at the carbon skeleton of the starting materials and products to see whether any carbon atoms have been lost or whether any carbon–carbon bonds have been formed.
- Consider whether the reaction is likely to be ionic, radical, or pericyclic.
- Look for likely nucleophiles and electrophiles; many reactions will simply be an interaction between these.
- Common reactivity patterns (for example the formation of a phosphorus–oxygen bond) are worth looking for.
- Always be aware of the thermodynamic driving force for a reaction.
- If there are not any obviously reactive species in a reaction mixture, can any of the species be made more reactive, for example by protonation?

- If more than one reaction step is possible, write down all possibilities and see which turns out to be more productive later.
- Always be aware of the pH of the reaction medium, and make sure all postulated species are compatible with it.

Problems

1. We saw in Section 3.2.3 that pyridines can be prepared by the Hantzsch pyridine synthesis, in which a β-dicarbonyl compound is treated with an aldehyde and an ammonium salt; an example is shown below. Identify all the possible nucleophilic and electrophilic reaction sites.

2. Write down a numbering system for the carbons in the product and match these with the carbons in the starting materials.
3. This numbering system should tell you which site must react with the aldehyde carbon. Is this consistent with the pattern of nucleophiles and electrophiles you identified in Problem 1. If not, what change could reasonably be made to one of the reagents so that this reaction could take place?
4. Suggest a plausible first step for the reaction. This type of reaction can give either one of two products, one formed by dehydration of the other. Which do you think is the more likely product here? Bear in mind what the next step of the reaction sequence is likely to be, and so what sort of functionality is required.
5. Write a complete mechanism for formation of the dihydropyridine.
6. A reaction for inverting the stereochemistry of a secondary alcohol is known as Mitsunobu inversion. The initial product of the reaction is an ester, which is then hydrolysed if the alcohol is required. The reagents used for this reaction are shown below. Identify all the nucleophiles and electrophiles in the reaction mixture.

7. Since the stereochemistry of the alcohol is inverted, it seems likely that the oxygen atom in the original alcohol is removed during the course of the reaction. Is there any atom to which you think it is particularly likely to become attached? Does the pattern of nucleophiles and electrophiles you have identified allow this to happen as a first step? If not, would appropriate reaction of the nucleophiles and electrophiles make it more likely for this bond to be formed?

8. Bearing in mind your answers to Problem 7, suggest a mechanism for the reaction.

9. Primary alcohols can be efficiently converted into aldehydes with the Swern oxidation. In this reaction, the alcohol is treated with dimethyl sulphoxide, oxalyl chloride, and a base, usually triethylamine. Identify all the nucleophiles and electrophiles.

10. Which of these nucleophiles is most likely to react with which electrophile? This reaction produces two new products, which then react with each other. Write mechanisms for these reactions. (Hint: carbon dioxide and carbon monoxide are side products of the reaction.)

11. When 1-butanol labelled with deuterium at the 1-position was subjected to this reaction, one of the products was CH_3SCH_2D. Bearing this in mind, write a mechanism for the rest of the reaction.

12. The McFadyen–Stevens reaction is another method for preparing aldehydes, this time by reduction of compounds at the carboxylic acid oxidation level. It is applicable only to carboxylic acids with no α-hydrogen, as the rather harsh basic conditions would lead to aldol reactions of the product. Suggest a mechanism for this reaction.

9

Case histories

If you have read the first eight chapters of the book, you will by now have learnt much about the basics of reaction mechanisms. However, in much the same way that no-one would read a book on how to play the violin and then immediately expect to be able to play the solo part in Beethoven's violin concerto, you may find it takes a little practice to become entirely proficient in writing mechanisms for reactions. As you proceed with your study of chemistry, it will not be difficult to obtain that practice, as you will encounter a great many reactions. If you think about the mechanism of each new reaction, you will soon find the process becomes more and more automatic. It is certainly easier to become a virtuoso with reaction mechanisms than with the violin!

To start you on your way, in this last chapter we will look at four reactions with interesting mechanisms. These have all had much work done on them to elucidate their mechanisms, and it will be instructive to see how the mechanisms were discovered. The discoveries were, of course, made from experiments, but the experiments themselves were designed from a knowledge of plausible mechanisms and knowing the right questions to ask.

As with Section 8.3, you will get more out of this chapter if you try to stay one step ahead of the text, and to think about what the mechanisms might be and what experiments might be done before they are described.

9.1 The formation of 9-benzylfluorene

To show that discovering reaction mechanisms is not necessarily the prerogative of eminent professors who have dedicated a lifetime to the art, most of the experiments I shall describe for this reaction were done by second-year undergraduates, and many of the experiments were also designed by them.

In this reaction, fluorene is treated with potassium hydroxide in refluxing benzyl alcohol, which leads to the formation of 9-benzylfluorene (Figure 9.1). It is not

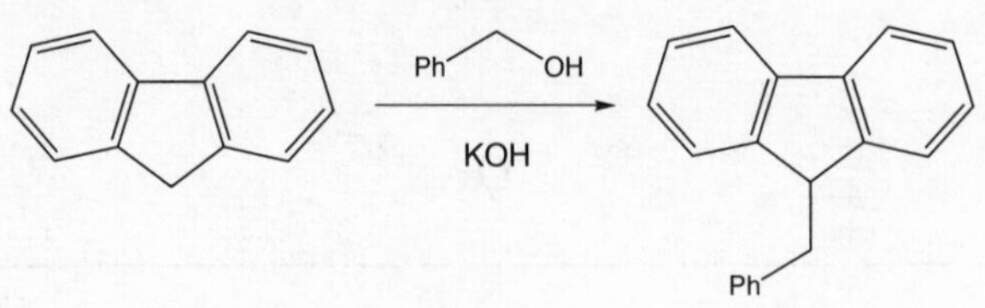

Figure 9.1

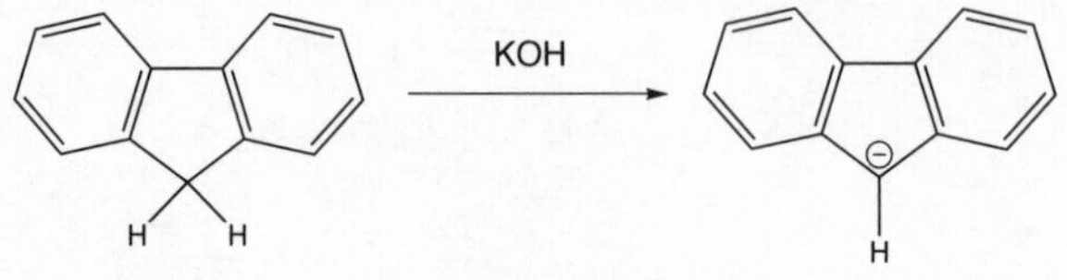

Figure 9.2

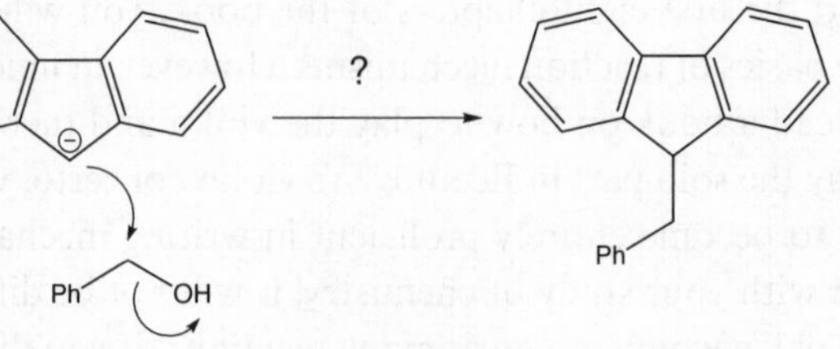

Figure 9.3

immediately obvious what the mechanism might be. However, a sensible first step would be deprotonation of fluorene (Figure 9.2). There are two reasons for thinking this likely. Firstly, the reaction is run in the presence of base (potassium hydroxide), and this position of the fluorene molecule is acidic, with a pK_a of about 25 (the resulting anion is stabilized by aromatic resonance, the molecule becoming a 14 π electron system). Secondly, some reaction must clearly take place at the 9-position of fluorene, and deprotonation seems most likely.

There is indeed evidence for the formation of this anion. The reaction mixture can be seen to have an orange colour, and the fluorene anion is orange. Obviously this is not proof of the presence of the anion, because many other things are also orange, but it does support the idea of the deprotonation.

One possible mechanism would be that this anion reacts with benzyl alcohol in an S_N2 displacement (Figure 9.3), although this seems barely plausible because OH$^-$ is such a poor leaving group and, owing to the basic conditions of the reaction, there is no possibility of turning it into a better leaving group by protonation.

Nonetheless, an experiment was done to rule out this possibility. Fluorene was treated with potassium hydroxide and benzyl phenyl ether under similar conditions to the original reaction and no reaction took place (Figure 9.4). If the reaction did

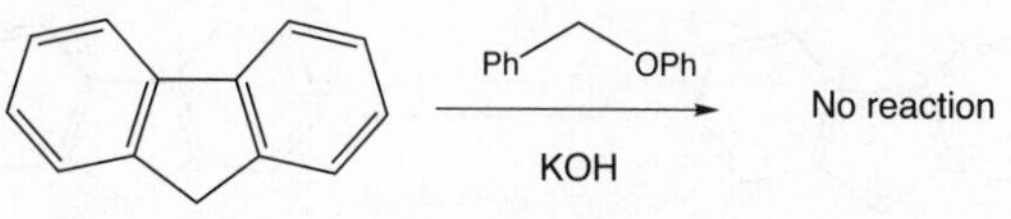

Figure 9.4

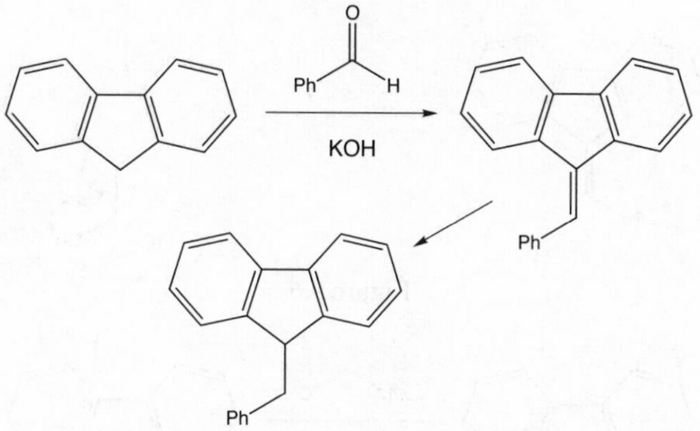

Figure 9.5

proceed by an S_N2 displacement, then this version should be even more efficient because the leaving group would now be PhO^-, which is a better leaving group than OH^-. We can therefore discount this mechanism.

The reaction was originally carried out in the context of an investigation of the reducing properties of solutions of potassium hydroxide in benzyl alcohol. A further possibility considered was therefore that something was reduced by benzyl alcohol to give the product. When benzyl alcohol acts as a reducing agent it is oxidized itself, yielding benzaldehyde.

Benzaldehyde would of course react readily with the fluorene anion, so a possible mechanism for the reaction is that the fluorene anion reacts with benzaldehyde to give 9-benzylidenefluorene, and that this is then reduced to 9-benzylfluorene (Figure 9.5). This is plausible, because more benzaldehyde is generated in the reduction step, so only a trace of benzaldehyde is required at the start. Benzaldehyde is often found in benzyl alcohol as an impurity, and benzyl alcohol would in any case be oxidized by oxygen in the air under the conditions of the reaction.

A number of experiments were carried out to test this idea. In one of these, fluorene was treated with potassium hydroxide and benzaldehyde, which led to the formation of 9-benzylidenefluorene under milder conditions than those required for the original reaction. This shows that any benzaldehyde present would indeed react readily with the fluorene anion. Note that there is a rather unusual dehydration step here, in that it takes place under basic conditions (Figure 9.6). It must therefore proceed by the E1cB mechanism (see Section 6.2.4).

249

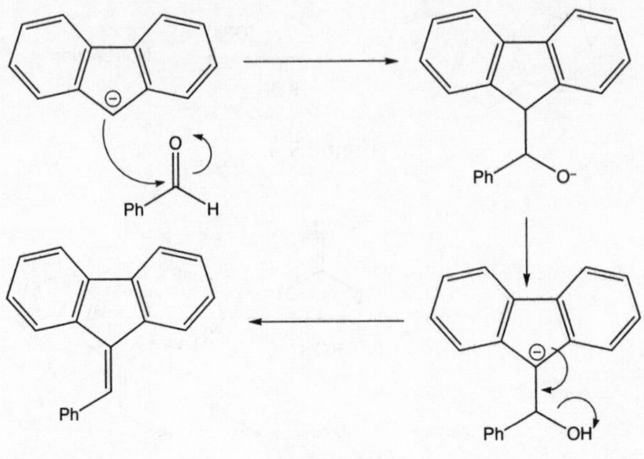

Figure 9.6

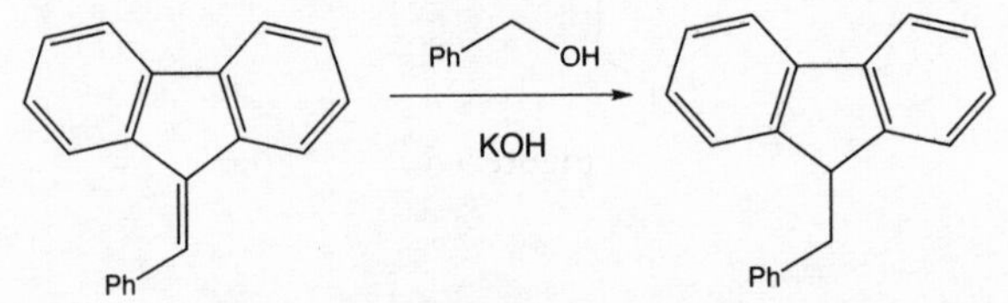

Figure 9.7

The product of this reaction, 9-benzylidenefluorene, was treated with a solution of potassium hydroxide in refluxing benzyl alcohol (the conditions of the original reaction), which reduced it to 9-benzylfluorene (Figure 9.7).

These two experiments show that both steps of the proposed mechanism are plausible. Further evidence is provided by the observation that addition of benzaldehyde to the original reaction increases the rate. Moreover, the reaction failed when it was carried out under an atmosphere of nitrogen with benzyl alcohol that had been carefully purified to remove the traces of benzaldehyde. Although we now have convincing evidence for the proposed mechanism, we should note that, as so often in organic chemistry, this is not the same as absolute proof.

If we accept that the reaction proceeds through 9-benzylidenefluorene, then we are still left with two possible mechanisms for the reduction step. The double bond may be reduced by hydride transfer from either benzyl alcohol (after deprotonation) or benzaldehyde (after addition of OH⁻) (Figure 9.8). If benzyl alcohol is the reducing agent, then this would regenerate benzaldehyde and so account for the requirement for only a trace of benzaldehyde in the reaction. However, it would be wrong to discount the possibility that benzaldehyde might be the reducing agent for this reason alone. The reaction is run in the presence of air, and it may well be that enough aerial oxidation takes place to provide the two equivalents of benzaldehyde needed for this mechanism (one for the formation of 9-benzylidenefluorene and one

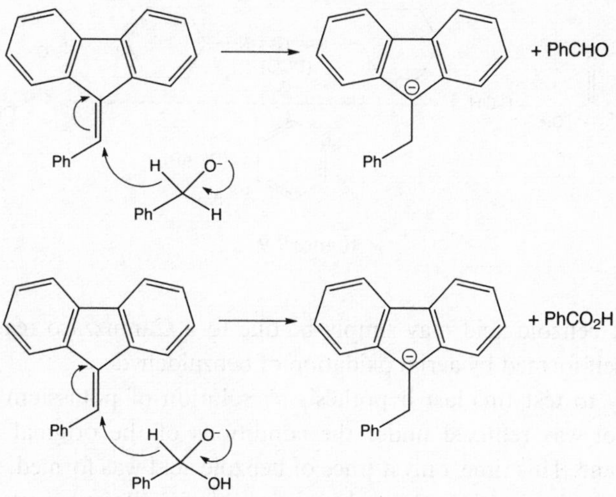

Figure 9.8

for its reduction). Moreover, there is good precedent for benzaldehyde acting as a reducing agent: this is exactly what happens in the Cannizzaro reaction (see Section 6.1.2).

We saw in Section 7.2 that we can often obtain useful information by looking at the side products of a reaction. A useful clue was given here by the finding that benzoic acid could be extracted from the aqueous washings of the reaction after acidification. If benzaldehyde is the reducing agent, then benzoic acid should be formed during the reaction. Since benzoic acid is formed, we have good evidence that benzaldehyde is the reducing agent.

If this is so, then we might expect that the reaction would fail if run under a nitrogen atmosphere, as it would then not be possible for the two equivalents of benzaldehyde required to be formed. As mentioned above, the reaction did indeed fail when carried out in this way with carefully purified benzyl alcohol. However, when the reaction was run under nitrogen with benzyl alcohol taken straight from the bottle, 9-benzylfluorene was formed in a similar yield to that obtained when the reaction was run in the presence of air. This time, no benzoic acid was formed.

It thus appears that when the reaction is carried out under nitrogen, the reducing agent must be benzyl alcohol. Under these conditions not enough benzaldehyde could be formed to act as the reducing agent, and if benzaldehyde were the reducing agent then we would expect to find benzoic acid.

This finding is somewhat at odds with the finding that benzoic acid is produced when the reaction is run in the presence of air. There are two possible explanations for this. Firstly, it may be that both benzaldehyde and benzyl alcohol can act as reducing agents, and that benzyl alcohol is dominant when the reaction is run under nitrogen, but when the reaction is run in air the benzaldehyde can also take part.

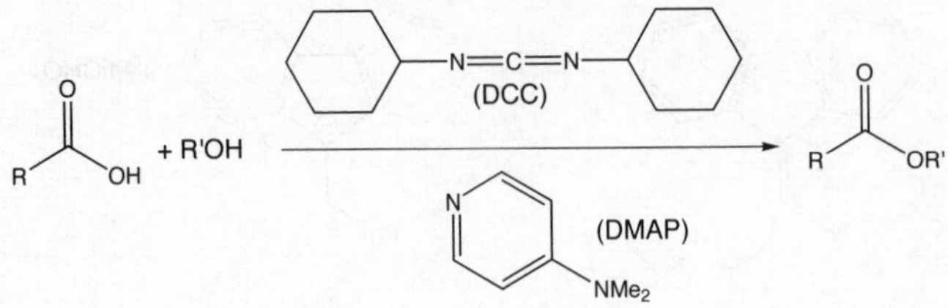

Figure 9.9

Secondly, the benzoic acid may simply be due to a Cannizzaro reaction of benzaldehyde, itself formed by aerial oxidation of benzaldehyde.

It was easy to test this last hypothesis. A solution of potassium hydroxide in benzyl alcohol was refluxed under the conditions of the original reaction, but without fluorene. This time, only a trace of benzoic acid was formed, which tells us that the Cannizzaro reaction cannot be responsible for all the benzoic acid formed in the original reaction.

Taking all this evidence together, we can see that the reduction can be effected by either benzaldehyde or benzyl alcohol. Benzyl alcohol is the reducing agent when the reaction is run under nitrogen, but benzaldehyde can compete effectively when the reaction is run in the presence of air.

9.2 DCC-mediated ester formation

One of the most useful reagents for coupling acids and alcohols to form esters is dicyclohexylcarbodiimide, or DCC. This is also commonly used to couple amino acids in peptide synthesis. A general coupling reaction is shown in Figure 9.9. The reaction is often catalysed by a base; pyridine is sometimes used, but 4-(dimethylamino)pyridine (DMAP) is more efficient. The mechanism of this reaction has received a great deal of attention. A vast amount of research has been done, and it will be possible to give only some edited highlights here.

A number of experiments were done on the reaction between the carboxylic acid and DCC in the absence of an added nucleophile. The product of this reaction is the anhydride: in other words, the acid couples with itself. As we saw in Section 7.2, it is important to be aware of all the products of a reaction, and the other products here were dicyclohexylurea, plus the N-acylurea in an amount varying with the conditions of the reaction (Figure 9.10). Any mechanism must therefore account for the formation of all of these products.

Compounds with cumulative double bonds (ketenes, isocyanates, etc.) are well known to undergo 1,2 addition reactions, and it would not be surprising to find carbodiimides reacting in this way. In fact, examples of their reacting like this were already known to researchers in this area. For example, when phenol is treated with DCC at 100 °C for 2 days, the 1,2 adduct shown in Figure 9.11 is formed.

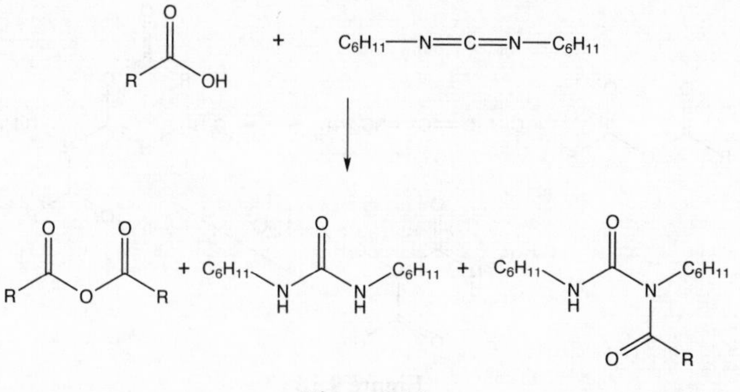

Figure 9.10

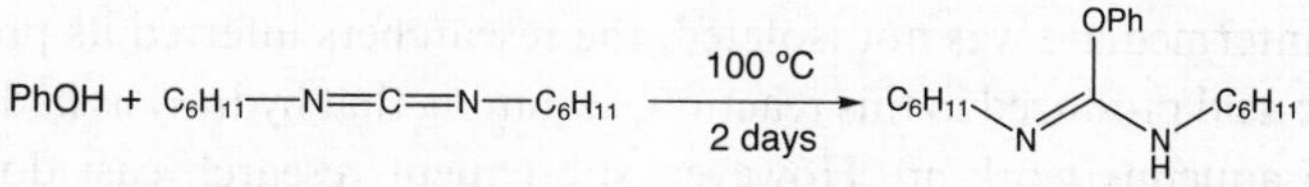

Figure 9.11

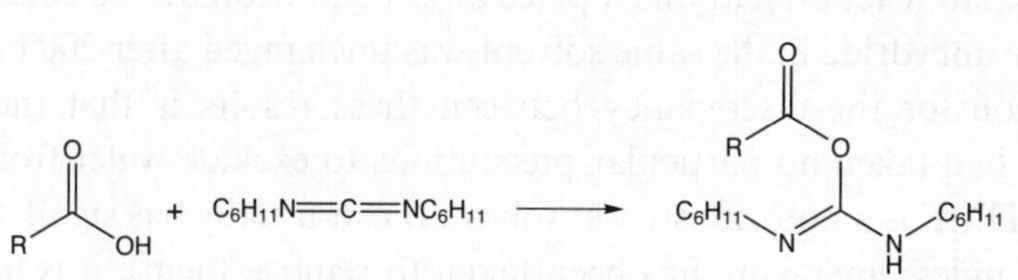

Figure 9.12

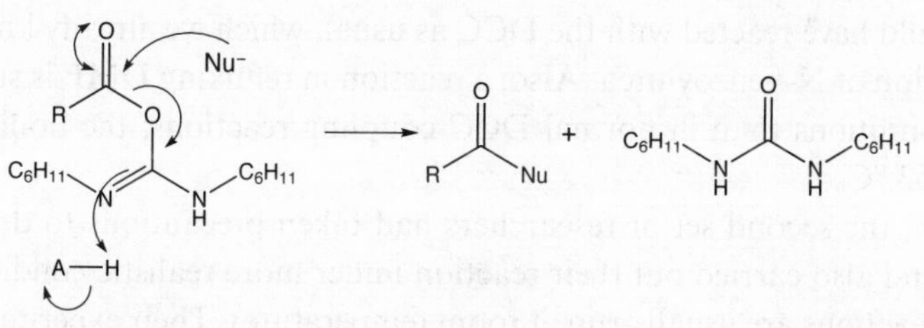

Figure 9.13

By analogy, the first step in the mechanism was postulated as a 1,2 addition of the carboxylic acid to the carbodiimide function, producing an O-acylisourea (Figure 9.12). This could then act as the acylating agent, as shown in Figure 9.13, or possibly undergo some other fate, as yet unspecified.

But what of the N-acylurea formed as a side product? An early suggestion was that it was formed by reaction between the anhydride and excess DCC to give the O,N-diacylisourea, followed by hydrolysis (Figure 9.14). The evidence that this was based on was that N-benzoylurea was formed in 97.5% yield when benzoic anhydride was treated with DCC and tosic acid in refluxing DMF for 2 h. Although the

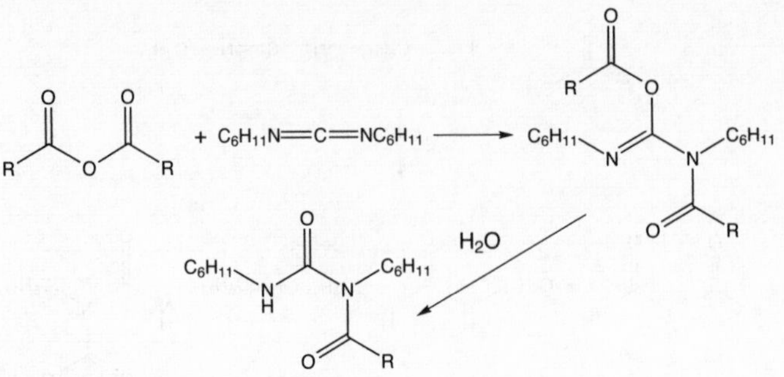

Figure 9.14

dibenzoyl intermediate was not isolated, the researchers inferred its presence from the benzoic acid produced in this reaction, assuming that hydrolysis had taken place during the aqueous work-up. However, subsequent research cast doubt on this hypothesis. When DCC was treated with acetic anhydride in acetonitrile at room temperature, no reaction had taken place after 65 h. Moreover, a solution of DCC and benzoic anhydride in the same solvent was unchanged after 200 h.

The reason for the discrepancy between these results is that the first set of researchers had taken no particular precautions to exclude water from their reaction. Since DMF is a notoriously wet solvent (i.e. it usually has significant traces of water in it, unless great care has been taken to remove them), it is likely that the benzoic anhydride would have been hydrolysed under the conditions of the reaction (the tosic acid present would have catalysed this reaction). The benzoic acid thus formed could have reacted with the DCC as usual, which we already know leads to the formation of N-benzoylurea. Also, a reaction in refluxing DMF is subject to far harsher conditions than in normal DCC coupling reactions; the boiling point of DMF is 153 °C.

However, the second set of researchers had taken precautions to dry their acetonitrile, and also carried out their reaction under more realistic conditions (DCC coupling reactions are usually run at room temperature). Their experiment is therefore more trustworthy. They did not stop there, however; they also investigated the effect of excess DCC on the reaction between DCC and acetic acid. Recall that the products of this reaction are acetic anhydride and N-acetylurea. If the N-acetylurea is formed by reaction between DCC and acetic anhydride, then excess DCC should increase the yield of N-acetylurea at the expense of the anhydride, but it had no effect on the product distribution. Furthermore, they investigated the effect of added acetic anhydride on the kinetics of the reaction, and found no evidence of any acceleration, as would have been expected if DCC were reacting with acetic anhydride.

The direct reaction between DCC and the anhydride can therefore be ruled out for several reasons. Since we seldom deal in certainties in organic chemistry, a

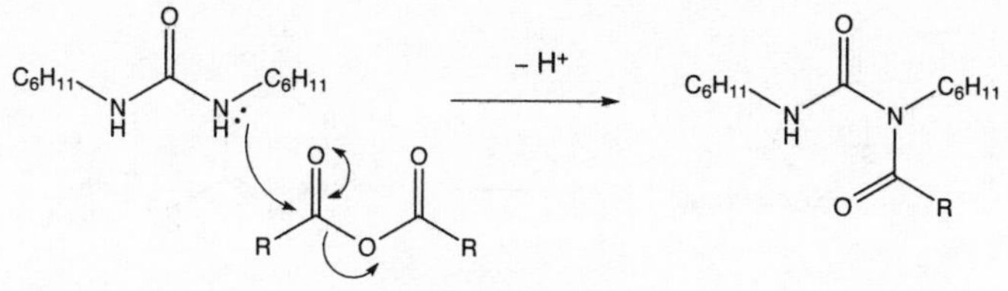

Figure 9.15

conclusion is always more convincing if it is supported by the results of several different experiments.

Another possibility for the formation of the N-acylurea is acylation of dicyclo-hexylurea by the anhydride (or another acylating agent), as shown in Figure 9.15. A number of experiments were carried out to test this hypothesis. Firstly, dicyclo-hexylurea was treated with acetic anhydride. No reaction occurred either in ace-tonitrile or chloroform, even with base catalysis. Furthermore, addition of acetic anhydride to reactions between acetic acid and DCC had no effect on the yield of N-acetylurea, and addition of diisopropylurea (this was used because of the poor solubility of dicyclohexylurea) also had no effect on the product distribution.

It is therefore safe to conclude that nothing in the reaction mixture is capable of acylating dicyclohexylurea. This leaves a third possibility, which is that the O-acylisourea initially formed undergoes a rearrangement to give the N-acylurea. Although there is no direct evidence to support this, it is generally assumed to be the correct mechanism because all plausible alternatives have been ruled out.

You may wonder why so much attention has been paid to the formation of a side product when it might be more interesting to consider the desired reaction. We do not study reaction mechanisms simply because it is intellectually satisfying, in the same way that a mathematician might devote years of his life to trying to prove some esoteric theorem. Knowledge of reaction mechanisms has practical use. Here, we are interested in the mechanisms of formation of the side product precisely because it is unwanted, and we would like to minimize its formation.

What we have learned is that the formation of the unwanted side product is due to rearrangement of an intermediate. Whilst we do not know the exact mechanism of this step, it seems likely that it is unimolecular, and would therefore have first-order kinetics. However, the step leading from this same intermediate to the desired reaction is presumably bimolecular, as we would expect that the intermediate would react with another species, either to give the product directly, or some acylating agent that later forms the product. We would therefore expect this step to have second-order kinetics.

The ratio of our desired product to the unwanted product therefore depends on the competition between these two reactions, one first-order and the other second-order. In general, second-order reactions compete more effectively with first-order reactions in more concentrated solution. For example, if the rate is proportional to

255

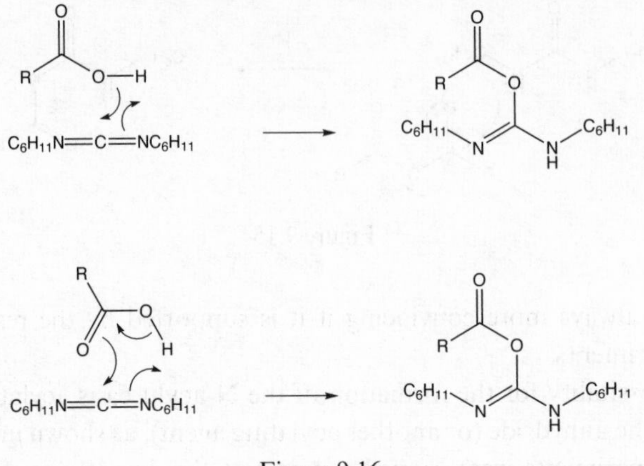

Figure 9.16

the concentration of two reactants, it will be 100 times faster if the solution is 10 times more concentrated, whereas the rate of a reaction proportional to only one of those will be only 10 times faster. We would therefore expect that the formation of the unwanted side product (the N-acylurea) would be minimized if the reaction is run in as concentrated solution as possible. This is indeed found to be so, and for this reason DCC coupling reactions are generally carried out in concentrated solutions. This is a lovely example of how well-conducted mechanistic investigations can give us information of real practical value.

Let us now return to the formation of the O-acylisourea. This could be formed either by a concerted reaction or by a reaction of two ions. If the reaction is concerted, it could proceed by either a four-centre or a six-centre mechanism (Figure 9.16), and if it is ionic, the ions could be present either as free ions or as a tightly associated ion pair.

The four-centre concerted mechanism seems unlikely, because this mechanism is also available to alcohols and phenols, but their reaction with carbodiimides is slower by many orders of magnitude. On balance, the evidence favours an ionic mechanism, since stronger acids react faster than weaker ones. The ionized forms would be expected to be present at a higher concentration for stronger acids, leading to a faster reaction.

However, it is unlikely that the reaction is between free ions, since when triethylammonium acetate was added, the rate fell. If free ions were involved, then increasing the concentration of one of them should increase the rate. It is thought that the reason why the rate is decreased is that the acetate forms a hydrogen-bonded complex with acetic acid, thus effectively reducing the concentration of the acid.

Acetic acid is known to form dimers in solution, and some attention was paid to whether the dimer or the monomer was the reacting species. The reaction proceeds 30 times faster in carbon tetrachloride than in acetonitrile. This suggests that the

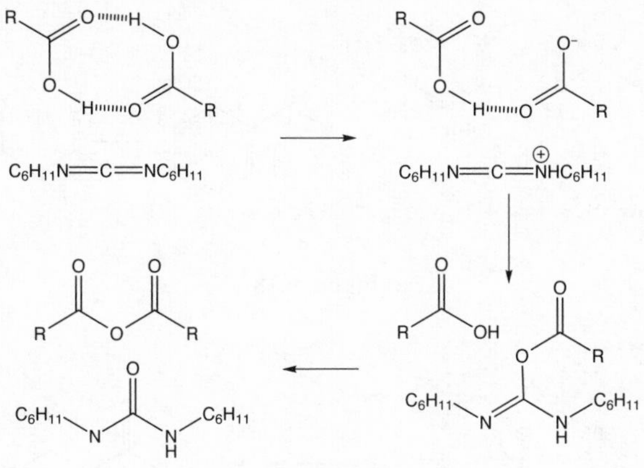

Figure 9.17

dimer reacts faster than the monomer, because acetic acid is almost completely dimerized in carbon tetrachloride, whereas in acetonitrile the dominant species is the monomer.

Moreover, when the kinetics of the reaction were measured, the reaction in acetonitrile was found to be somewhat higher than first order in acetic acid. As the concentration of acetic acid rises, the proportion of the dimer increases. This increases the rate faster than would be expected just from the increase in concentration, which is why the reaction has higher than first-order kinetics.

A related observation was that the yield of anhydride increased under conditions favouring formation of the dimer. Thus the yield was greater in carbon tetrachloride than in acetonitrile, and the yield in acetonitrile increased with increasing concentration of acetic acid. When the dimer reacts, a second molecule of acetic acid is already within the solvent cage to react with the O-acylisourea (Figure 9.17), so anhydride formation is able to compete more effectively with rearrangement to the N-acylurea. Since the yield is sensitive to the proportion of dimer in this way, it shows that the monomer can react if the concentration of dimer is low, but the dimer does appear to be the preferred reactant. If the monomer were unable to take part in the reaction, then the proportion of dimer would have no effect on the yield, just on the rate.

All the discussion of this reaction so far has to some extent been peripheral, useful and interesting though it has been, as the main feature of the reaction is that it allows acylation of alcohols and amines, and we have not yet looked at the acylation step.

When the reaction is used as a way of acylating an alcohol, say, the question arises of whether the acylating agent is the O-acylisourea itself, the anhydride, or some other species. The simplest answer would be that it is the O-acylisourea, but we already know that carboxylic acids can form anhydrides in the presence of DCC, and anhydrides are good acylating agents. We should, of course, always also consider any other possibilities.

257

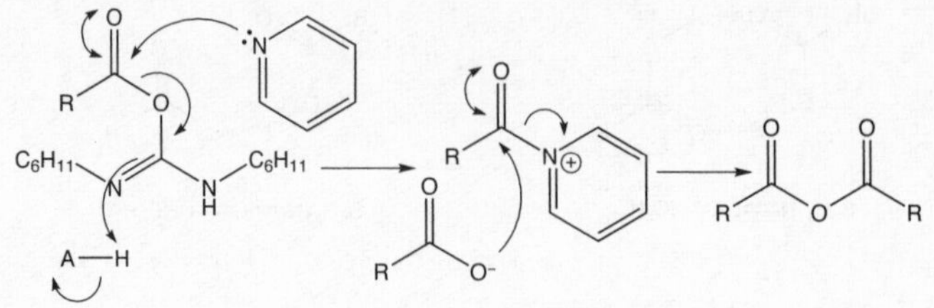

Figure 9.18

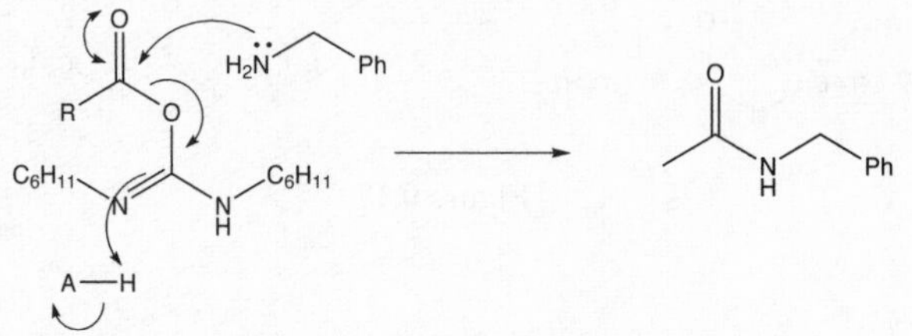

Figure 9.19

Coupling reactions between acids and alcohols using DCC are often carried out in the presence of bases, so the effect of bases on the reaction was investigated. Returning to the reaction between acetic acid and DCC with no added nucleophile present, it was found that triethylamine decreased the yield of anhydride, increasing the yield of N-acylurea, whereas pyridine had the opposite effect.

There are two possible explanations for why triethylamine has this effect. It may be that by converting acetic acid into acetate, it reduces the concentration of the acid, which we have already seen decreases the yield of anhydride. It may also be that in some way it catalyses the rearrangement leading to the N-acylurea. From this evidence, it is not possible to say which is more important. However, any effect pyridine may have as a base is obviously masked by some other more important mechanism. The most likely explanation is that it is acting as a nucleophilic catalyst (Figure 9.18). There is good evidence for its ability to do this: acylpyridinium salts have occasionally been isolated and are excellent acylating agents (see Section 6.3.5).

When benzylamine is added to the reaction the results are most instructive. This time, the yield of N-acylurea is about the same as if no base is added. Of course, no acetic anhydride is formed here, since it reacts rapidly with benzylamine. Since benzylamine is a base, we would expect it to have a similar effect to that of triethylamine, namely that it would increase the yield of N-acylurea. If the anhydride were the acylating agent, this should be the only effect, as the course of the reaction would be the same as with triethylamine until the anhydride is formed. Since benzylamine does not increase the yield of N-acylurea, it must be participating in the reaction other than as a base before the anhydride is formed. This suggests that it is reacting directly with the O-acylisourea (Figure 9.19). We may therefore conclude that the

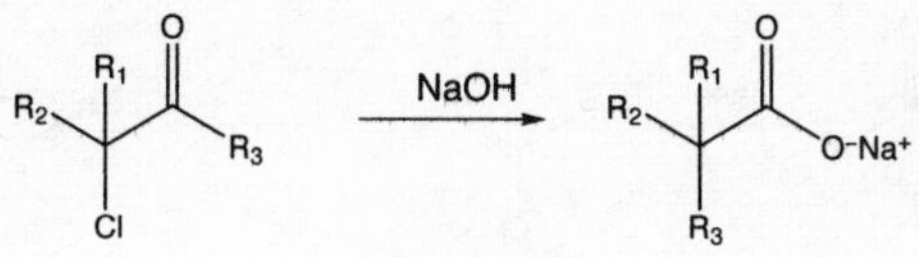

Figure 9.20

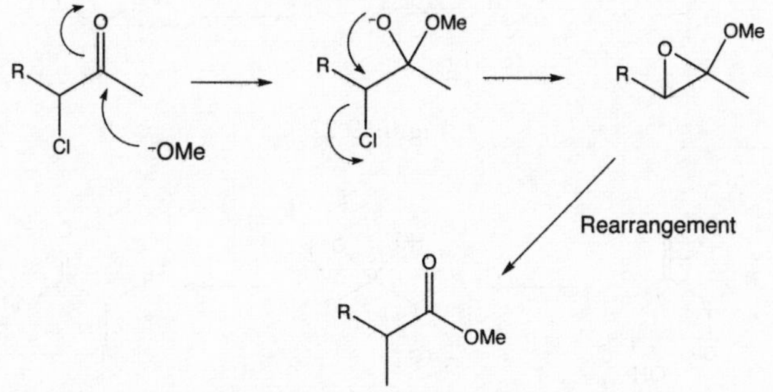

Figure 9.21

O-acylisourea acts as the acylating agent. Of course, this does not completely rule out participation of the anhydride, but it does show that it is not the major pathway. When pyridine is present, it appears that the acylating agent is the acylpyridinium species.

The nature of the acylating agent may, of course, change if the reaction is carried out under different conditions. The reaction has been widely investigated, and most of the evidence points to the O-acylisourea as the acylating agent, with the acylpyridinium species taking over if pyridine (or a substituted pyridine) is present. However, there are some circumstances in which anhydride formation appears to be more important, notably during solid-phase peptide synthesis.

As with many questions in organic chemistry, the answer is not simple.

9.3 The Favorskii rearrangement

When α-halo ketones are treated with base, they often rearrange to a carboxylate derivative (Figure 9.20). This is known as the Favorskii rearrangement. It is reasonably general, although the yields are variable. The nature of the carboxylate function depends on the base used: hydroxide leads to a carboxylate salt, alkoxides lead to esters, and amines give amides. The reaction has also been carried out with α,β-epoxyketones.

A number of mechanisms were initially proposed for this reaction. One of these was that base attack on the carbonyl group leads to formation of an epoxide, which rearranges to give the product (Figure 9.21). Another mechanism goes through a ketene (Figure 9.22) in a reaction similar to the Curtius rearrangement (see Section

259

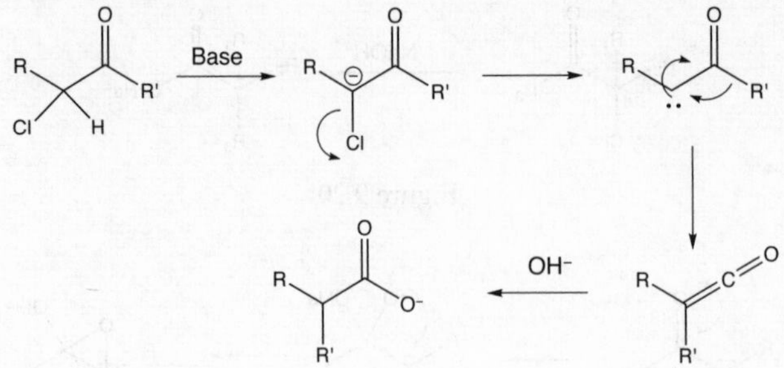

Figure 9.22

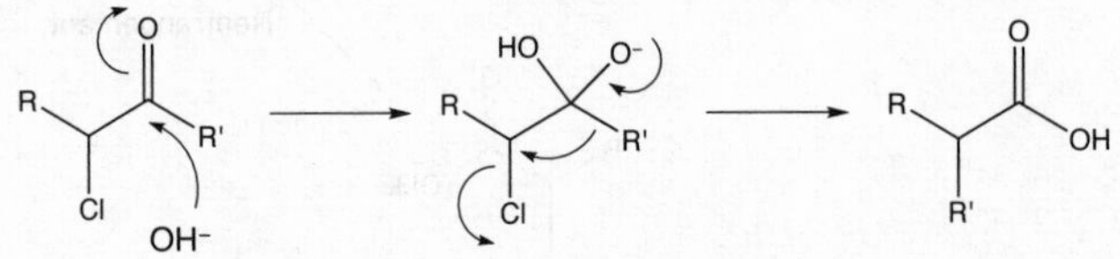

Figure 9.23

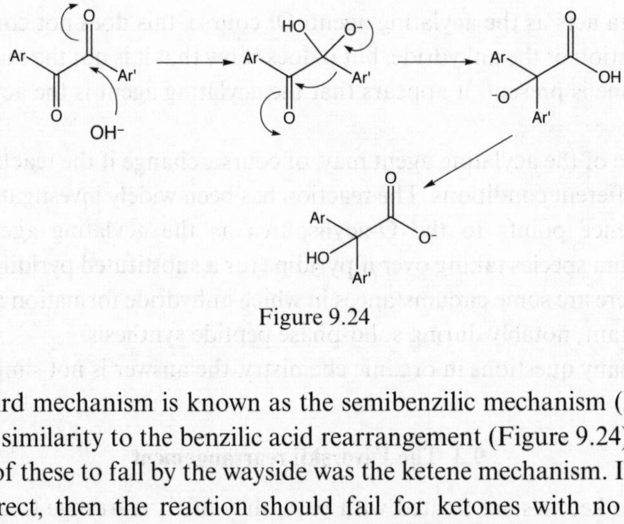

Figure 9.24

5.2.5). A third mechanism is known as the semibenzilic mechanism (Figure 9.23), owing to its similarity to the benzilic acid rearrangement (Figure 9.24).

The first of these to fall by the wayside was the ketene mechanism. If this mechanism is correct, then the reaction should fail for ketones with no α-hydrogen. However, the reaction has been carried out successfully with many compounds of this type.

The other two mechanisms began to run into trouble when it was found that the two isomeric chloroketones shown in Figure 9.25 gave the same product. This led to the postulation of another mechanism, with a cyclopropane intermediate (Figure 9.26). This cyclopropane mechanism accounts for the formation of a single product from the two chloroketones in Figure 9.25 because they will both give the same cyclopropane, which will open preferentially to give the more stable carbanion, in other words with the negative charge adjacent to the aromatic ring.

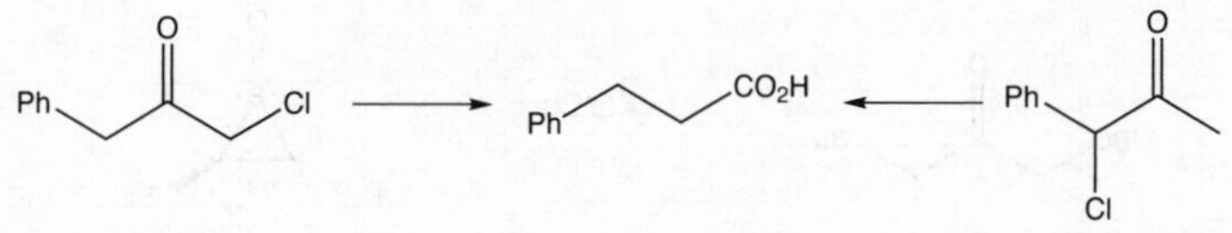

Figure 9.25

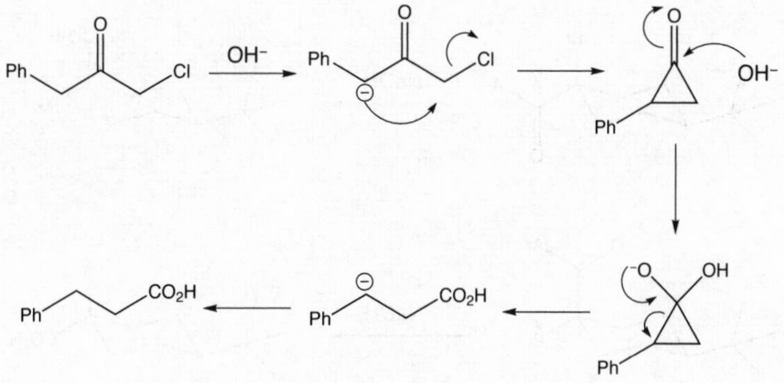

Figure 9.26

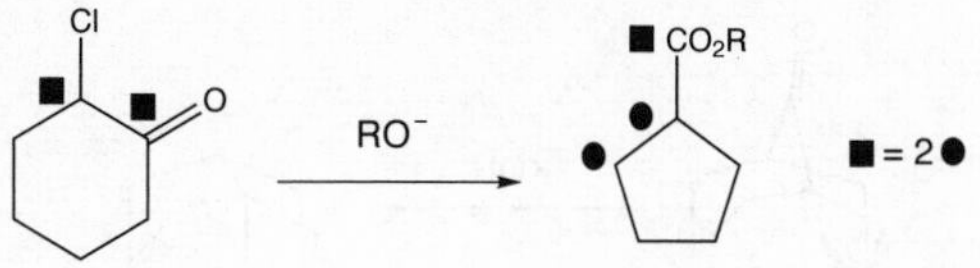

Figure 9.27

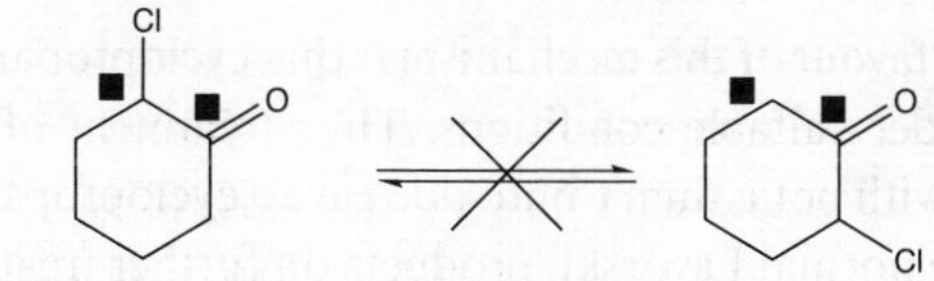

Figure 9.28

An elegant experiment using radiolabelling was done to test this hypothesis. α-Chlorocyclohexanone labelled equally with ^{14}C at the carbonyl and α-positions was synthesized, and then treated with alkoxide under typical Favorskii conditions. This led to the expected product, with half of the radioactivity in the carbonyl carbon. The remaining half was distributed equally between the α- and β-positions (Figure 9.27). Moreover, the starting material was recovered and was found to be labelled in the same way as at the start of the reaction, thus ruling out the possibility of equilibration of the chlorine before the Favorskii reaction (Figure 9.28). This showed unequivocally that there must be a symmetrical intermediate. A cyclopropane is the obvious choice.

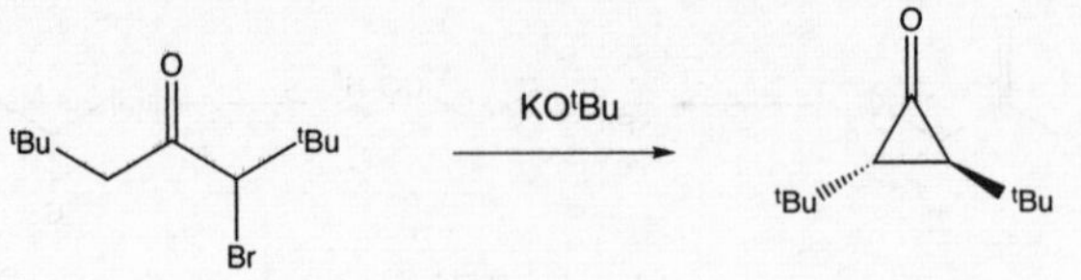

Figure 9.29

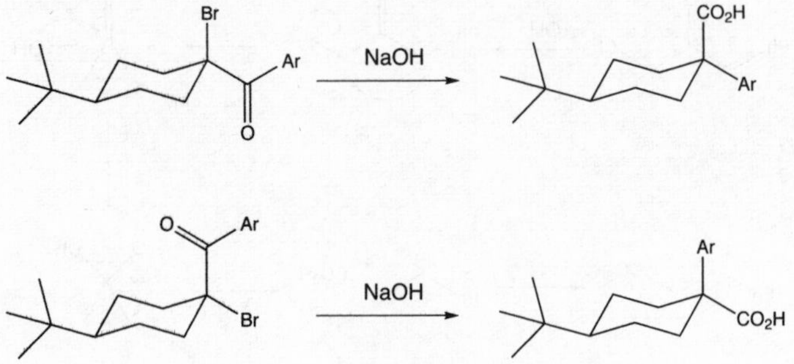

Figure 9.30

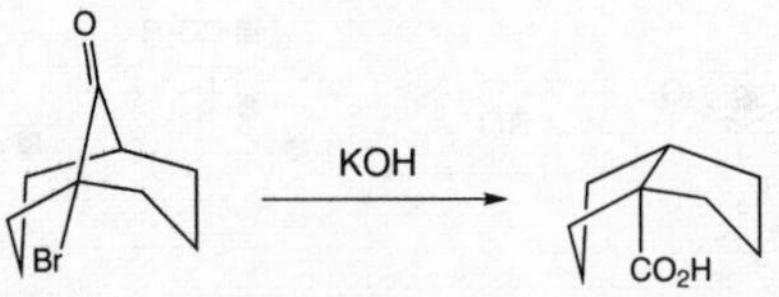

Figure 9.31

Further evidence in favour of this mechanism is that cyclopropanes may be isolated from the reaction under suitable conditions. Thus treatment of the bromoketone shown in Figure 9.29 with potassium t-butoxide gave a cyclopropanone that could be isolated. This gave the normal Favorskii products on further treatment with base.

However, there is a problem even with this mechanism. It requires that there be a proton at the α'-position, and many α-haloketones with no proton at this position still undergo the reaction. Since the evidence for the cyclopropane mechanism is so good, the most likely explanation for this discrepancy is that a different mechanism operates in compounds with no α'-hydrogen.

It is thought that the semibenzilic mechanism is at work in such molecules. One piece of evidence for this comes from examination of the stereochemistry. When the bromoketones shown in Figure 9.30 were treated with sodium hydroxide, they gave the acids with inversion of configuration at the carbon originally bearing the bromine.

The semibenzilic mechanism may also operate in some compounds with an α'-hydrogen. For example, the bicyclic bromide in Figure 9.31 is prevented from enolizing by its geometry, and so cannot react via the cyclopropane mechanism, but still gives a Favorskii rearrangement.

262

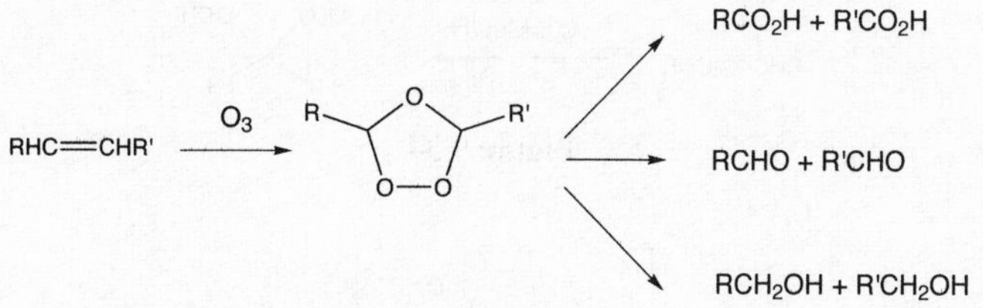

Figure 9.32

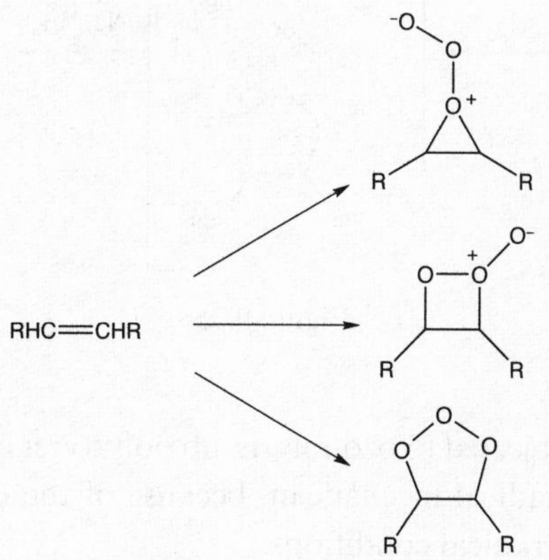

Figure 9.33

9.4 Ozonolysis of olefins

Treatment of olefins with ozone leads to the formation of ozonides (Figure 9.32). These ozonides are not usually isolated, owing to their explosive nature, but are instead cleaved either oxidatively to give carboxylic acids, or reductively to give either aldehydes or alcohols, depending on the reducing agent used.

The reaction is very general and is of great synthetic importance. Historically, it was used extensively in structural elucidation. Cleavage of compounds containing double bonds would lead to simpler fragments, and the structure of the original molecule could be deduced from the structure of the fragments. Nowadays, this use is all but extinct, since structures can now be determined far more easily by spectroscopic means.

A great deal of work has been done on the steps leading to the formation of the ozonide, and this is what we will consider here. An observation bearing on the initial step of the reaction is that electron-rich olefins react faster than electron-deficient ones. Ozone is an electrophilic reagent, and so it seems likely that the first step of the reaction is an addition of ozone to the double bond. There are a number of ways in which ozone might add to the double bond. An olefin–ozone adduct (a so-called primary ozonide) could be formed with a three-, four-, or five-membered ring (Figure 9.33). This could take place by an ionic, pericyclic, or radical process.

263

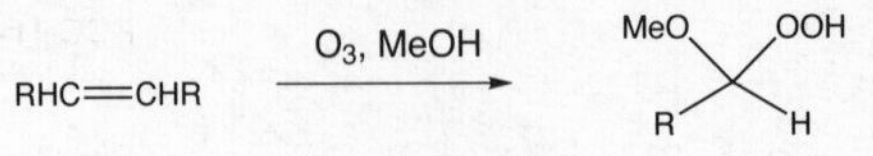

Figure 9.34

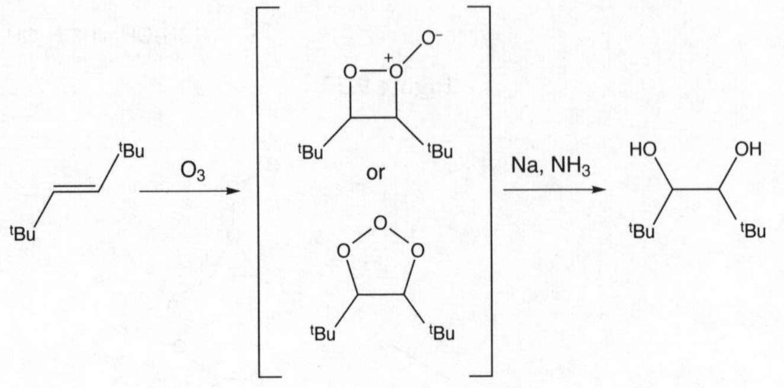

Figure 9.35

When styrene was subjected to ozonolysis, no polystyrene could be detected. This appears to rule out a radical mechanism, because of the ease with which styrene polymerizes under free radical conditions.

The primary ozonide is presumably very unstable, as it has been very difficult to isolate. However, it has been detected spectroscopically, and has also been trapped. Treatment of trans-di-t-butylethylene with ozone at low temperatures leads to a crystalline precipitate that decomposes on warming to room temperature. This decomposition was found to be exothermic, with a heat of reaction about half the value of the heat of a typical ozonolysis, which suggests that it is an intermediate.

Further evidence that this precipitate is an intermediate comes from addition of methanol. Ozonolysis in the presence of methanol usually results in methoxyhydroperoxides rather than ozonides (Figure 9.34). In this experiment, if methanol is added before warming to room temperature the methoxyhydroperoxide is formed, whereas it is not if methanol is added only after warming to room temperature. This shows that the crystalline precipitate formed at low temperatures is not the ozonide, but an intermediate on the way to the ozonide that can be diverted into an alternative reaction pathway in the presence of methanol.

Perhaps the best piece of evidence that this precipitate is the primary ozonide comes from its reaction with sodium in liquid ammonia, which reduced it to the 1,2-diol (Figure 9.35). This is consistent with either the four-membered or five-membered structure, but not the three-membered one, which would not be expected to give a diol on reduction.

It later became possible to obtain an NMR spectrum of this primary ozonide. This showed that the methine protons were equivalent, thus ruling out the four-

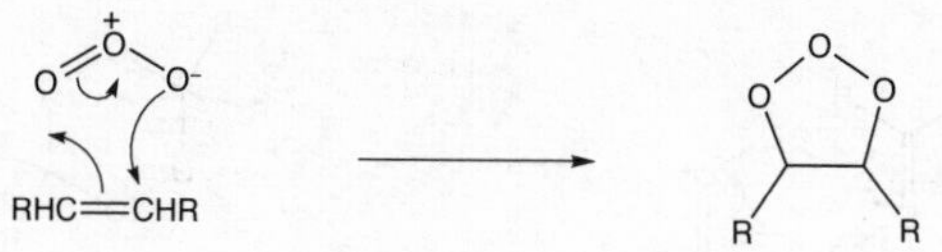

Figure 9.36

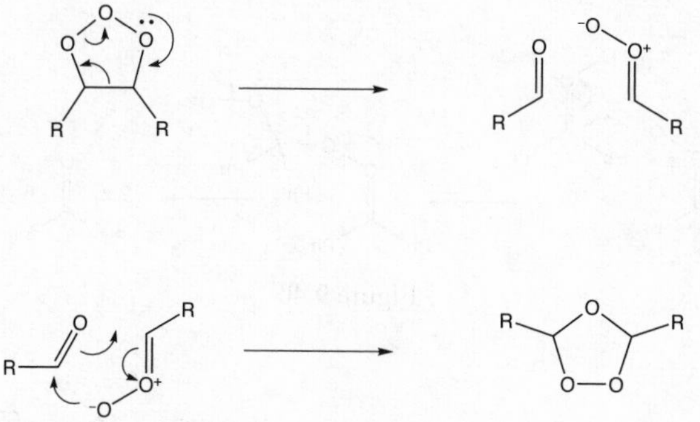

Figure 9.37

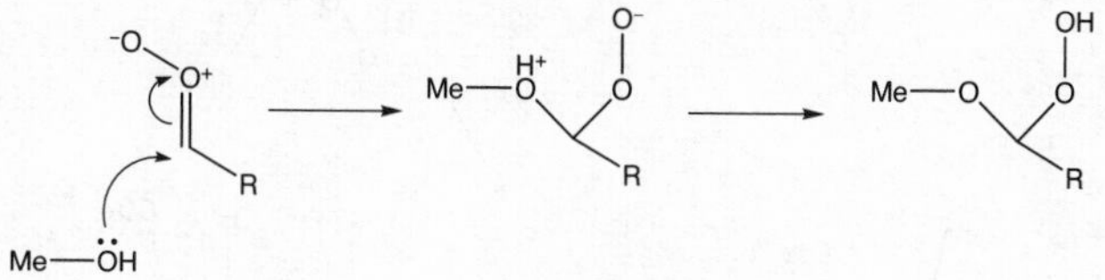

Figure 9.38

membered structure. The primary ozonide thus appears to be the 1,2,3-trioxolane, presumably formed by a 1,3-dipolar cycloaddition reaction (Figure 9.36).

The primary ozonide must somehow be converted into the ozonide. This process must include cleavage of the carbon–carbon bond. A mechanism was proposed for this step in which the primary ozonide splits into two constituents by a cycloreversion reaction. The species thus formed are an aldehyde and a carbonyl oxide. These can then recombine in another cycloaddition to give the ozonide (Figure 9.37).

There is much evidence for this proposal. We have already mentioned that methoxyhydroperoxides are formed when olefins are treated with ozone in the presence of methanol. This is entirely consistent with the presence of the carbonyl oxide, which would add methanol as shown in Figure 9.38.

In general, tetrasubstituted olefins do not give ozonides. This is consistent with the proposed mechanism, in that cleavage of the primary ozonide would lead to a carbonyl oxide and a ketone. Ketones are considerably less reactive than aldehydes, and so we should not be surprised that they do not react in the same way. The products that are formed in these reactions have been identified as the expected ketone,

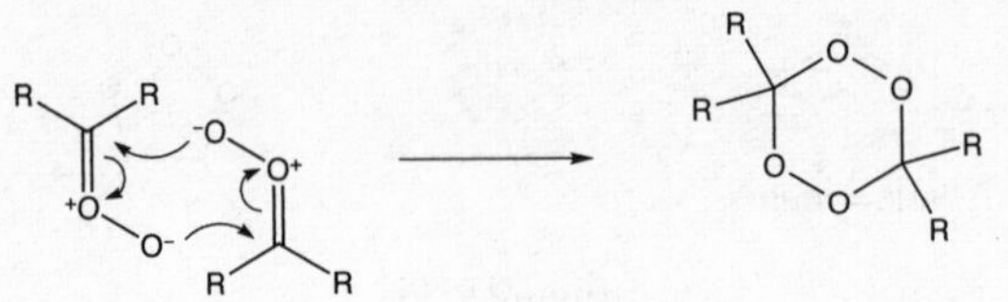

Figure 9.39

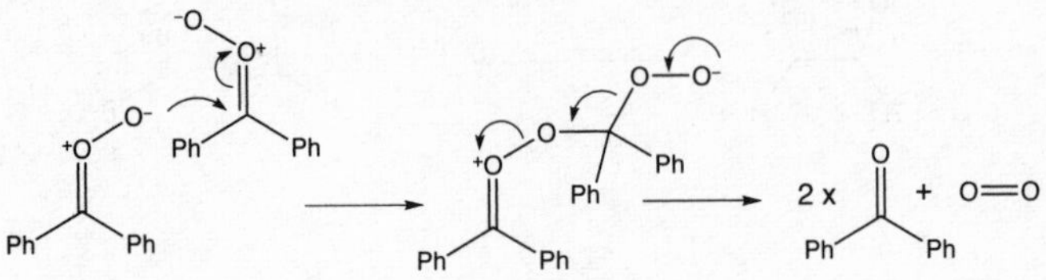

Figure 9.40

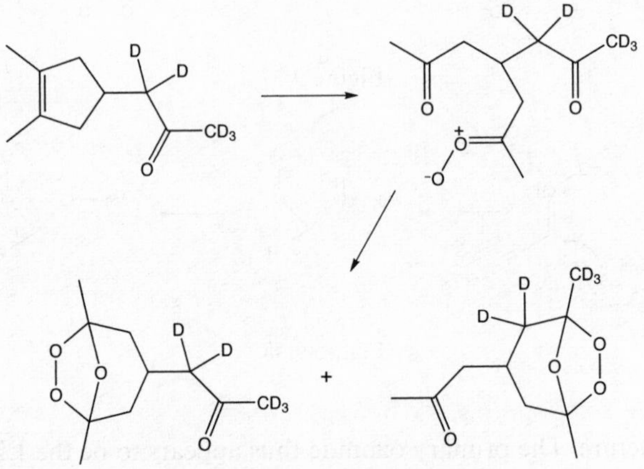

Figure 9.41

plus other products consistent with the intermediacy of a carbonyl oxide. These include the peroxide dimers shown in Figure 9.39.

Careful measurement of the yields of various products in the ozonolysis of tetra-phenylethylene showed that over half of the product was benzophenone. This shows that some of the carbonyl oxide must eventually be converted into the ketone. A plausible mechanism is shown in Figure 9.40.

Whilst ketones do not normally react with carbonyl oxides, they can if they are particularly reactive or if the reaction is intramolecular. A particularly neat experiment made use of this intramolecular reaction. The deuteriated molecule shown in Figure 9.41 was treated with ozone. If the postulated mechanism is correct, then a symmetrical intermediate should be formed, and the label should be scrambled in the product. This was indeed so, the two different labelling patterns of the product

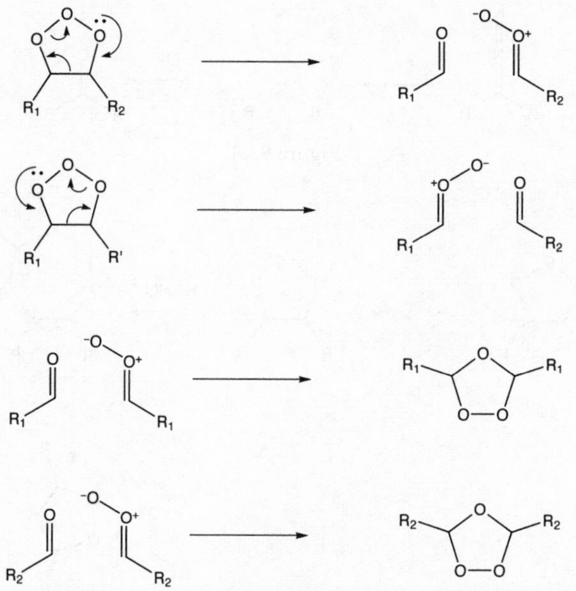

Figure 9.42

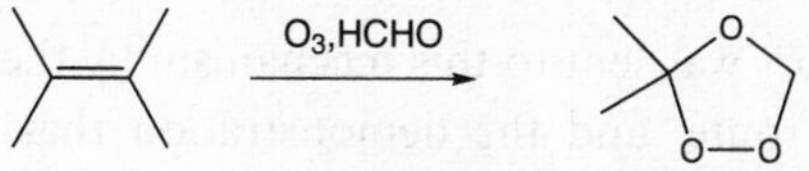

Figure 9.43

being found in equal amounts. This shows that the two ends of the original double bond must become separated, and that the carbonyl oxide must have sufficient lifetime to react with either ketone group, rather than reacting preferentially with the one to which it is originally closer.

One of the most convincing pieces of evidence in favour of this mechanism is the formation of cross products. If an unsymmetrical olefin is ozonolysed, we can envisage the formation of symmetrical products in addition to the expected one, as a result of the primary ozonide cleaving in different ways (Figure 9.42). These have been found for a number of olefins. The yield of cross product tends to increase as the solution becomes more concentrated, which suggests that the two halves of the molecule are initially held together in a solvent cage, but that this cage effect becomes less pronounced in more concentrated solution. More polar solvents also favour formation of the cross products, as they tend to solvate the intermediates more closely, and so are less likely to hold them together within a cage.

Cross ozonides can also be formed if an aldehyde is added to the reaction mixture. For example, ozonolysis of tetramethylethylene in the presence of formaldehyde leads to the unsymmetrical ozonide shown in Figure 9.43.

267

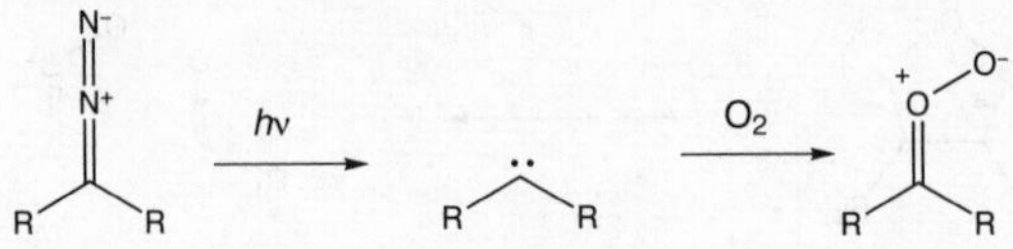

Figure 9.44

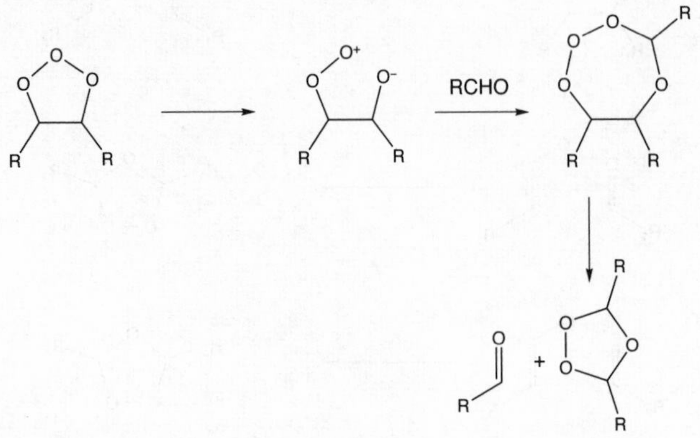

Figure 9.45

Considerable support was lent to this mechanism by the synthesis of a carbonyl oxide by a different route, and the demonstration that it reacted as expected. Photolysis of diazoalkanes leads to carbenes, as mentioned in Section 4.4.2. If this reaction is done in the presence of oxygen, the carbene may react with the oxygen to form a carbonyl oxide (Figure 9.44). When diphenyldiazomethane was photolysed in the presence of oxygen and various aldehydes the expected ozonides were formed. Formation of ozonides without any ozone present is particularly convincing evidence for the proposed mechanism.

Despite the successes of this mechanism in explaining a great deal of experimental data, it did run into trouble. When the stereochemistry of the ozonides was investigated, it turned out that the *cis/trans* ratio was often a function of the geometry of the original olefin. This is not consistent with the mechanism we have been discussing, since it predicts that the information of the initial double bond geometry will be lost during the reaction.

Another mechanism was therefore proposed in an attempt to explain the stereochemistry of the reaction, in which the primary ozonide reacted with an aldehyde to give a seven-membered intermediate, which then broke down to give the ozonide (Figure 9.45). It was suggested that the preferred conformation of the intermediate would be a function of the initial olefin geometry, and that different conformations of the intermediate would lead to different *cis/trans* ratios in the product.

However, this mechanism did not stand up to scrutiny. A prediction of this mech-

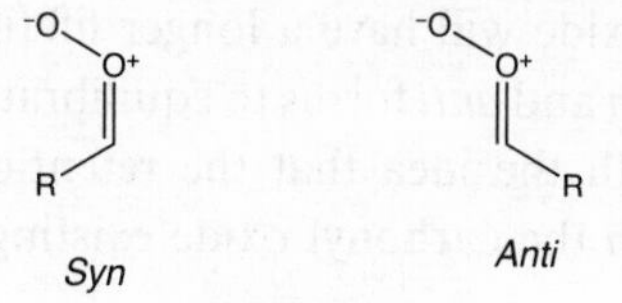

Syn   *Anti*

Figure 9.46

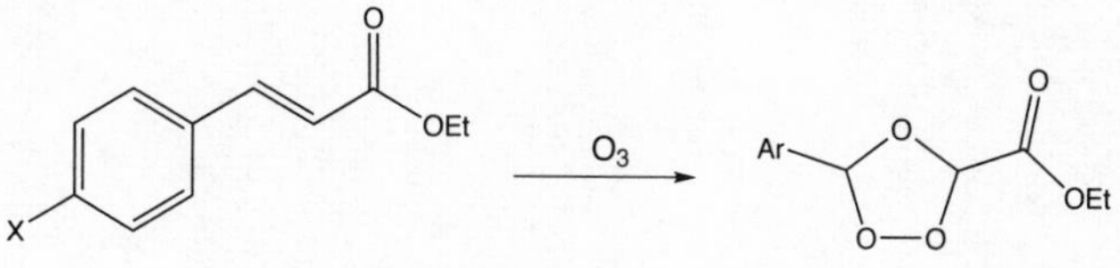

Figure 9.47

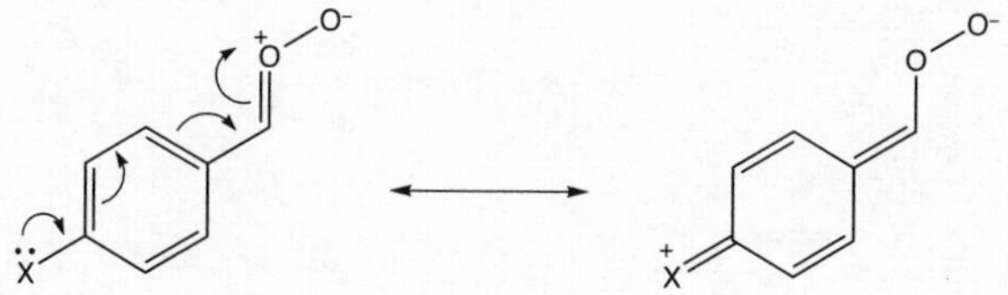

Figure 9.48

anism is that the *cis/trans* ratio of the product should be affected by added aldehyde, but no examples of this effect were found. Secondly, this mechanism can be distinguished from our earlier mechanism by oxygen-labelling experiments. If this new mechanism is correct, then the oxygen from an added aldehyde should end up in the peroxy bridge of the ozonide, whereas the earlier mechanism will place it in the ether linkage. When stilbene was ozonolysed in the presence of ^{18}O-labelled benzaldehyde, the product was found to contain the label in the ether position.

With this new mechanism ruled out, an attempt was made to refine the original mechanism to take account of the stereochemistry. In this refinement, the carbonyl oxides were suggested to exist in *syn* and *anti* forms (Figure 9.46), their relative amounts depending on the geometry of the original olefin. Careful consideration of the conformation of the primary ozonide supports this idea. Similar considerations of conformations of reactants leading to the ozonide suggest that the *cis/trans* ratio will be different depending on whether the *syn* or *anti* carbonyl oxide is the dominant species.

A neat piece of evidence in support of this was provided by the ozonolysis of a series of *para*-substituted methyl cinnamates (Figure 9.47). As the *para*-substituent became more electron-donating, the *cis/trans* ratio of the ozonide became less of a function of the initial geometry. An electron-donating substituent will tend to stabilize the carbonyl oxide, as we can see from the resonance structures in Figure 9.48.

269

A more stable carbonyl oxide will have a longer lifetime, and will therefore have more of a chance for its *syn* and *anti* forms to equilibrate before it reacts. This result is therefore in accord with the idea that the retention of stereochemical information is dependent upon the carbonyl oxide existing in a particular conformation.

Glossary

α-elimination	the removal of two groups from the same atom
β-elimination	the removal of two groups from adjacent atoms
π bond	a covalent bond between two atoms in which the electron density is above and below the bond, but not along the internuclear axis
σ bond	a covalent bond between two atoms in which the electron density is at a maximum along the internuclear axis
activation energy	the energy a molecule or collection of molecules must obtain in order to undergo a reaction; it corresponds to the difference in free energy between the starting materials and the transition state
alkane	a hydrocarbon with no double bonds or functional groups
alkene	a hydrocarbon containing at least one double bond
alkyne	a hydrocarbon containing at least one triple bond
alpha effect	enhancement of nucleophilicity when a nucleophile is adjacent to a pair of electrons
ambident electrophile	an electrophile that can react at two different sites, such as an α,β-unsaturated carbonyl compound, which may react at the carbonyl carbon or at the β-position
ambident nucleophile	a nucleophile that can react at two different

	sites, such as an enolate, which may react at the carbon or oxygen
antarafacial	a term applied mainly to pericyclic reactions, meaning that a π system reacts on opposite faces at each end
antiaromatic	a cyclic polyene with $4n$ π electrons, not showing aromatic stability
antibonding orbital	a molecular orbital in which there is reduced electron density between two atoms, therefore exposing the two positively-charged nuclei to each other and so tending to push them apart
antiperiplanar	a configuration in which two groups are in the same plane as each other and the atoms that join them, but are on opposite sides of the bond in between them
aprotic	normally applied to solvents, meaning devoid of acidic protons
aromatic	a cyclic polyene with $(4n+2)$ π electrons, with particular stability
aromatic resonance	the energetic advantage of an aromatic system, due to the delocalization of the electrons
asymmetric	refers to a compound that is not superimposable on its mirror image, normally because it contains a carbon atom with four different groups attached; not to be confused with unsymmetrical
atomic orbital	a possible arrangement of electrons within an atom; each orbital can contain two electrons
autoxidation	the oxidation of compounds, normally hydrocarbons, on exposure to air
bimolecular	refers to a step of a reaction in which two molecules react
bonding orbital	a molecular orbital in which there is increased electron density between two atoms, therefore shielding the two positively-charged nuclei from each other and so tending to keep them together
Brønsted acid	a substance that can give up a proton
Brønsted base	a substance that can accept a proton
canonical form	a structure that contributes to a resonance hybrid of a molecule with delocalized

	electrons, written as though the electrons were localized
carbanion	a species with a negatively-charged carbon atom
carbene	a divalent carbon species; carbenes have only six electrons in their valence shell instead of the normal eight
carbenium ion	a rare term for a trivalent carbocation
carbenoid	a species that behaves like a carbene in its reactions, although not actually a carbene
carbocation	a species with a positively-charged carbon atom
carbonium ion	an alternative term for carbocation, although it is probably best avoided when talking about trivalent carbocations because other 'onium' ions are hypervalent
chiral	asymmetric
concerted reaction	a reaction in which more than one process takes place at a time, for example an E2 elimination, in which the proton is removed at the same time as the leaving group leaves
conjugate acid	the conjugate acid of a base is formed by protonation of the base
conjugate addition	addition taking place over an extended π system, normally applied to addition to α,β-unsaturated carbonyl compounds in which the nucleophile attacks at the β position
conjugate base	the conjugate base of an acid is formed by deprotonation of the acid
covalent bond	a bond between two atoms in which electrons are shared between them
curly arrows	a representation of the movement of electrons in chemical reactions or the imaginary movement of electrons between resonance structures
cycloaddition reaction	a type of pericyclic reaction in which two compounds add to each other through their π systems to give a cyclic compound; an example is the Diels–Alder reaction
degenerate	applied to orbitals; degenerate orbitals have precisely the same energy because of the

	symmetry properties of the atom or molecule
delocalized	delocalized electrons are not localized in particular bonds, but are spread out over a number of bonds through a conjugated π system
diastereoisomer	diastereoisomers (the less correct term diastereomers is sometimes heard) are compounds that differ from each other in the configuration about some of their chiral centres, although they are not mirror images of each other
dienophile	a compound, normally an olefin, that reacts with a diene in the Diels–Alder reaction
dipole moment	a compound with a dipole moment has some degree of charge separation, in other words one end is at least partially negatively charged and the other end is at least partially positively charged
driving force	the thermodynamic push behind a reaction, analogous to gravity being the driving force for apples falling off trees
E1 reaction	an elimination reaction in which the leaving group leaves before the proton is removed
E1cB reaction	an elimination reaction in which the proton is removed before the leaving group leaves
E2 reaction	an elimination reaction in which the proton is removed at the same time as the leaving group leaves
Ei reaction	an internal elimination reaction, in which a part of the molecule attached to the leaving group acts as a base, removing the proton
electrocyclic reaction	pericyclic ring-opening or cyclization reaction
electrofugal leaving group	a leaving group that leaves without its electron pair; the most common example is the proton
electron sink	an atom capable of accepting a pair of electrons; this is where curly arrows end up in a drawing of a reaction mechanism
electron source	an atom capable of providing a pair of electrons; this is where curly arrows start in a drawing of a reaction mechanism

electronegative	an electronegative atom has a tendency to attract electrons
electrophile	an electron-deficient species, either with a positive charge or with a vacant orbital capable of accepting electrons
electropositive	an electropositive atom does not hold its electrons very tightly and easily forms positive charges
elimination	a type of reaction in which two groups leave a molecule, usually resulting in the formation of a double bond
endothermic	an endothermic reaction has a positive enthalpy of reaction, in other words it requires an input of heat
energy-level diagram	a diagram showing the energy of the orbitals of an atom or molecule
enthalpy	the heat given out or absorbed during a chemical reaction
entropy	a measure of disorder; reactions that create entropy (i.e. those which lead to more disorder) are favourable
exothermic	an exothermic reaction has a negative enthalpy of reaction, in other words it gives out heat
extrusion	a type of elimination in which a group is taken from inside a molecule and the two ends thus freed join together
free energy	the combination of enthalpy and entropy that determines how favourable a reaction is
free radical	a species with an unpaired electron
frontier orbital	either an orbital containing the electrons that take part in a reaction, or an empty orbital that accepts electrons during a reaction
functional group interconversion	a type of reaction in which the carbon skeleton of a molecule remains unchanged but the functional groups are altered
hard	a term applied to acids, bases, nucleophiles and electrophiles; hard species are usually charged and are not easily polarized; ionic interactions are more important than frontier orbital interactions in their reactions
heteroatom	any atom other than carbon or hydrogen

275

heterolytic cleavage	the cleavage of a bond to produce a cation and an anion
Hofmann elimination	elimination in which the double bond goes to the least highly substituted position
HOMO	highest occupied molecular orbital; a type of frontier orbital; the HOMO contains the electrons of highest energy in a molecule; these are the electrons that take part in reactions in nucleophiles
homolytic cleavage	the cleavage of a bond to produce two radicals
Hückel rule	cyclic polyenes with $(4n+2)$ π electrons are aromatic, whereas those with $4n$ π electrons are antiaromatic
hybridization	the combination of atomic orbitals within the same atom to give atomic orbitals of different shapes; this a mathematical convenience rather than a physical process
hydrophilic	literally 'water loving'; hydrophilic molecules are often charged and are readily soluble in water
hydrophobic	literally 'water hating'; hydrophobic molecules are non-polar and dissolve in organic solvents more easily than in water
hyperconjugation	a form of conjugation in which one or more canonical form has a broken σ bond
inductive effect	an electronic effect acting through σ bonds; it may be either negative (as in halogens) or positive (as in alkyl groups)
initiation	a part of a radical reaction in which radicals are first formed
intermediate	a species formed during a chemical reaction that takes place in more than one step; intermediates are genuine chemical entities with finite lifetimes, although if they are high in energy these lifetimes may be very short
intermolecular	describes a process that takes place between at least two separate molecules
intramolecular	describes a process that takes place within a single molecule
ionic bond	a bond formed by the electrostatic attraction of a positive and a negative ion
leaving group	a group that becomes detached from the rest

	of a molecule during a reaction; leaving groups may be either nucleofugal or electrofugal, but the term is usually used to mean a nucleofugal leaving group
levelling effect of water	the effect that water has of attenuating the strength of very strong acids or bases by becoming either deprotonated or protonated
Lewis acid	a species that can accept a pair of electrons, usually some sort of metal ion
Lewis base	a species that can donate a pair of electrons; although the definition is slightly different, Lewis bases are the same as Brønsted bases
LUMO	lowest unoccupied molecular orbital; a type of frontier orbital; the LUMO is the lowest in energy of the empty orbitals of a molecule; this is the orbital that takes part in reactions in electrophiles
Markovnikov addition	addition to a double bond such that the end of the double bond that had more protons to begin with gains another
mesomeric effect	a type of electronic effect acting through π bonds
Michael addition	conjugate addition to an α,β-unsaturated carbonyl compound
molecular orbital	a possible arrangement of electrons within a molecule, often considered as being formed from a combination of atomic orbitals; each orbital can contain two electrons
molecularity	refers to the number of species taking part in a given step of a reaction
neighbouring group participation	the assistance in a reaction by an adjacent group; for example, an internal S_N2 displacement by an adjacent nucleophile, giving a product that can then react with an external nucleophile
nitrene	a monovalent nitrogen species, analogous to a carbene; nitrenes have only six electrons in their valence shell instead of the normal eight
nucleofugal leaving group	a type of leaving group that leaves with its electron pair, such as a halide ion
nucleophile	an electron-rich species, either with a negative

	charge or with a lone pair capable of donating electrons
olefin	an alternative term for alkene
pericyclic	a pericyclic reaction is one in which the electrons move in a circuit in a concerted process
photochemical	a photochemical reaction is one that requires irradiation with light, usually in the ultraviolet region of the spectrum
primary kinetic isotope effect	the slowing of the rate of reaction due to replacement of an atom by a heavier isotope when a bond to this atom is broken in the rate-determining step; this is most commonly observed when hydrogen is replaced by deuterium
propagation	in radical reaction, a propagation step is one in which a radical reacts to create another radical
protecting group	a group that is added to a molecule before carrying out some synthetic step or steps, to avoid unwanted reactions in susceptible parts of the molecule; it is removed after the transformations have been carried out
pyrolysis	breakdown of a molecule on heating
quantum number	a parameter that describes an atomic orbital; each atomic orbital is specified by a set of four quantum numbers, n, l, m, and s
racemization	conversion of an optically active (chiral) compound into a mixture of enantiomers, so that it loses its optical activity
radical	an alternative term for free radical
radical chain	a series of reactions involving radicals, in which there are many propagation steps
radical inhibitor	a substance that readily reacts with radicals to produce stable molecules, thus inhibiting the propagation of radical chains
radical initiator	a substance that readily forms free radicals
rate-determining step	the slowest step of a reaction; it is called rate-determining because the overall rate of a reaction cannot exceed the rate of the slowest step
reaction order	an experimentally determined kinetic property

	of a reaction, equal to the sum of the powers of the concentrations appearing in the rate equation
reaction profile	a diagram showing the energy of the reactants during the course of a reaction; minima on this diagram correspond to intermediates and maxima correspond to transition states
rearrangement	a reaction in which the arrangement of atoms within a molecule changes, so that the product is an isomer of the starting material
resonance energy	the energy gained by π bonds coming into conjugation with each other, applied particularly to aromatic systems
resonance hybrid	a term applying to molecules with delocalized electrons; the resonance hybrid is the combination of the imaginary structures with localized electrons (canonical forms) that represents the true distribution of electrons in the molecule, although it is not as easy to draw as canonical forms
resonance structures	another term for canonical forms
ring strain	the energy disadvantage in small rings due to the deviation of bond angles from their ideal values
S_N1 reaction	a substitution reaction in which the leaving group leaves to give a carbocation, which then combines with a nucleophile
S_N2 reaction	a substitution reaction in which the incoming nucleophile expels the leaving group in a concerted process
$S_N i$ reaction	a substitution reaction in which the nucleophile is derived from the leaving group
Saytzeff elimination	elimination in which the double bond goes to the more highly substituted position
Schrödinger equation	an equation whose solutions are wavefunctions, in other words the mathematical description of the behaviour of electrons in atoms and molecules
secondary kinetic isotope effect	an isotope effect that arises for reasons other than in a primary kinetic isotope effect
sigmatropic rearrangement	a type of pericyclic reaction in which a σ

	bond moves across a π system to a new position
singlet carbene	a carbene in which the two non-bonding electrons are of opposing spins; in other words the carbene has one lone pair orbital and one empty orbital
soft	a term applied to acids, bases, nucleophiles and electrophiles; soft species are easily polarized and do not usually have a high charge density; frontier orbital interactions are more important than ionic interactions in their reactions
solvent cage	an enclosure formed by solvent molecules around reactants, holding them together
SOMO	singly occupied molecular orbital; the frontier orbital of a radical, containing the unpaired electron
stereoelectronic effect	an effect due to alignment of orbitals
steric	to do with the physical size of molecules or parts of molecules
substitution	a reaction in which one group is replaced by another
suprafacial	a term applied mainly to pericyclic reactions, meaning that a π system reacts on the same face at each end
synperiplanar	a configuration in which two groups are in the same plane as each other and the atoms that join them, and are on the same side of the bond in between them
termination	in radical chains, termination reactions are the combination of two radicals to form a stable molecule
transition state	the point of highest energy in a reaction step; the transition state does not have a finite lifetime
triplet carbene	a carbene in which the two non-bonding electrons are of the same spin; in other words the carbene has two singly occupied orbitals and behaves as a diradical
unimolecular	refers to a step of a reaction in which only one molecule takes part
unsymmetrical	not symmetrical, although not in the same

	sense as asymmetric (for example an olefin with different substituents at each end of the double bond)
valence orbital	an orbital in the valence shell of an atom
valence shell	the outermost occupied shell (collection of atomic orbitals with a common value of the quantum number n) of an atom, containing the highest energy electrons; it is these electrons that take part in reactions
wavefunction	a mathematical description of the behaviour of electrons in atoms or molecules; wavefunctions have their physical counterparts in orbitals
Wheland intermediate	an intermediate in an aromatic substitution reaction, in which two groups are attached to one carbon atom of the ring and a positive charge is spread over the rest of the ring
work-up	a term used to describe operations done when the main reaction is complete and before the product can be isolated, such as addition of acid or extraction between water and organic solvents
ylide	a molecule with positive and negative charges on adjacent atoms
zero-point energy	the vibrational energy of a bond in its ground state; quantum mechanical calculations show that this must always be more than zero
zwitterion	a molecule that contains both positively and negatively charged groups, such as an amino acid

Answers to problems

Chapter 1

1. a: sp$_2$; b: sp$_2$; c: sp$_3$; d: sp.
2. Imaginary movement in resonance structures: a, c. Real movement in reactions: b, d.

a   b   c   d

3.

4. a: oxygen atom; b: lone pair on hydroxyl oxygen; c: carbon–carbon double bond; d: lone pair on nitrogen atom.

5. a: oxygen atom; b: nitrogen atom; c: bromine atom.
   Note: this is not the same as the site of reaction of the molecule with a nucleophile, but is the final home of the electrons involved in any reaction.

6.

7.

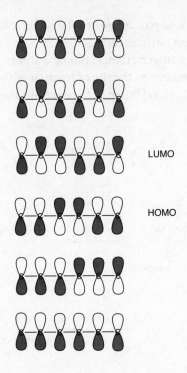

LUMO

HOMO

8. a, c.

9. The cation. This is an aromatic system, with 2 π electrons ($4n+2$, with $n=0$).

10. The charge will arise because the following resonance structure contributes substantially to the resonance hybrid, as both rings are aromatic in this structure.

Chapter 2

1. Nucleophiles: c, d, e; electrophiles: a, b, f.

2. Strongest to weakest: a>g>d>f>e>c>b.

284

3. Fastest to slowest: e>d>a>b>c.
   The thiolate ion (e) is the fastest, despite being a weaker base than d, because it is a very soft nucleophile and so reacts efficiently with the soft electrophile ethyl iodide. The reason why b reacts faster than c is because of the greater steric bulk of c; this has nothing to do with their basicities.
4. Hydroxide is more basic, hydroperoxide is more nucleophilic. The electron-withdrawing effect of the adjacent oxygen stabilizes the negative charge in hydroperoxide, reducing the basicity, but the alpha effect makes it more nucleophilic.
5. Propyl bromide will give more O-alkylation because it is a harder electrophile.
6.

7. Best leaving group to worst: f>a>d>e>c>b.

Chapter 3

1. The equilibrium lies in favour of C+D.
2. A: kinetic product; B: transition state; C: intermediate; D: thermodynamic product.
3. Reactions shown with intermediates: a, d.
   Reactions shown with transition states: b, c.
4. The reaction is under kinetic control, because it is irreversible; the base will remove the most kinetically accessible proton, this determines which product forms.
5. The alternative four-membered lactone does not form because four-membered rings are thermodynamically less stable than four-membered rings. Under acidic conditions, the formation of lactones is reversible, so the thermodynamic product forms.

6. This is because the dissolution is an endothermic process, in other words the bonds between the ions within the solid salt are stronger than those between the ions and water, so heat is absorbed from the medium when the bonds between the ions are broken. The reason why this process nevertheless occurs is because there is a large increase in entropy when a solid dissolves, which can often compensate for the unfavourable enthalpy change.
7. Fastest to slowest: c>d>a>b>e>f.

8. On treatment with base, a will give an allene and b an alkyne. In b, the bromine is fixed in an antiperiplanar conformation to a hydrogen across the double bond, so this elimination will be efficient. In a, elimination across the double bond must take place from a synperiplanar conformation, which is less efficient, so elimination across the single bond to give the allene will predominate.

Chapter 4

1. Most stable to least stable: b>g>f>d>e>c>a.
2. Butyllithium is nucleophilic and would therefore react directly with the carbonyl group in an addition reaction. The steric bulk of LDA renders it much less nucleophilic.
3. c will react only by protonating the Grignard reagent to give benzene. a will not react. b, d, and e will react as follows:

4.

5. Most stable to least stable: e>g>f>b>a>d>c.
6.

7. Nucleophilic attack on the carbonyl group by water is hindered by the steric bulk of the ester group. The ester can leave the t-butyl group in an S_N1 process because of the stability of the tertiary carbocation.

286

8. Most stable to least stable: c>e>d>a>b.

9.

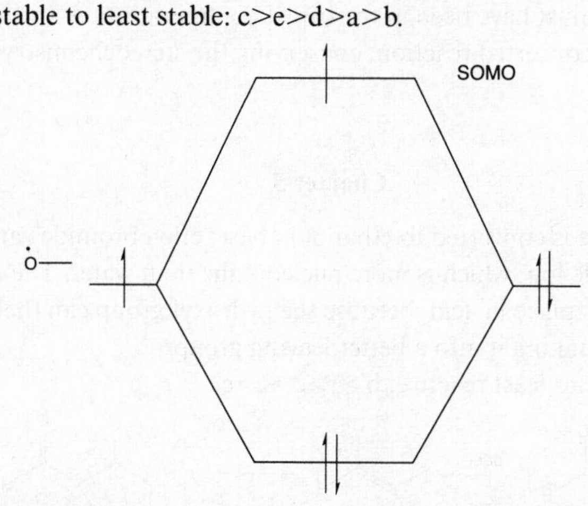

The radical is nucleophilic, because the energy of its SOMO is raised.

10. The mechanism is shown below. In the sequence leading to the other isomer, the methoxy group would be adjacent to the negative charge in the intermediates, which is unfavourable because of the interaction between the negative charge and the oxygen lone pair.

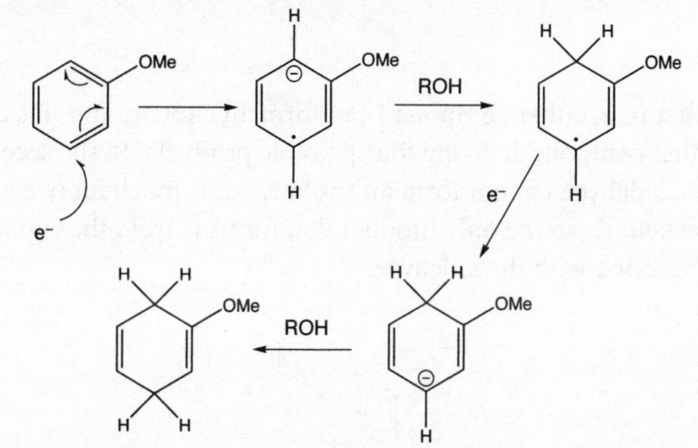

11.

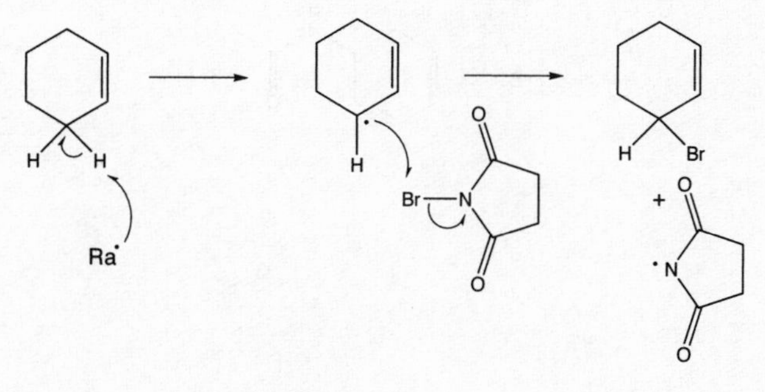

12. The carbene must have been in the triplet configuration. A singlet carbene would give a concerted reaction, conserving the stereochemistry of the olefin.

Chapter 5

1. Ethyl bromide is converted to ethanol in base; ethyl bromide can then react with hydroxide ion, which is more nucleophilic than water. The reverse reaction takes place in acid, because the hydroxyl group can then be protonated, making it into a better leaving group.

2. Most reactive to least reactive: b>a>d>e>c.

3.

4. In the first pair, either compound can form an enolate, and this can react with either compound, giving four possible products. In the second pair, the aromatic aldehyde cannot form an enolate, but is much more electrophilic than the ketone, so the only product that forms is from the enolate of the ketone reacting with the aldehyde.

5.

A:

B:

6.

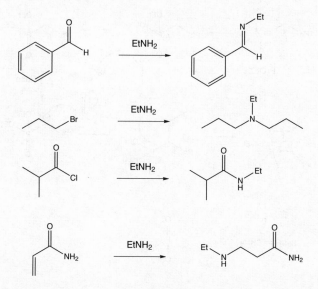

7. Hydrazine is oxidized to diimide by oxygen, with copper sulphate acting as a catalyst. The diimide reduces the double bond.

8.

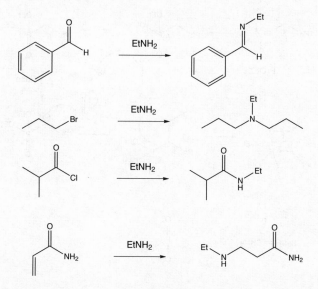

9.

10.

11. Phosphomonoesters and phosphodiesters have a negative charge under basic conditions, and so are not easily attacked by hydroxide, because the negative charges repel each other. Phosphotriesters are neutral molecules in base, and so no such barrier exists to their hydrolysis.

12. It acts as a Lewis acid, helping the halide to leave.

Chapter 6

1. The bromonium ion can be written as the following combination of resonance structures. The resonance form with the secondary carbocation will contribute more to the resonance hybrid than the form with the primary carbocation, so the secondary position will carry the greater positive charge, and hence will react more readily with the chloride.

2. The stereochemistry of the intermediate is determined by the *trans* addition of bromine. This intermediate must adopt a conformation in which a bromine is antiperiplanar to a hydrogen to eliminate HBr, leading to the geometry of the product as shown.

3.

4.

5. The intermediate α-lactone can be opened intramolecularly by the hydroxyl group to give a protonated epoxide that can be opened by chloride at either end.

6. The first step in this reaction is rate-determining, and is faster with fluoride than with other halogens, because fluoride is more electron-withdrawing, thus making the ring more reactive towards nucleophiles. Although the second step is slower with fluoride than with other halogens, this is immaterial because it is not rate-determining.

7. The weak oxygen–oxygen bond in the peracid is broken and replaced with stronger bonds.

8.

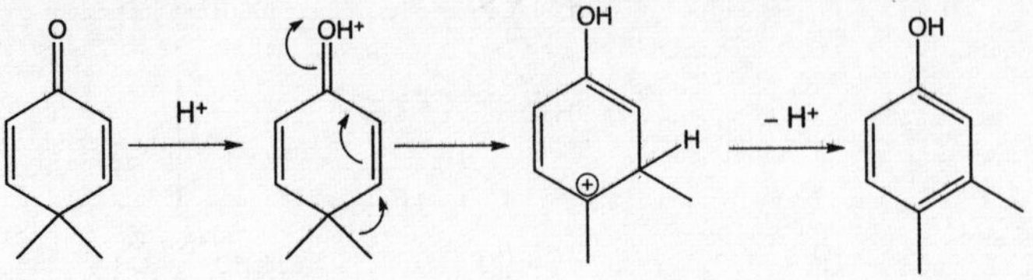

9.

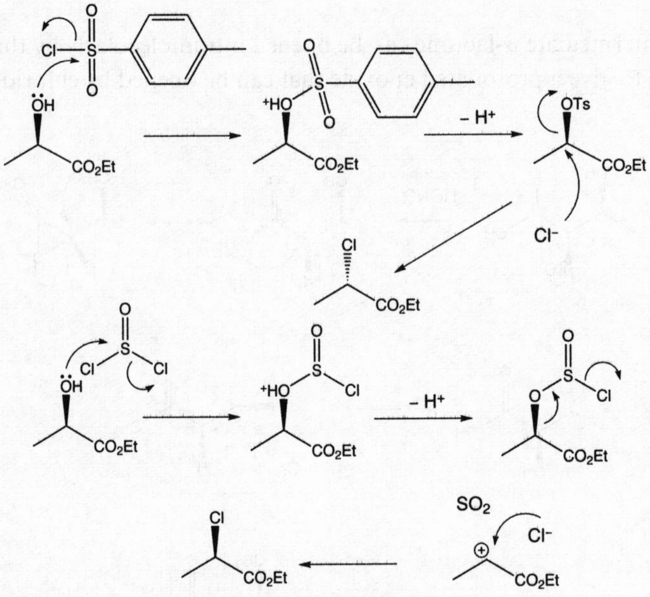

The products have opposite optical rotations.

10.

a

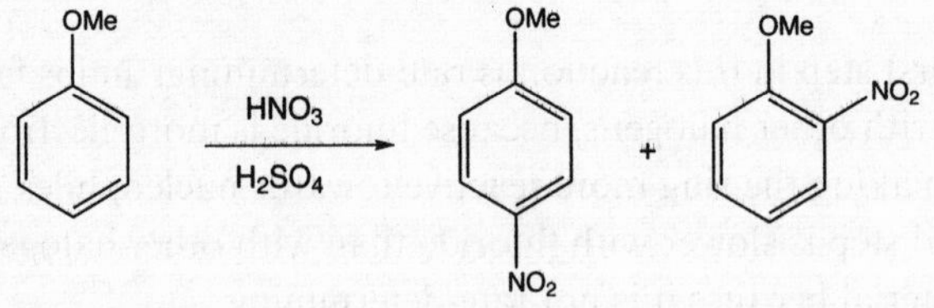

b

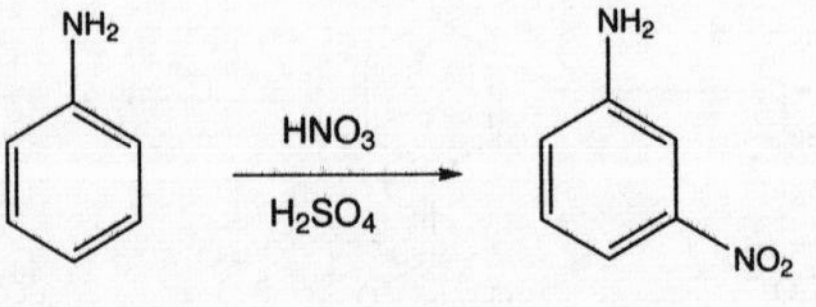

11. a will give more elimination, because of the enhanced acidity of the proton α to the carbonyl group. The use of potassium t-butoxide would increase the proportion of elimination from both compounds.

12. Fastest to slowest: c>d>e>b>a.

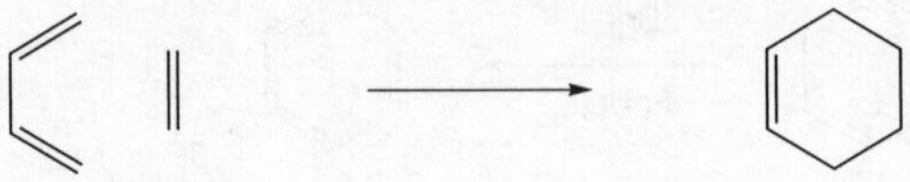

a

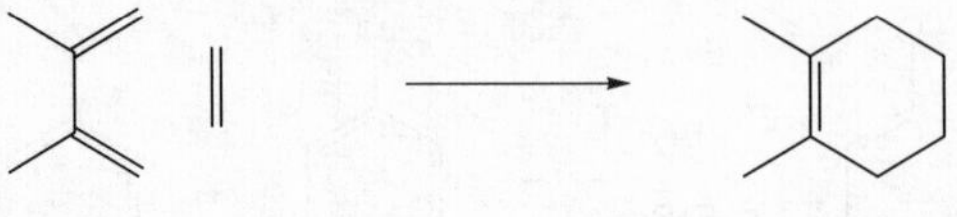

b

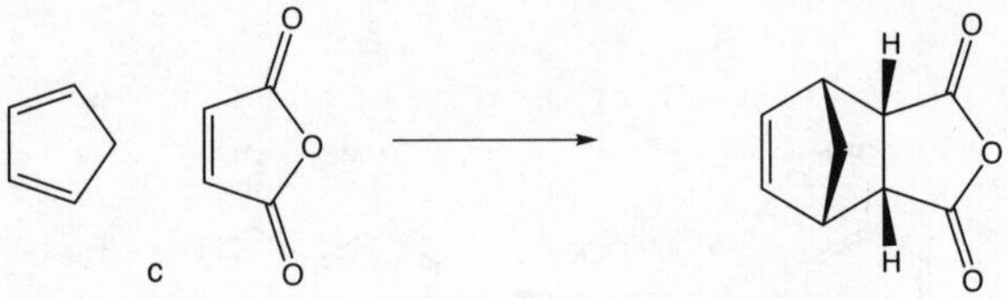

c

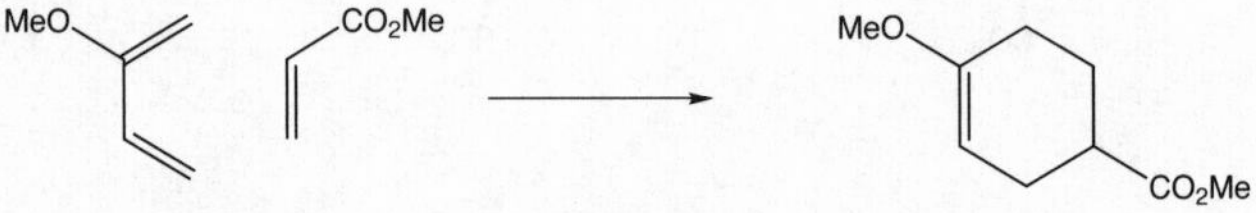

d

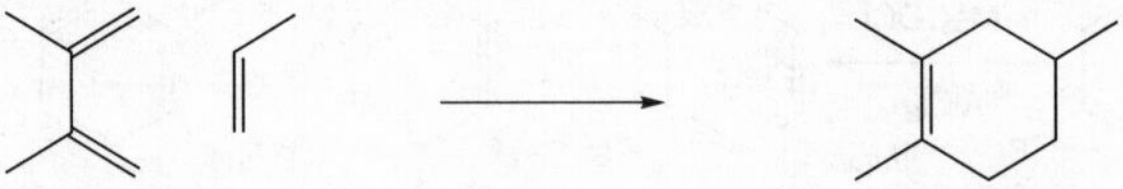

e

Chapter 7

1. The hydrolysis reaction could be carried out in ^{18}O-labelled water. The label would end up in the carboxylic acid if the acyl–oxygen bond were cleaved (as it normally is) and in the alcohol if the alkyl–oxygen bond were cleaved.

2. Although the rate laws are different, in practice it would not be possible to distinguish between the mechanisms directly. The rate law for normal acid hydrolysis is $r=k$ [H$^+$][Ester][H$_2$O], and that for hydrolysis of a tertiary ester with cleavage of the alkyl–oxygen bond is $r=k$ [H$^+$][Ester]. Since water is present in large excess in such reactions, we will not see the effect of it when measuring the rate. Another piece of evidence that could be used to verify this mechanism is to look at the stereochemistry if a chiral tertiary alkyl

group is used. If the alkyl–oxygen bond remains intact, then the stereochemistry of this group must be conserved. If the reaction takes place with inversion or racemization at this group, then this shows that the alkyl–oxygen bond must be cleaved at some point during the reaction.

3. The products of the reaction could be subjected to the reaction conditions for extended periods of time. If the ratio of products is unchanged, then this would show that the hydrazino compound could not be an intermediate in the formation of the reduced product.

4. The most likely contender is diimide, as this would readily reduce the N–N double bond in azobenzene.

5. This is not consistent with straightforward displacement of the tosyl group. The explanation is neighbouring-group participation by the aromatic ring. The product is racemic because the symmetrical cyclopropane can open on either side.

6. This tells us that the rate-determining step (RDS) does not involve bromine, but that acetone is deprotonated in this step. We can postulate a mechanism in which the RDS is enolization, which is followed by fast reaction with bromine.

Chapter 8

1. Electrophilic sites: all three carbonyl carbons. Nucleophiles: acetate anion, ammonia (in equilibrium with ammonium ion).
2. A possible numbering system, but not the only valid one, is as follows.

3. This tells us that the carbon labelled 3, 5 must react with the aldehyde carbon. This is not a site that we have identified as nucleophilic, but deprotonation of the β-ketoester will give a stable enolate, in which this site is nucleophilic.
4. The dehydrated product is more likely here; the next step can be Michael addition of another molecule of enolate to the double bond we have just formed.

5.

6. Nucleophiles: triphenylphosphine, hydroxyl groups of alcohol and carboxylic acid.

Electrophiles: carbonyl groups and nitrogens of diethyl azodicarboxylate.

7. The oxygen is most likely to become attached to the phosphorus atom. We know that phosphorus forms strong bonds with oxygen and this can often be the driving force for reactions. This is not possible initially, since both the hydroxyl group and the phosphorus are nucleophiles. However, the phosphorus can be made into an electrophile by reaction with the diethyl azodicarboxylate.

299

8.

9. Nucleophiles: hydroxyl group of alcohol, oxygen in dimethyl sulphoxide, triethylamine. Electrophiles: sulphur in dimethyl sulphoxide, carbonyl groups in oxalyl chloride.

10. The most reactive electrophile is oxalyl chloride, the most reactive nucleophile is the oxygen of the dimethyl sulphoxide.

11.

12. There are two possible mechanisms for this reaction; the evidence seems to favour the nitrene mechanism (mechanism b), but this has not been proved conclusively.

Mechanism a:

Mechanism b:

then as for mechanism a

Index

302